A. Baudisch, D. Seese, H.-P. Tuschik, M. Weese

# Decidability and Generalized Quantifiers

**Mathematical Research** **Mathematische Forschung**

Wissenschaftliche Beiträge
herausgegeben von der
Akademie der Wissenschaften der DDR
Zentralinstitut für Mathematik und Mechanik

---

**Band 3**

# Decidability and Generalized Quantifiers

von

A. Baudisch
D. Seese
H.-P. Tuschik
M. Weese

# Decidability and Generalized Quantifiers

von

Andreas Baudisch
Detlef Seese
Hans-Peter Tuschik
Martin Weese

Akademie-Verlag · Berlin 1980

Verfasser:

Dr. Andreas Baudisch
Dr. Detlef Seese

Zentralinstitut für Mathematik und Mechanik der
Akademie der Wissenschaften der DDR, Berlin

Dr. Hans-Peter Tuschik
Dr. Martin Weese

Humboldt-Universität, Berlin

Erschienen im Akademie-Verlag, DDR-108 Berlin, Leipziger Str. 3—4
Lektor: Dr. Reinhard Höppner

Lizenznummer: 202 · 100/416/80
Umschlag: Rolf Kunze
Gesamtherstellung: VEB Druckerei „Thomas Müntzer", 582 Bad Langensalza
Bestellnummer: 762 808 1 (2182/3) · LSV 1025
Printed in GDR
DDR 31,— M

# PREFACE

The decision problem for a given formal theory T consists in deciding whether the set of theorems of T is recursive. Or in other words, is there an algorithm which for every sentence $\varphi$ of the language of T decides whether $\varphi$ is a theorem or not? The existence of algorithms is one of the main problems in mathematical logic (and mathematics in general). The proof of the decidability of a theory T gives a deep insight into the structure of the definable predicates of T.

Examples of decidable theories are the elementary theories of algebraically closed fields, real-closed fields, p-adic fields, Abelian groups, Boolean algebras, and linear orderings. Undecidable theories are, for example, the elementary theories of the arithmetic of natural numbers, groups, rings, fields, and graphs. However, the decidable theories mentioned above become undecidable if the first-order language is extended by allowing quantification over sets.

Yet it is desirable to have stronger languages than first-order logic to be able to formalize mathematics as much as possible.

In 1956 A. Mostowski raised the possibility of adding new quantifiers to first-order logic. He defined cardinality quantifiers $Q_\alpha$ with the intended meaning "there exist $\omega_\alpha$-many". These quantifiers can be regarded as generalizations of the existential quantifier $\exists$. Since that time many logicians (e.g. Fuhrken, Vaught, Keisler) have been engaged in generalized quantifiers and have got a lot of results. Some important results are: the recursive enumerability of the logic with the quantifier "there exist uncountably many" (Vaught), the explicite axiomatization and the completeness theorem for this logic (Keisler).

The extended first-order logic obtained by adding cardinality quantifiers is more expressive than elementary logic. But the quantifiers $Q_\alpha$ are weak enough to have a workable model theory. It is natural to search for decidable theories in the extended languages. Especially, whenever K is an elementary class with a decidable elementary theory, then we ask whether the theory of K in the extended language remains decidable.

In this monograph we study the most interesting elementary classes which have a decidable theory in $L(Q_\alpha)$. The results are based on the research work of the authors and their common interest in generalized quantifiers. Their contributions vary with the chapters. In detail, the content of the single chapters is based on their work as cited in the following: Baudisch (3), Seese (4, 7), Tuschik (2, 4, 5) and Weese (6).

The material is presented in such a way that each of the Chapters 2 - 7 can be read independent of the others. This should help those readers who are only interested in special structures.

We hope that this monograph is not only of interest for mathematicians working on decidability or generalized quantifiers but also for those who are interested in structure theoretical investigations of Abelian groups, Boolean algebras, linear orderings, graphs, or pseudo-trees. It is not necessary that the reader has a deep knowledge of logic. However, it is useful to know some of the basic facts from mathematical logic and set theory.

In the following let us summarize the content of the single chapters. In Chapter 1 basic notions and results from the model theory of cardinality quantifiers are introduced. Moreover, it contains further definitions and notational conventions.

In Chapter 2 the elimination of the quantifiers $Q_\alpha$ is investigated. It is proved that $Q_\alpha$ is eliminable in an elementary theory T if and only if $Q_\alpha$ is definable in T. The most interesting result is the definability of $Q_1$ for theories which have no Vaughtian pairs. Examples of such theories are the theories of algebraically closed fields of characteristic p (p prime or zero) and the theory of real closed fields.

Abelian groups are studied in Chapter 3. First we remember the elementary case and the results of Szmielew. Then these methods are extended to the language with cardinality quantifiers by introducing new simple core sentences (e.g. "there exist $\omega_\alpha$-many x which are divisible by m"). This leads to a complete characterization of the expressive power of the logic with $Q_\alpha$ for Abelian groups. The decidability of the $Q_\alpha$-theory of Abelian groups and Abelian p-groups is deduced using these results.

Chapters 4 - 7 have in common that the same method of investigation for different classes of structures is used. This method can be described as follows. Let K be any class of structures of the same signature and let L be the corresponding language. A subclass N of K is said to be dense for K with respect to L if each sentence of L which is satisfiable in K is also satisfiable in N. Now the theory of K in the language L is decidable if there exists an algorithm deciding for every $\mathfrak{A} \in N$ and each $\varphi \in L$ whether $\mathfrak{A} \models \varphi$. This method is successful if the models from the dense system have a simple structure. In all four chapters the models of the corresponding dense systems are represented as terms. The most difficult part in the proof is to construct a sufficiently simple class of structures and to prove its density. To do this we use the generalization of Ehrenfeucht's game to the logic with $Q_\alpha$, developed by Lipner, Brown and Vinner. In each case simplified versions of this game are used which are adapted to the corresponding structures.

Graphs and forests are investigated in Chapter 4. A graph is a relational structure with a binary, irreflexive and symmetric relation. In 1965 Hanf introduced the notion of n-spheres and gave a sufficient condition for the elementary equivalence of two graphs of finite valency. This criterion may be regarded as a special form of Ehrenfeucht's game for graphs. A generalization of this theorem is given for graphs of any valency and languages with the quantifier $Q_1$. A forest is a graph without circles. The connected forests are called trees. A dense system for the $Q_1$-theory of trees is constructed using simple graph theoretical operations. Then it is proved that each tree of this system has a decidable $Q_1$-theory. From this the decidability of the $Q_1$-theory of forests is deduced.

Linear orderings are regarded in Chapter 5. Contrary to the other chapters the extended language contains m cardinality quantifiers $Q_0, Q_{\alpha_1}, \ldots, Q_{\alpha_{m-1}}$, where $\alpha_1, \ldots, \alpha_{m-1}$ have to be successor ordinals. The corresponding game for finitely many additional cardinality quantifiers is a straightforward generalization of the game for only one quantifier. According to the method of Läuchli and Leonard we split linear orderings by monotone mappings. Iterated splitting is coded as terms which will form a dense system. As a special case we regard the subclass of well-orderings. A general product of linear orderings is introduced. It generalizes all known operations on ordered sets. All considerations strongly depend on the axioms of set theory. For simplicity we suppose the general continuum hypothesis, other convenient axioms of set theory are possible.

In Chapter 6 the decidability of the $Q_\alpha$-theory $Th_\alpha(BA)$ of Boolean algebras is proved. For each ordinal $\alpha$, $Th_\alpha(BA)$ can be reduced to one of the following three theories: $Th_0(BA)$, $Th_1(BA)$ or $Th_\omega(BA)$. The necessary background definitions and results for Boolean algebras are given in § 1. In § 2 a new product for Boolean algebras, which generalizes the strong and the weak direct product is introduced. This new product plays an essential role in the construction of the dense system. The special form of the Lipner-game for Boolean algebras is introduced in § 3. This game is denoted as $H^2$-game. Dense systems for $Th_0(BA)$, $Th_1(BA)$ and $Th_\omega(BA)$ are given in § 4. In § 5 it is proved that these systems are uniformly decidable. The proofs of both paragraphs use $H^2$-games. The chapter ends with a characterization of all $Th_\alpha(BA)$ without using GCH.

In Chapter 7 pseudo-trees are investigated. A pseudo-tree is a partial ordering $(A,\leq)$ which has the property that $\{b: b \leq a\}$ is linearly ordered for each $a \in A$. $(A,\leq)$ is said to be an $\omega$-tree if $\{b: b \leq a\}$ is finite for each $a \in A$. A set $C \subseteq A$ is said to be bounded if there exists a natural number n so that $card(\{b: b \leq c\}) \leq n$ for all $c \in C$. The main result of Chapter 7 is the decidability of the theory of $\omega$-trees in a language which allows quantification on bounded sets of cardinality $\geq \omega_1$. This result is quite useful since the language is more powerful than an extension of the elementary logic by $Q_1$ alone. Let $(\mathfrak{A}_i)_{i\in\omega}$ be a sequence of elementary n-equivalent structures and let $\mathfrak{B}_i$ be a substructure of $\mathfrak{A}_i$. The chapter starts with a criterion to deduce the elementary n-equivalence of the union of the structures $\mathfrak{B}_i$. This criterion is used to prove that a certain class of $\omega$-trees is dense for the corresponding theory. The dense system is closely connected with the system of Chapter 4, and the decidability of the $Q_1$-theories of forests and trees can be deduced as a corollary from the results of Chapter 7.

There are a lot of various cardinality quantifiers different from the quantifiers $Q_\alpha$. Some of them are mentioned in Chapter 8. Adding these quantifiers to the elementary logic some of the regarded theories lead to undecidable extensions while others have decidable extensions and are nevertheless stronger than $Q_\alpha$. Moreover, it is interesting that the methods presented in the monograph could also be used in these cases to prove the decidability of the regarded theories. This was one reason for confining ourselves to quantifiers $Q_\alpha$ for a detailed presentation.

Finally we are grateful to Dr. H.-J. Pohl for preparing and correcting our manuscript for typing. We owe special thanks to Mrs. E. Pohl for her expert typing of the manuscript. Special credit is due to Professor L. Budach and Dr. H.-J. Hoehnke for their interest and encouragement. The authors also acknowledge the helpful cooperation with Dr. R. Höppner from the Akademie Verlag.

Berlin, February 25th, 1979

A. B. D. G. S.
H. P. T. M. W.

# TABLE OF CONTENTS

# CHAPTER 1

# DEFINITIONS AND BASIC FACTS

## § 1. Notations

We assume that the reader is familiar with some of the basic notions of mathematical logic as they can be found for instance in the first chapters of [Shoenfield, 1967]. Most of our notations are standard. We shall give some basic notations and conventions in this paragraph.

Throughout the whole book we will use the axiom of choice without mentioning it. Natural numbers will be denoted by k, l, m, n, i, j, r, p, q. Each ordinal $\alpha$ will be identified with the set of all smaller ordinals, $\{\beta : \beta < \alpha\}$. Therefore the empty set can be denoted by 0 and the ordering relation $<$ for ordinals by $\in$. The set of all natural numbers is denoted by $\omega$ and the class of all ordinal numbers by On, respectively. Ordinal numbers and cardinal numbers are denoted by $\alpha$, $\beta$, $\gamma$ and $\varkappa$, $\lambda$, respectively. Let card(A) be the cardinality of A, and pow(A) be the powerset of A. $A^{<\omega}$ is the set of all finite sequences of elements from A. Let $\bar{b}$ be an element of $A^{<\omega}$, then $lh(\bar{b})$ is the length of $\bar{b}$. We can write $\bar{b} = \langle \bar{b}(0), \ldots, \bar{b}(lh(\bar{b}) - 1)\rangle$. $\emptyset$ is the empty sequence. To indicate finite sequences we make a bar above the letter, e.g. $\bar{a}$. "if and only if" is shortened to "iff". The end of a proof will be indicated by "Q.E.D.". The symbol ":=" has the meaning "is defined to be".

Let L be an elementary first order language with equality "=", i.e. L contains:

existential and universal quantifiers $\exists$, $\forall$ (there exists, for all),
logical connectives $\neg$, $\vee$, $\wedge$, $\longrightarrow$, $\longleftrightarrow$ (not, or, and, implies, iff),
individual variables $v_1, v_2, \ldots, v_n, \ldots$,
finitely many nonlogical constants, relation symbols, function symbols, and individual constants depending on the signature,
punctuation ( ).

The set of terms is defined inductively as follows. Every variable and every constant is a term. If $t_1,\ldots,t_n$ are terms and f is an n-ary function symbol, then $f(t_1,\ldots,t_n)$ is a term. Let R be an n-ary relation symbol of L and $t_1,\ldots,t_n$ be terms of L, then the expressions $R(t_1,\ldots,t_n)$ and $t_1 = t_2$ are called atomic formulas.

Now the set of all formulas is the smallest set that contains all atomic formulas and is closed under the following formation rules:

> If $\varphi$ and $\psi$ are formulas, then $(\neg\varphi)$, $(\varphi\wedge\psi)$, $(\varphi\vee\psi)$, $(\varphi\longrightarrow\psi)$, $(\varphi\longleftrightarrow\psi)$, $(\exists v_i\varphi)$, and $(\forall v_i\varphi)$ are also formulas.

Sentences are formulas without free variables. For instance

$$\varphi = (\forall v_1(\forall v_2(v_1 + v_2 = v_2 + v_1)))$$

is a sentence while

$$\psi = (\forall v_1(\exists v_2(v_1 + v_2 = 0 \wedge v_3 < 0)))$$

is not since $v_3$ is free in $\psi$. An exact definition of freeness of variables in formulas is given in [Shoenfield, 1967].

If we do not have in mind some fixed individual variables. then x, y, z are used as meta-variables for individual variables. Sequences of variables are denoted by $\bar{x}$, $\bar{y}$, or $\bar{z}$. Formulas are denoted by $\varphi$, $\psi$, $\chi$, or $\theta$. $\varphi(\bar{x})$ means that the formula $\varphi$ contains at most the variables of the sequence $\bar{x}$ as free variables. We write $\exists\bar{x}\varphi(\bar{x})$, $\forall\bar{x}\varphi(\bar{x})$, $\bigvee_{i<n}\varphi_i$, and $\bigwedge_{i<n}\varphi_i$ instead of $\exists x_0\ldots\exists x_{lh(\bar{x})-1}\varphi(\bar{x})$, $\forall x_0\ldots\forall x_{lh(\bar{x})-1}\varphi(\bar{x})$, $\varphi_0\vee\ldots\vee\varphi_{n-1}$, and $\varphi_0\wedge\ldots\wedge\varphi_{n-1}$, respectively. Moreover we shall omit a lot of parentheses by using the usual preference rules.

Structures for L will be denoted by $\mathfrak{A}$, $\mathfrak{B}$, $\mathfrak{C}$, $\mathfrak{M}$, or $\mathfrak{N}$. The domain of a structure $\mathfrak{A}$ will be denoted by $|\mathfrak{A}|$. If $\varphi(\bar{x})$ is a formula and $\bar{a}$ is a sequence from $|\mathfrak{A}|^{<\omega}$ of the same length as $\bar{x}$, then $\mathfrak{A}\models\varphi[\bar{a}]$ means that the formula $\varphi$ is satisfied in $\mathfrak{A}$ if the variables of $\bar{x}$ are interpreted by the corresponding elements of $\bar{a}$. The elementary theory of $\mathfrak{A}$, denoted by $Th(\mathfrak{A})$, is defined to be the set of all sentences $\varphi$ which are valid in $\mathfrak{A}$. For a class K of structures we define

$$Th(K) := \bigcap_{\mathfrak{A}\in K} Th(\mathfrak{A}).$$

A theory T is a set of sentences in a certain language. If this language is elementary, then T is said to be an elementary or first order theory. A structure $\mathfrak{A}$ is a model of T if all sentences of T are valid in $\mathfrak{A}$. Mod(T) denotes the class of all models of T. We write $T \models \varphi$ if $\mathfrak{A} \models \varphi$ for every $\mathfrak{A} \in$ Mod(T).

A monadic second order language L' is an expansion of an elementary language L by adding:

set variables $V_1, V_2, \ldots, V_n, \ldots,$
the membership relation "$\in$".

To the atomic formulas of L we add $v_i \in V_j$ as new atomic formulas. The set of all formulas of L' is the smallest set that contains the atomic formulas and is closed by the usual formation rules.

Satisfaction for sentences of the monadic second order language L' is defined as usual by adding the clauses

$\mathfrak{A} \models \exists X\varphi(X)$ iff there exists a set $A \subseteq |\mathfrak{A}|$ such that $\mathfrak{A} \models \varphi[A]$.

$\mathfrak{A} \models \forall X\varphi(X)$ iff for all sets $A \subseteq |\mathfrak{A}|$ it holds that $\mathfrak{A} \models \varphi[A]$.

For each structure $\mathfrak{A}$ the monadic second order theory of $\mathfrak{A}$, denoted by $Th_{II}(\mathfrak{A})$, is defined to be the set of all sentences of the corresponding monadic second order language which are valid in $\mathfrak{A}$. If the set variables vary over finite sets in the definition of $\mathfrak{A} \models \varphi$ only, then we get the so-called weak monadic second order logic. The corresponding weak monadic second order theory of $\mathfrak{A}$ is denoted by $Th_w(\mathfrak{A})$. As in the elementary case $Th_{II}(K)$ and $Th_w(K)$ are defined for a class K of structures.

## § 2. Generalized quantifiers

For each ordinal $\alpha$ we take a new quantifier symbol $Q_\alpha$. Let L be a formal language which does not contain $Q_\alpha$ among its symbols. Now we extend L by the quantifier $Q_\alpha$ and call the resulting language $L(Q_\alpha)$. Starting with the formulas of L the formulas of $L(Q_\alpha)$ are built by the rules of formation from L (extended to $L(Q_\alpha)$) and the new rule

If $\varphi$ is a formula of $L(Q_\alpha)$, then $(Q_\alpha x\varphi)$ is also a formula of $L(Q_\alpha)$.

The following formulas are examples of formulas of $L(Q_1)$:

$$(Q_1 v_2(v_2 = v_2)), \quad (\exists v_1(Q_1 v_2(v_2 < v_1))).$$

The satisfaction relation $\models$ for L is extended to $L(Q_\alpha)$ by the following clause:

$$\mathfrak{A} \models Q_\alpha x\varphi(x) \quad \text{iff} \quad \mathrm{card}(\{a: a \in |\mathfrak{A}| \text{ and } \mathfrak{A} \models \varphi[a]\}) \geq \omega_\alpha.$$

Regarding our above examples we see that the first formula is valid in a structure $\mathfrak{A}$ iff the domain of this structure contains at least $\omega_1$ many elements. The second sentence is for instance not valid in the structure $\langle \omega_1, < \rangle$; but it is valid in each structure $\langle \beta, < \rangle$ with $\omega_1 < \beta$.

If L is a first order language, then $Th_\alpha(\mathfrak{A})$ denotes the set of all sentences $\varphi$ of $L(Q_\alpha)$ such that $\mathfrak{A} \models \varphi$. For a class $K$ of structures we define similarly as above:

$$Th_\alpha(K) := \bigcap_{\mathfrak{A} \in K} Th_\alpha(\mathfrak{A}) \quad \text{and} \quad Th_\alpha^+(K) := Th_\alpha(K) \cup \{Q_\alpha x(x = x)\}.$$

If K is the class of models of an elementary theory T, then we shall write $Th_\alpha(T)$ instead of $Th_\alpha(K)$ sometimes.

It is important to know how $Th_\alpha(T)$ changes if $\alpha$ varies. To investigate this we introduce the following notion.

For each ordinal $\alpha$ and each sentence $\varphi$ from $L(Q_\alpha)$ $^\bullet\varphi^\bullet$ is the sentence obtained from $\varphi$ by omitting $\alpha$ in each occurrence of $Q_\alpha$. In this case we assume that Q is not a symbol of L. If S is a set of sentences of $L(Q_\alpha)$, then $^\bullet S^\bullet$ is defined to be the set $\{^\bullet\varphi^\bullet : \varphi \in S\}$.

A partial answer to the question above is given by the following theorem.

Theorem 1 [Fuhrken, 1965], [Bell, Slomson, 1969], [Slomson, 1976].
(i) For all $\alpha$, $^\bullet Th_\alpha(T)^\bullet \subseteq {}^\bullet Th_0(T)^\bullet$.
(ii) For all $\alpha > 0$ with $\omega_\alpha$ regular, $^\bullet Th_1(T)^\bullet \subseteq {}^\bullet Th_\alpha(T)^\bullet$.
(iii) (G.C.H.) For all $\alpha$ with $\omega_\alpha$ regular, $^\bullet Th_{\alpha+1}(T)^\bullet \subseteq {}^\bullet Th_1(T)^\bullet$.
(iv) (V = L) For all $\alpha$, $^\bullet Th_{\alpha+1}(T)^\bullet \subseteq {}^\bullet Th_1(T)^\bullet$.
(v) For all $\alpha$, $\beta$ with $\omega_\alpha$ a strong limit cardinal and $\omega_\beta$ singular, $^\bullet Th_\alpha(T)^\bullet \subseteq {}^\bullet Th_\beta(T)^\bullet$.

For a proof of these results the reader should consult [Fuhrken, 1965] and [Bell, Slomson, 1969]. Some interesting hints can be found in [Slomson, 1976]. From Theorem 1 it follows that some strong assumptions such as

V = L + "there are no inaccessible cardinals"

imply that for a given first order theory T there are at most three distinct theories, namely '$Th_0(T)$', '$Th_1(T)$', '$Th_\omega(T)$'. However, for a lot of interesting theories this result holds without further set-theoretical assumptions. The following theorems are also useful.

Theorem 2 [Mostowski, 1957]. For all $\alpha$ and for all sentences $\varphi$ from $L(Q_\alpha)$ and for all models $\mathfrak{A}$ of $\varphi$ with $card(|\mathfrak{A}|) > \omega_\alpha$ there exists a model $\mathfrak{B}$ of $\varphi$ with $card(|\mathfrak{B}|) = \omega_\alpha$.

Theorem 3 [Vaught, 1964], [Keisler, 1970, 1968]. If T is a recursively enumerable theory, then $Th_1(T)$ and $Th_\omega(T)$ are recursively enumerable, too.

For $Th_1(T)$ this was proved by [Vaught, 1964]. An explicite axiomatization for $L(Q_1)$ was given by [Keisler, 1970]. The axiomatizability of $L(Q_\omega)$ was proved by [Keisler, 1968]. In contrast to Theorem 3 $L(Q_0)$ is not axiomatizable (see [Mostowski, 1957]). This follows from Gödel's Incompleteness Theorem since with $Q_0$ we obtain a categorical recursive set of axioms for the standard model of arithmetic.

For some technical reasons we now define for each fixed $\alpha$ the quantifier rank qr of a formula:

$$qr : \{\varphi : \varphi \text{ is a formula of } L(Q_\alpha)\} \longrightarrow \omega;$$

$$qr(\varphi) := 0 \text{ if } \varphi \text{ is atomic},$$

$$qr((\neg\varphi)) := qr(\varphi),$$

$$qr((\varphi \wedge \psi)) := qr((\varphi \vee \psi)) := qr((\varphi \rightarrow \psi)) := qr((\varphi \leftrightarrow \psi)) := \max(\{qr(\varphi), qr(\psi)\}),$$

$$qr((\exists x\varphi)) := qr((\forall x\varphi)) := qr((Q_\alpha x\varphi)) := qr(\varphi) + 1.$$

Let $\mathfrak{A}$ and $\mathfrak{B}$ be structures for L. For each $n \in \omega$, $\alpha \in On$, and $\beta \in On$

we define:

$\mathfrak{A} \equiv_n \mathfrak{B} \Leftrightarrow_{df}$ for each sentence $\varphi$ from L with $qr(\varphi) \leq n$ it holds that $\mathfrak{A} \models \varphi$ iff $\mathfrak{B} \models \varphi$;

$\mathfrak{A} \equiv \mathfrak{B} \Leftrightarrow_{df}$ for all $k \in \omega$ it holds that $\mathfrak{A} \equiv_k \mathfrak{B}$;

$\mathfrak{A} \,{}^{\alpha}{\equiv}_n^{\beta}\, \mathfrak{B} \Leftrightarrow_{df}$ for all sentences $\varphi$ from $L(Q_\alpha)$, $\psi$ from $L(Q_\beta)$ with $qr(\varphi) \leq n$ and $'\varphi' = '\psi'$ it holds that $\mathfrak{A} \models \varphi$ iff $\mathfrak{B} \models \psi$;

$\mathfrak{A} \,{}^{\alpha}{\equiv}^{\beta}\, \mathfrak{B} \Leftrightarrow_{df}$ for all $k \in \omega$ it holds that $\mathfrak{A} \,{}^{\alpha}{\equiv}_k^{\beta}\, \mathfrak{B}$.

Let $\mathfrak{A}$ be a structure for L, T be a theory for L, n be a natural number, and $\alpha$, $\beta$ be ordinals. We define:

$typ(n,\mathfrak{A}) := \{\varphi: \varphi$ is a sentence of L with $qr(\varphi) \leq n$ and $\mathfrak{A} \models \varphi\}$;

$typ(n,\alpha,\mathfrak{A}) := \{\varphi: \varphi$ is a sentence of $L(Q_\alpha)$ with $qr(\varphi) \leq n$ and $\mathfrak{A} \models \varphi\}$.

If $\mathfrak{A}$ and $\mathfrak{B}$ are structures for L and $\mathfrak{A} \equiv_n \mathfrak{B}$, then $typ(n,\mathfrak{A}) = typ(n,\mathfrak{B})$; if $\mathfrak{A} \,{}^{\alpha}{\equiv}_n^{\beta}\, \mathfrak{B}$, then $'typ(n,\alpha,\mathfrak{A})' = 'typ(n,\beta,\mathfrak{B})'$. Furthermore, set

$typ(n,T) := \{typ(n,\mathfrak{A}): \mathfrak{A} \in Mod(T)\}$,

$typ(n,\alpha,T) := \{typ(n,\alpha,\mathfrak{A}): \mathfrak{A} \in Mod(T)$ and $card(\mathfrak{A}) \geq \omega_\alpha\}$.

In case $\alpha = \beta$, we write $\mathfrak{A} \equiv_n^{\beta} \mathfrak{B}$ and $\mathfrak{A} \equiv^{\beta} \mathfrak{B}$ instead of $\mathfrak{A} \,{}^{\alpha}{\equiv}_n^{\beta}\, \mathfrak{B}$ and $\mathfrak{A} \,{}^{\alpha}{\equiv}^{\beta}\, \mathfrak{B}$, respectively.

A very useful method to prove the elementary n-equivalence, $\mathfrak{A} \equiv_n \mathfrak{B}$, of two structures $\mathfrak{A}$ and $\mathfrak{B}$ was introduced by [Ehrenfeucht, 1961] and [Fraïssé, 1954]. Ehrenfeucht, extending the work of Fraïssé, showed that $\mathfrak{A} \equiv_n \mathfrak{B}$ can be characterised in terms of a two-person game. Similarly extended games were defined to characterise $\mathfrak{A} \,{}^{\alpha}{\equiv}_n^{\beta}\, \mathfrak{B}$. This was shown independently by [Lipner, 1970] and [Brown, 1971] in game-theoretical terms and by [Vinner, 1972] in terms of partial isomorphisms.

The game $G(n,\alpha,\beta,\mathfrak{A},\mathfrak{B})$ is a two-person game over n rounds. In each round player I begins and decides whether a Q-round or an $\exists$-round is to be played. If he decides to play a Q-round, then he chooses a subset $C \subseteq |\mathfrak{A}|$ ($C \subseteq |\mathfrak{B}|$) such that $card(C) \geq \omega_\alpha$ ($card(C) \geq \omega_\beta$, respectively). Now player II chooses a subset D of $|\mathfrak{B}|$ ($D \subseteq |\mathfrak{A}|$) such that $card(D) \geq \omega_\beta$ ($card(D) \geq \omega_\alpha$, respectively). Then player I continues by choosing an element $d \in D$, after this player II chooses an element $c \in C$. The result of the round are the elements c and d. If player I decides to play an $\exists$-round, then he chooses one of the structures $\mathfrak{A}$ and $\mathfrak{B}$, too. From the domain of the chosen structure he takes an element c. Then player II takes an element d from the domain of the other structure. The result of the round are the elements c and d. Player II wins the game if $\{(a_i, b_i): i < n\}$ is a partial isomorphism between $\mathfrak{A}$ and $\mathfrak{B}$, where $(a_i, b_i)$ is the result of the i-th round; otherwise player I wins. The statement that player II has a winning strategy in the game described above is shortened by $\mathfrak{A} \;{}^{\alpha}\!\!\sim_n^{\beta} \mathfrak{B}$. In case $\alpha = \beta$ we shall simply write $\mathfrak{A} \sim_n^{\alpha} \mathfrak{B}$.

The proof of the following useful theorem can be found in [Lipner, 1970] and [Vinner, 1972].

Theorem 4. Let L be an elementary language and T be a theory in L. Moreover, let $\mathfrak{A}$ and $\mathfrak{B}$ be models of T, $\alpha, \beta$ be ordinals, and $n \in \omega$. Then:
(i) $\mathfrak{A} \;{}^{\alpha}\!\!\sim_n^{\beta} \mathfrak{B}$ implies $\mathfrak{A} \;{}^{\alpha}\!\!\equiv_n^{\beta} \mathfrak{B}$.
(ii) If there are, up to logical equivalence with respect to T, only finitely many sentences $\varphi$ of $L(Q_\alpha)$ and of $L(Q_\beta)$ with $qr(\varphi) \leq n$, then

$$\mathfrak{A} \;{}^{\alpha}\!\!\sim_n^{\beta} \mathfrak{B} \quad \text{iff} \quad \mathfrak{A} \;{}^{\alpha}\!\!\equiv_n^{\beta} \mathfrak{B}.$$

Sometimes we need the following theorem.

Theorem 5 [Wojciechowska, 1969]. Let L be an elementary language, $n \in \omega$, $\alpha, \beta \in On$, and $\mathfrak{A}$, $\mathfrak{A}'$, $\mathfrak{B}$, $\mathfrak{B}'$ be structures for L with $\mathfrak{A} \;{}^{\alpha}\!\!\equiv_n^{\beta} \mathfrak{A}'$ and $\mathfrak{B} \;{}^{\alpha}\!\!\equiv_n^{\beta} \mathfrak{B}'$. Then $\mathfrak{A} \times \mathfrak{B} \;{}^{\alpha}\!\!\equiv_n^{\beta} \mathfrak{A}' \times \mathfrak{B}'$ holds.

This theorem generalizes a corresponding result of [Feferman, Vaught, 1959] for elementary logic to languages with generalized quantifiers $Q_\alpha$.

## § 3. Decidability

A theory is decidable if there is an effective method with the help of which it is possible to decide for T whether $\varphi \in T$ or not in a finite number of steps for every sentence $\varphi$ of the language L. In the following L need not necessarily be an elementary language.

To give this definition a mathematical sense we assume that we have an arithmetization for the language L. The method of arithmetization of the syntax of an elementary theory is due to K. Gödel, hence sometimes called "Gödelization". The method consists in adjoining a natural number $\gamma(\varphi)$ to every formula $\varphi$ of a language L, sometimes written as $\ulcorner\varphi\urcorner$ such that every natural number which is in the range of $\gamma$ is uniquely "readable" as a formula of L, i.e. has a unique grammatical structure. The definition of $\gamma$ is by induction on the number of symbols in $\varphi \in$ L (i.e. the complexity of $\varphi$) and is carried out in detail in [Feferman, 1960]. Now a theory T is said to be decidable if $\{\ulcorner\varphi\urcorner : T \models \varphi\}$ is a recursive set of natural numbers (see [Shoenfield, 1967]). T is said to be undecidable if this set is not recursive.

Some examples of undecidable theories are:

the Peano arithmetic of natural numbers [Gödel, 1931];
the elementary theories of groups, rings (Tarski in [Tarski, Mostowski, Robinson, 1953], fields [Robinson, 1949], lattices [Tarski, 1949 a], one binary relation [Kalmar, 1936];
(using CH) the monadic theory of linear orderings [Shelah, 1975].

Examples of decidable theories are:

elementary theory of Abelian groups [Szmielew, 1955],
of Boolean algebras [Tarski, 1949], [Ershov, 1964],
of real closed fields [Tarski, 1949 b],
of algebraically closed fields [Tarski, 1949 b],
of p-adic fields [Ax, Kochen, 1965/66], [Ershov, 1965];
monadic second order theory of $\langle\omega_1, <\rangle$ [Büchi, 1973];
monadic second order theory of two successor functions [Rabin, 1969].

A good survey of theories proved to be decidable or undecidable can be found in [Ershov, Lavrov, Taimanov, Taitslin, 1965].

Let L and L' be two languages and T, T' be theories in L, L', respectively. T is said to be recursive (or interpretable in T') if there exists an effective mapping g from the set of sentences of L into the set of sentences of L' such that for all sentences $\varphi$ of L:

$$T \models \varphi \quad \text{iff} \quad T' \models g(\varphi)$$

With this it is effectively meant that

$$\{(\ulcorner\varphi\urcorner, \ulcorner g(\varphi)\urcorner) : \varphi \text{ is a sentence of } L\}$$

is recursive. Obviously the following lemma holds.

<u>Lemma 6.</u> Let L, L', T, T' be as above. If T is recursive in T', then the decidability of T' implies the decidability of T.

There are some refined methods of interpretability. The reader should consult [Tarski, Mostowski, Robinson, 1953] and [Rabin, 1964].

A theory T in a language L admits elimination of quantifiers if for each formula $\varphi(\bar{x})$ of L there is a formula $\psi(\bar{x})$ of L without quantifiers such that $T \models \forall\bar{x}(\varphi(\bar{x}) \leftrightarrow \psi(\bar{x}))$. The theories of algebraically closed fields, real closed fields (see [Tarski, McKinsey, 1948]), and Abelian groups (see [Szmielew, 1955]) admit the elimination of quantifiers.

Let T be a theory in a language L. A class K of structures for L is said to be dense for T if:

$K \subseteq \mathrm{Mod}(T)$ and
for each sentence $\varphi$ of L which is valid in some model of T there exists an $\mathfrak{A} \in K$ such that $\mathfrak{A} \models \varphi$.

<u>Theorem 7</u> [Ershov, 1963]. Let T be a recursively enumerable theory in a language L. Moreover, let $\{\mathfrak{A}_i : i \in \omega\}$ be a class of structures which is dense for T and let

$$\{(i, \ulcorner\varphi\urcorner) : i \in \omega \text{ and } \varphi \text{ is a sentence of L and } \mathfrak{A}_i \models \varphi\}$$

be recursively enumerable. Then T is decidable.

Sometimes it is not necessary to assume the axiomatizability of T.

<u>Theorem 8.</u> Let T be a theory in a language L and let $\{\mathfrak{A}_i : i \in \omega\}$ be dense for T. Assume that

$$\{(i, \ulcorner\varphi\urcorner) : i \in \omega \text{ and } \varphi \text{ a sentence of L and } \mathfrak{A}_i \models \varphi\}$$

is recursively enumerable and that there is a recursive function

$g : \omega \to \omega$ such that for all sentences of L with $T \nvdash \neg\varphi$ there is a $k < g(\ulcorner\varphi\urcorner)$ with $\mathfrak{A}_k \models \varphi$. Then T is decidable.

Let us finish this paragraph with the following trivial remark.

Remark. Let T be a decidable elementary theory, $\alpha$ be an ordinal. Then $Th_\alpha(T)$ is decidable iff $Th^+_\alpha(T)$ is decidable.

# CHAPTER 2

# ELIMINATION OF GENERALIZED QUANTIFIERS

Elementary theories with additional generalized quantifiers have in general a more expressive power than elementary theories. In this chapter we study elementary theories which have eliminable cardinality quantifiers, i.e. elementary theories in which we have a definition of "infinity" and "uncountability". It turns out that some interesting theories have eliminable cardinality quantifiers, e.g. the theories of algebraically closed fields of characteristic p or 0 and the theory of real-closed fields. As a corollary we get that these theories are decidable with various additional cardinality quantifiers.

## § 1. Elimination of $Q_\alpha$

We give the definition of elimination of the quantifiers $Q_\alpha$ and then we show that the elimination can be done in a uniform way.

Definition 1. Let T be an elementary theory in L, then $Q_\alpha$ is eliminable in T iff for every formula $\varphi(y,x_1,\dots,x_n)$ of L there is a formula $\psi(x_1,\dots,x_n)$ of L such that $Th^+_\alpha(T) \models Q_\alpha y\varphi \leftrightarrow \psi$.

Definition 2. Let T be an elementary theory in L, then $Q_\alpha$ is definable in T iff for every formula $\varphi(y,x_1,\dots,x_n)$ of L there is a natural number m such that $Th^+_\alpha(T) \models Q_\alpha y\varphi \longleftrightarrow \exists y^{\geq m}\varphi$, where $\exists y^{\geq m}$ means "there are at least m many".

Lemma 1. Let T be an elementary theory in L and $\alpha$ be an ordinal. Then $Q_\alpha$ is eliminable in T iff $Q_\alpha$ is definable in T.

Proof. One direction is obvious.

Now assume that $Q_\alpha$ is eliminable and some formula $\varphi(y,x_1,\dots,x_n)$ is given. By the elimination property there is some $\psi(x_1,\dots,x_n)$ such that

$$Th^+_\alpha(T) \models Q_\alpha y\varphi(y,x_1,\dots,x_n) \longleftrightarrow \psi(x_1,\dots,x_n).$$

This implies that the following theory is inconsistent:

(1) axioms of T,

(2) $\{c_i \neq c_j: i \neq j \text{ and } i,j < \omega_\alpha\}$,

(3) $\{\varphi(c_i,d_1,\dots,d_n): i < \omega_\alpha\}$,

(4) $\neg\psi(d_1,\dots,d_n)$,

where $d_1,\dots,d_n$ and $c_1,c_2,\dots$ are new constants. By the Compactness Theorem there is a finite set of sentences from (2) - (4) which is inconsistent with T. But this means that there is some natural number m such that

$$T \models \exists y^{\geq m}\varphi(y,x_1,\dots,x_n) \longrightarrow \psi(x_1,\dots,x_n).$$

By the elimination property again we get

$$Th^+_\alpha(T) \models Q_\alpha y\varphi(y,x_1,\dots,x_n) \longleftrightarrow \exists y^{\geq m}\varphi(y,x_1,\dots,x_n),$$

i.e. $Q_\alpha$ is definable in T.

Q.E.D.

Recalling the content of the lemma we can say: If some cardinality quantifier $Q_\alpha$ is eliminable, then it is eliminable in the strong form described in Definition 2.

<u>Lemma 2.</u> Let T be an elementary theory. Then $Q_1$ is eliminable in T iff $Q_0$ is eliminable in T and $\cdot Th_0(T)\cdot = \cdot Th_1(T)\cdot$.

<u>Proof.</u> One direction is obvious.

Let $\mathfrak{A}$ be a countable model of T. For every formula $\varphi(y,x_1,\dots,x_n)$ there is a natural number m such that

$$Th^+_1(T) \models Q_1 y\varphi(y,x_1,\dots,x_n) \longleftrightarrow \exists y^{\geq m}\varphi(y,x_1,\dots,x_n).$$

Let $\mathfrak{B}$ be an uncountable elementary extension of $\mathfrak{A}$. Assume for some elements $a_1,\dots,a_n$ of $\mathfrak{A}$ that $\mathfrak{A} \models \exists y^{\geq m}\varphi(y,a_1,\dots,a_n)$. Then $\varphi(y,a_1,\dots,a_n)$ defines an uncountable set in $\mathfrak{B}$. Hence it defines an infinite set in $\mathfrak{A}$. Thus we have

$$\mathrm{Th}_0^+(T) \models Q_0y\varphi(y,x_1,\dots,x_n) \longleftrightarrow \exists y^{\geq m}\varphi(y,x_1,\dots,x_n).$$

Q.E.D.

Let $\mathfrak{A}$ be a structure and $\varphi(x,x_1,\dots,x_n)$ be a formula of L. Suppose $\varphi(x,a_1,\dots,a_n)$ defines an infinite set in $\mathfrak{A}$ for some elements $a_1,\dots,a_n$ of $\mathfrak{A}$. Assume $\mathfrak{A}$ has a proper elementary extension $\mathfrak{B}$ such that the set defined by $\varphi(x,a_1,\dots,a_n)$ remains fixed, i.e. $\varphi(x,a_1,\dots,a_n)$ defines the same set in $\mathfrak{B}$ as in $\mathfrak{A}$, $\varphi(\mathfrak{B},a_1,\dots,a_n) = \varphi(\mathfrak{A},a_1,\dots,a_n)$. In this situation $(\mathfrak{A},\mathfrak{B})$ is called a Vaughtian pair for $\varphi(x,a_1,\dots,a_n)$. The theory T is said to have Vaughtian pairs if there are a formula $\varphi$ and two models $\mathfrak{A}$ and $\mathfrak{B}$ of T such that $(\mathfrak{A},\mathfrak{B})$ is a Vaughtian pair with respect to $\varphi(x,a_1,\dots,a_n)$ for some elements $a_1,\dots,a_n$ of $\mathfrak{A}$. Surprisingly the property to have Vaughtian pairs is strongly connected with the $Q_1$-elimination property as the following theorem will show.

<u>Theorem 3.</u> Let T be an elementary theory. Then $Q_1$ is eliminable in T iff T possesses no Vaughtian pairs.

<u>Proof.</u> Assume $Q_1$ is eliminable in T. Suppose T has a Vaughtian pair, then there is a countable model $\mathfrak{A}$ of T, and an uncountable elementary extension $\mathfrak{B}$ of $\mathfrak{A}$, and a formula $\varphi(y,x_1,\dots,x_n)$ and elements $a_1,\dots,a_n$ of $\mathfrak{A}$ such that $\varphi(y,a_1,\dots,a_n)$ defines an infinite set in $\mathfrak{A}$ and every element of $\mathfrak{B}$ which satisfies $\varphi(y,a_1,\dots,a_n)$ belongs to $\mathfrak{A}$. Thus $\varphi(y,a_1,\dots,a_n)$ defines a countable set in $\mathfrak{B}$. This contradicts the $Q_1$-elimination.

Suppose now T possesses no Vaughtian pairs. We extend the elementary language L by adding a new predicate P and new constants $d_1,\dots,d_n,d$ and $c_1,c_2,\dots$ Then the following theory T' is inconsistent since every of its models would be a Vaughtian pair for T. T' consists of the following axioms:

(1) T,

(2) P defines an elementary substructure,

(3) $\forall y[\varphi(y,d_1,\ldots,d_n) \longrightarrow P(y)]$,

(4) $\{\varphi(c_i,d_1,\ldots,d_n): i < \omega\}$,

(5) $\{c_i \neq c_j: i,j < \omega$ and $i \neq j\}$,

(6) $\neg P(d)$.

By the Compactness Theorem there is a finite set of axioms which is inconsistent. Assume that such a finite inconsistent set of sentences contains at most the constants $d,d_1,\ldots,d_n$, $c_1,\ldots,c_m$. Let $\mathfrak{A}$ be a model of T and $a_1,\ldots,a_n$ be elements such that $\mathfrak{A} \models \exists y^{\geq m}\varphi(y,a_1,\ldots,a_n)$. Assume $\varphi(y,a_1,\ldots,a_n)$ defines a finite set in $\mathfrak{A}$. Let $\mathfrak{B}$ be a proper elementary extension of $\mathfrak{A}$. Then we can expand the model $\mathfrak{B}$ to a model of that part of T' which contains only the constants $d,d_1,\ldots,d_n$, $c_1,\ldots,c_m$ as follows. Interpret the predicate P by the elementary substructure $\mathfrak{A}$, let d be any element of $\mathfrak{B}$ which is not in $\mathfrak{A}$, $d_1,\ldots,d_n$ are interpreted by the elements $a_1,\ldots,a_n$, $c_1,\ldots,c_m$ are interpreted by any m elements of $\varphi(y,a_1,\ldots,a_n)$; by the hypothesis there are at least m elements in $\varphi(y,a_1,\ldots,a_n)$. By the assumption $\varphi(y,a_1,\ldots,a_n)$ defines a finite set, thus (3) is satisfied. However, this contradicts the inconsistency of that part of T'. Hence the following

$$Th_0^+(T) \models Q_0 y\varphi(y,x_1,\ldots,x_n) \longleftrightarrow \exists y^{\geq m}\varphi(y,x_1,\ldots,x_n)$$

is valid. Now let $\mathfrak{A}$ be an uncountable model of T and $a_1,\ldots,a_n$ be elements of $\mathfrak{A}$ such that $\mathfrak{A} \models \exists y^{\geq m}\varphi(y,a_1,\ldots,a_n)$. Hence by the property above $\varphi(y,a_1,\ldots,a_n)$ defines an infinite set in $\mathfrak{A}$. Let $\mathfrak{B}$ be an elementary substructure of $\mathfrak{A}$ such that $\mathfrak{B}$ contains all elements of $\mathfrak{A}$ which satisfy the formula $\varphi(y,a_1,\ldots,a_n)$, the elements $a_1,\ldots,a_n$, and which has the same cardinality as the set defined by $\varphi(y,a_1,\ldots,a_n)$ in $\mathfrak{A}$. By the Löwenheim-Skolem-Tarski-Theorem such a model exists. Assume $\mathfrak{B}$ were a proper elementary substructure, then $(\mathfrak{A},\mathfrak{B})$ would be a Vaughtian pair for T. Hence $\mathfrak{B} = \mathfrak{A}$. Thus $\varphi(y,a_1,\ldots,a_n)$ defines an uncountable subset in $\mathfrak{A}$. Hence we have proved that

$$Th_1^+(T) \models Q_1 y\varphi(y,x_1,\ldots,x_n) \longleftrightarrow \exists y^{\geq m}\varphi(y,x_1,\ldots,x_n),$$

thus $Q_1$ is definable in T.

Q.E.D.

Corollary 4. An elementary theory T is $\omega_1$-categorical iff T is $\omega$-stable and $Q_1$ is eliminable in T.

Proof. By the theorem the $Q_1$-elimination is equivalent to the property having no Vaughtian pairs. But T is $\omega_1$-categorical iff T is $\omega$-stable and has no Vaughtian pairs. This follows from the works of [M. Morley, 1965] and [J. Baldwin, and A. Lachlan, 1971].

Q.E.D.

From Theorem 1 in Chapter 1 we get the following corollary about the elimination of $Q_\alpha$.

Corollary 5. Let T be an elementary theory and $\alpha$ be an ordinal. Then the following holds:

(1) if $Q_\alpha$ is eliminable in T, then $Q_0$ is eliminable,
(2) if $Q_1$ is eliminable in T, then $Q_\alpha$ is eliminable in T,
(3) (G.C.H.) if $Q_{\alpha+1}$ is eliminable in T and $\omega_\alpha$ is regular, then $Q_1$ is eliminable,
(4) (V = L) if $Q_{\alpha+1}$ is eliminable in T, then $Q_1$ is eliminable,
(5) if $Q_\alpha$ is eliminable in T and $\omega_\alpha$ is a strong limit cardinal, then $Q_\beta$ is eliminable for all $\beta$ such that $\omega_\beta$ is singular.

(2) follows from Theorem 3 since T has no Vaughtian pairs if $Q_\alpha$ is eliminable for $\alpha > 0$.

For stable theories we get stronger results.

Theorem 6. Let T be an elementary stable theory with $Q_\alpha$ eliminable in T for some $\alpha > 0$. Then $Q_1$ is eliminable in T.

Proof. Assume $Q_1$ is not eliminable in T. By Theorem 3 T has a Vaughtian pair. We can assume that it is countable. Let $\varphi(y,a_1,\dots,a_n)$ be the unary predicate which does not increase in the proper elementary extension. By Lachlan's theorem [Lachlan, 1972] we can build an elementary chain of length $\omega_\alpha$ beginning with the Vaughtian pair such that $\varphi(y,a_1,\dots,a_n)$ does not increase in the chain. The union of the chain has cardinality at least $\omega_\alpha$, but $\varphi(y,a_1,\dots,a_n)$ defines a countable subset. This contradicts the $Q_\alpha$ elimination.

Q.E.D.

## § 2. Decidability

Since the quantifier $Q_1$ is recursively axiomatized some decidability results about T transfer to $Th_1^+(T)$. This is expressed in Theorem 3 of Chapter 1.

Theorem 7. Let T be an elementary decidable theory and $Q_1$ be eliminable in T. Then $Th_\alpha^+(T)$ is decidable for all ordinals $\alpha$.

Proof. For every $\varphi(y,x_1,\ldots,x_n)$ of L there is some natural number m such that $Q_1y\varphi(y,x_1,\ldots,x_n) \longleftrightarrow \exists y^{\geq m}\varphi(y,x_1,\ldots,x_n)$ is a theorem of $Th_1^+(T)$. This follows from the elimination property. Since $Th_1^+(T)$ is recursively enumerable, we can find m effectively. Thus for every formula $\Phi(x_1,\ldots,x_n)$ of $L(Q_1)$ we can effectively find a formula $\Psi(x_1,\ldots,x_n)$ of L such that $\Phi \longleftrightarrow \Psi$ is a theorem of $Th_1^+(T)$. But T is decidable by the hypothesis, hence $Th_1^+(T)$ is so, too. From Corollary 5 (2) it follows that for all $\alpha$, $^\bullet Th_1(T)^\bullet = {}^\bullet Th_\alpha(T)^\bullet$. Thus for all $\alpha$ $Th_\alpha^+(T)$ is decidable.

Q.E.D.

Theorem 8. The following elementary theories are decidable with any additional quantifier $Q_\alpha$:

(1) $ACF_p$, the theory of algebraically closed fields of characteristic p, p a prime or zero,

(2) RCF, the theory of real closed fields.

# CHAPTER 3

# ABELIAN GROUPS

Let AG be the class of Abelian groups. In this chapter we will describe the expressive power of the quantifier $Q_\alpha$ in connection with AG and prove the decidability of $Th_\alpha(AG)$. Contrary to the other parts of this book we will use the method of elimination of quantifiers. We extend the classical results of Szmielew. By a group we always understand an additively written Abelian group.

## § 1. The results of Szmielew

In this chapter we use p,q to denote primes. Let $\mathfrak{Z}(p^n)$ be the cyclic group of order $p^n$, $\mathfrak{Z}(p^\infty)$ be the group of the type $p^\infty$, and $\mathfrak{Z}_p$ be the additive subgroup of all rationals a/b with L.C.M.(b,p) = 1. $\oplus$ denotes the direct sum and $(\oplus\mathfrak{A})^\lambda$ the $\lambda$-fold direct sum of $\mathfrak{A}$. Let $k\mathfrak{A}$ be the subgroup $\{ka: a \in \mathfrak{A}\}$ of $\mathfrak{A}$ and $[k]\mathfrak{A}$ be the subgroup $\{a: a \in \mathfrak{A}$ and $ka = 0\}$. We use $\{0\}$ to denote the trivial group (or subgroup) containing "0" only. Let $\mathfrak{A}_t$ be the torsion subgroup of $\mathfrak{A}$ and $\mathfrak{A}_p$ be the maximal p-subgroup of $\mathfrak{A}$. By a p-group we denote a group in which the orders of the elements are powers of one and the same prime p. As well-known $\mathfrak{A}_t = \bigoplus_p \mathfrak{A}_p$. For $a \in \mathfrak{A}$ and $\mathfrak{B} \subseteq \mathfrak{A}$ let be $a + \mathfrak{B} = \{c: c = a + b,\ b \in \mathfrak{B}\}$.

Th(AG) is formulated in the elementary language L with the nonlogical signs "+", "-", "0", and "$p^n|$" for every p and n. $p^n|x$ means $\exists y(p^n y = x)$. We use ky to abbreviate the term y + ... + y, where y occurs k times. Szmielew proved that every formula of L is equivalent with respect to Th(AG) to a Boolean combination of open formulas and certain core sentences. She defined the elementary core sentences $\zeta_i(p,n,k)$ and $\zeta_4(m)$ for $1 \leq i \leq 3$ and every p,n,m,k > 0 such that:

$\mathfrak{A} \models \zeta_1(p,n,k)$ iff there is a subgroup of $\mathfrak{A}$ isomorphic to $(\oplus\mathfrak{Z}(p^n))^k$,

$\mathfrak{A} \models \zeta_2(p,n,k)$ iff $\mathfrak{A}/p^n\mathfrak{A} \models \zeta_1(p,n,k)$,

$\mathfrak{A} \models \zeta_3(p,n,k)$ iff there is a direct summand of $\mathfrak{A}$ isomorphic to $(\bigoplus \mathfrak{Z}(p^n))^k$,

$\mathfrak{A} \models \zeta_4(m)$ iff $ma = 0$ for every $a \in \mathfrak{A}$.

Furthermore define

$$\zeta_i(p,n,\mathfrak{A}) = \sup(\{k \in \omega : \mathfrak{A} \models \zeta_i(p,n,k)\} \cup \{0\}) \quad \text{for } 1 \leq i \leq 3.$$

Theorem 1 [Szmielew, 1955].
(i) Every formula of L is equivalent with respect to Th(AG) to a Boolean combination of core sentences and atomic formulas.
(ii) Every sentence $\varphi$ of L that is true in Abelian group is fulfilled in a direct sum of finitely many groups $\mathfrak{Z}(p^n)$, $\mathfrak{Z}(p^\infty)$, and $\mathfrak{Z}_p$.
(iii) AG is decidable.

The following useful lemma you can find in [Szmielew, 1955], too:

Lemma 2. $\zeta_i(p,n,\mathfrak{A}) = \zeta_i(p,m,\mathfrak{A}) + \sum_{j=n}^{m-1} \zeta_3(p,j,\mathfrak{A})$ for $m \geq n$ and $i \in \{1,2\}$.

## § 2. Pure and basic subgroups

In this paragraph we summarize some group-theoretical results we need. You can find these theorems in [Fuchs, 1966].

Theorem A (Prüfer). If $m\mathfrak{A} = \{0\}$, then $\mathfrak{A}$ is a direct sum of cyclic groups.

Definition. $\mathfrak{B}$ is a pure subgroup of $\mathfrak{A}$ iff for every n, $n\mathfrak{A} \cap \mathfrak{B} = n\mathfrak{B}$.

Corollary 3 (Szmielew). $\mathfrak{B}$ is an elementary subgroup of $\mathfrak{A}$ iff $\mathfrak{A} \equiv \mathfrak{B}$ and $\mathfrak{B}$ is pure in $\mathfrak{A}$.

Theorem B (Szele). Let $\mathfrak{B} \subseteq \mathfrak{A}$ and $n\mathfrak{B} = \{0\}$. Then $\mathfrak{B}$ is a direct summand of $\mathfrak{A}$ if and only if $\mathfrak{B}$ is a pure subgroup of $\mathfrak{A}$.

Definition. $\mathfrak{A}$ is divisible iff for every $a \in \mathfrak{A}$ and every p and n, $p^n | a$.

The basic subgroup of p-groups is one of the most fundamental notions for working with Abelian p-groups. This notion is due to Kulikov. Let $\mathfrak{A}$, $\mathfrak{B}$ be p-groups up to the end of this paragraph.

Definition. A subgroup $\mathcal{B}$ of $\mathfrak{A}$ is a basic subgroup of $\mathfrak{A}$ iff
(i) $\mathcal{B}$ is a direct sum of cyclic groups,
(ii) $\mathcal{B}$ is pure in $\mathfrak{A}$,
(iii) $\mathfrak{A}/\mathcal{B}$ is divisible.

Theorem C (Kulikov). Every p-group $\mathfrak{A}$ contains a basic subgroup.

Theorem D (Baer). Assume that $\mathcal{B}$ is a subgroup of the p-group $\mathfrak{A}$ and $\mathcal{B} = \bigoplus_{n<\omega} \mathcal{B}_n$, where $\mathcal{B}_n$ is a direct sum of cyclic groups of order $p^n$.
Then $\mathcal{B}$ is a basic subgroup of $\mathfrak{A}$ if and only if
$\mathfrak{A} = \mathcal{B}_1 \oplus \ldots \oplus \mathcal{B}_n \oplus \{\mathcal{B}_n^*, p^n\mathfrak{A}\}$ for every n, where $\mathcal{B}_n^* = \mathcal{B}_{n+1} \oplus \mathcal{B}_{n+2} \oplus \ldots$,
and $\{\mathcal{B}_n^*, p^n\mathfrak{A}\}$ is the smallest subgroup of $\mathfrak{A}$ containing $\mathcal{B}_n^*$ and $p^n\mathfrak{A}$.

## § 3. Elimination of quantifiers

Lemma 4. Let $\mathcal{B}$ be a subgroup of $\mathfrak{A}$ and a,b be elements of $\mathfrak{A}$. R(x) should be a relation either of the form $k(a + x) \neq b$ or $n/(k(a + x) + b)$. Assume $\mathfrak{A} \models R(0)$. If $\text{card}(\{x \in \mathcal{B} : \mathfrak{A} \models \neg R(x)\}) \geq \omega_\alpha$, then there exists a subgroup $\mathcal{L}$ of $\mathcal{B}$ such that $\text{card}(\mathcal{L}) \geq \omega_\alpha$ and $\mathfrak{A} \models R(c)$ for every $c \in \mathcal{L}$.

Proof. Take $\mathcal{L}$ to be the subgroup generated by the differences of the elements of $\mathcal{B}$ fulfilling $\neg R(x)$.

Q.E.D.

Lemma 5. If $\text{card}(\mathfrak{A}) \geq \omega_\alpha$ and $\text{card}(k\mathfrak{A}) < \omega_\alpha$, then there exists some $p|k$ with $\text{card}([p]\mathfrak{A}) = \text{card}(\mathfrak{A}) \geq \omega_\alpha$.

Proof. For every $a \in k\mathfrak{A}$ let V(a) be $\{b \in \mathcal{B} : kb = a\}$ and $b_a$ be an element with $kb_a = a$. Then $\mathfrak{A} = \dot{\bigcup}_{a \in k\mathfrak{A}} V(a)$, where we use $\dot{\cup}$ to denote the disjoint union. Furthermore, $V(a) = \{b_a + c: c \in [k]\mathfrak{A}\}$. By Theorem A you get $\text{card}(V(a)) = \max_{p|k} \{\text{card}([p]\mathfrak{A})\}$ or $\text{card}(V(a)) < \omega$. Therefore there exists a prime p with $p|k$ and $\text{card}(\mathfrak{A}) = \text{card}(k\mathfrak{A})\text{card}([p]\mathfrak{A})$.
This implies $\text{card}([p]\mathfrak{A}) = \text{card}(\mathfrak{A}) \geq \omega_\alpha$.

Q.E.D.

Lemma 6. If $\mathfrak{A}$ is a p-group and $\text{card}(\mathfrak{A}) > \omega$, then $\text{card}(\mathfrak{A}) = \text{card}([p]\mathfrak{A})$.

Proof. Since $\mathfrak{A}$ is a p-group, $\mathfrak{A} = \bigcup_{0<n<\omega} [p^n]\mathfrak{A}$. By Theorem A $\text{card}([p^n]\mathfrak{A}) \leq \text{card}([p]\mathfrak{A}) + \omega$. $\text{card}(\mathfrak{A}) > \omega$ implies the assertion.

Q.E.D.

To extend the Szmielew elimination procedure we add the new core sentences $Q_\alpha x(px = 0 \wedge p^{n-1}|x)$ and $Q_\alpha x(n|x)$ for every p and $n \geq 1$. If $\varphi(\bar{x},x)$ is a conjunction of atomatic formulas and negated atomic formulas, then we use $\varphi^+(\bar{x},x)$ to denote the conjunction of all atomic formulas in $\varphi(\bar{x},x)$ which are not negated. Lemma 4 implies

Corollary 7. For every conjunction $\varphi(\bar{x},x)$ of atomatic formulas and negated atomatic formulas $Th_\alpha(AG) \models Q_\alpha x\varphi(\bar{x},x) \leftrightarrow \exists x\varphi(\bar{x},x) \wedge Q_\alpha x\varphi^+(\bar{x},x)$.

Theorem 8. For every formula $\gamma(\bar{x})$ of $L(Q_\alpha)$ you can effectively construct a Boolean combination $\chi(\bar{x})$ of core sentences and atomatic formulas such that $\gamma$ and $\chi$ are equivalent with respect to $Th_\alpha(AG)$.

Proof. It is sufficient to eliminate "$Q_\alpha x$" for formulas $Q_\alpha x\varphi(\bar{x},x)$, where $\varphi(\bar{x},x)$ is a Boolean combination of core sentences and atomatic formulas. The case "$\exists x$" will be handled using the Szmielew method for the elementary case (Theorem 1). Since we have in general $\models Q_\alpha x \bigvee_i \varphi_i(\bar{x},x) \longleftrightarrow \bigvee_i Q_\alpha x\varphi_i(\bar{x},x)$ let us confine to a conjunction $\varphi(\bar{x},x)$ of unnegated and negated atomic formulas and core sentences:

$$\varphi(\bar{x},x) = \bigwedge_i k_i x = t_i(\bar{x}) \wedge \bigwedge_i k_i'x \neq t_i'(\bar{x}) \wedge \bigwedge_i q_i^{r_i}|(f_i(\bar{x}) + g_i x) \wedge \bigwedge_i q_i'^{s_i}\nmid(f_i'(\bar{x}) + g_i'x) \wedge \varphi'(\bar{x}),$$

where x does not occur free in $\varphi'$, $t_i(\bar{x})$, $t_i'(\bar{x})$, $f_i(\bar{x})$, and $f_i'(\bar{x})$ are terms in "0", "+", and "−" over the variables of $\bar{x}$, $k_i,k_i',g_i,g_i',r_i,s_i$ are natural numbers different from 0, and $q_i,q_i'$ are primes. It is possible that some of the conjunctions in $\varphi(\bar{x},x)$ are empty. Finally we can cancel $\varphi'$.

First suppose that the condition $\bigwedge_i k_i x = t_i(\bar{x})$ is not empty. Let k be the greatest common measure of the $k_i$. Then

(1) $Th(AG) \models \bigwedge_i k_i x{=}t_i(\bar{x}) \wedge \bigwedge_i k_i y{=}t_i(\bar{x}) \longleftrightarrow \bigwedge_i k_i x{=}t_i(\bar{x}) \wedge k(x{-}y){=}0.$

If $k = 1$, by (1) $Q_\alpha x\varphi(\bar{x},x)$ is false in every Abelian group. Assume $k > 1$ and let k be $p_1^{v_1}p_2^{v_2}\dots p_n^{v_n}$, where the $p_i$ are different primes. Define $w_i = \max(\{r_j: p_i = q_j \text{ and } q_j\nmid g_j\} \cup \{0\})$. Then

(2) $Th_\alpha(AG) \models Q_\alpha x\varphi(\bar{x},x) \longleftrightarrow \exists x\varphi(\bar{x},x) \wedge (\bigvee_{1\leq i\leq n} Q_\alpha x(p_i x = 0 \wedge p_i^{w_i}|x)).$

The implication from the left to the right follows from (1). To show the other direction we assume w.l.o.g.

$$\mathfrak{A} \models \varphi(\bar{a},a) \wedge Q_\alpha x(p_1 x = 0 \wedge p_1^{w_1} | x).$$

Then one has $\mathfrak{A} \models \varphi^+(\bar{a},a + c)$ for every $c \in [p_1](p_1^{w_1}\mathfrak{A})$. The assertion follows from Corollary 7.

Now assume that the condition $\bigwedge_i k_i x = t_i(x)$ is empty. If $\varphi^+(\bar{x},x)$ is completely empty, Corollary 7 implies

(3) $\mathrm{Th}_\alpha(\mathrm{AG}) \models Q_\alpha x\varphi(\bar{x},x) \longleftrightarrow \exists x\varphi(\bar{x},x) \wedge Q_\alpha x(1|x)$.

Otherwise let k' be the lowest common multiple of all $q_i^{r_i}$ with $q_i \nmid g_i$ in $\varphi(\bar{x},x)$ and let k be the lowest common multiple of all $q_i^{r_i}$ in $\varphi(\bar{x},x)$. Then

(4) $\mathrm{Th}_\alpha(\mathrm{AG}) \models Q_\alpha x\varphi(\bar{x},x) \longleftrightarrow \exists x\varphi(\bar{x},x) \wedge Q_\alpha x(k'|x)$.

To prove the implication from the left to the right use that the difference of two elements fulfilling $\varphi(\bar{x},x)$ is divisible by k'. To show the other direction assume $\mathfrak{A} \models \varphi(\bar{a},a) \wedge Q_\alpha x(k'|x)$. If $\mathrm{card}(k\mathfrak{A}) \geq \omega_\alpha$, then Corollary 7 implies (4). If $\mathrm{card}(k\mathfrak{A}) < \omega_\alpha$, apply Lemma 5 to the group $k'\mathfrak{A}$. Then you get some p such that $p|(k/k')$ and

$$\mathrm{card}([p](k'\mathfrak{A})) = \mathrm{card}(k'\mathfrak{A}) \geq \omega_\alpha.$$

This implies $\mathfrak{A} \models \varphi^+(\bar{a},a + c)$ for every $c \in [p](k'\mathfrak{A})$. You get (4) from Corollary 7.

Q.E.D.

## § 4. The new core sentences

The next lemma gives an algebraic characterisation of the new core sentences $Q_\alpha x(n|x)$.

Lemma 9. $\mathfrak{A} \models \neg Q_\alpha x(n|x)$ iff there are subgroups $\mathfrak{B}$ and $\mathfrak{C}$ such that $\mathfrak{A} = \mathfrak{B} \oplus \mathfrak{C}$, $\mathrm{card}(\mathfrak{B}) < \omega_\alpha$, and $n\mathfrak{C} = 0$.

<u>**Proof.**</u> First assume $\mathfrak{A}$ has the desired form $\mathfrak{B} \oplus \mathfrak{L}$ with $\mathrm{card}(\mathfrak{B}) < \omega_\alpha$ and $n\mathfrak{L} = 0$. $\mathfrak{A} \models Q_\alpha x(n|x)$ would imply the existence of $\omega_\alpha$-many $c_\gamma \in \mathfrak{L}$ and some $b \in \mathfrak{B}$ such that $n|(b + c_\gamma)$. But then it would follow that $n|(c_{\gamma_1} - c_{\gamma_2})$ for every pair $c_{\gamma_1}, c_{\gamma_2}$ of such $c_\gamma$, and this is a contradiction.

Now suppose $\mathfrak{A} \models \neg Q_\alpha x(n|x)$. Using Zorn's Lemma you get some maximal direct summand $\mathfrak{L}$ of $\mathfrak{A}$ with $n\mathfrak{L} = 0$. To prove the chain condition apply that a group $\vartheta$ with $n\vartheta = 0$ is a direct summand iff it is pure (Theorem B, Szele). Let $\mathfrak{A} = \mathfrak{B} \oplus \mathfrak{L}$. We show that $\mathrm{card}(\mathfrak{B}) < \omega_\alpha$. $\mathfrak{A} \models \neg Q_\alpha x(n|x)$ implies

(5) $\mathrm{card}(\mathfrak{A}/\mathfrak{A}_t) < \omega_\alpha$.

Furthermore it is $\mathfrak{A}_t = \mathfrak{L} \oplus (\mathfrak{A}_t \cap \mathfrak{B})$. By the First Isomorphism Theorem it follows that

(6) $\mathfrak{A}/\mathfrak{A}_t \cong \mathfrak{B}/\mathfrak{A}_t \cap \mathfrak{B}$.

If now $\mathrm{card}(\mathfrak{B}) \geq \omega_\alpha$, it follows from (5) and (6) that $\mathrm{card}(\mathfrak{A}_t \cap \mathfrak{B}) \geq \omega_\alpha$. Let $\vartheta$ be $\mathfrak{A}_t \cap \mathfrak{B}$. Then $\mathfrak{A} \models \neg Q_\alpha x(n|x)$ implies $\mathrm{card}(n\vartheta) < \omega_\alpha$. By Lemma 5 there is some $p|n$ such that $\mathrm{card}([p]\vartheta) \geq \omega_\alpha$. We get the desired contradiction proving $\mathfrak{A} \models n|b$ for every $b \in [p]\vartheta$. Otherwise there is some $b \in [p]\vartheta$ such that $\mathfrak{A} \models n\nmid b$. Then there exist some $j$ and some $b_1 \in \mathfrak{A}$ such that $p^j b_1 = b$, where $j$ is chosen maximal. $p^j|n$ follows. Let $\{b_1\}$ be the subgroup generated by $b_1$. The maximality of $j$ implies that $\{b_1\}$ is a pure subgroup of $\mathfrak{A}$. By Theorem B (Szele) $\{b_1\}$ is a direct summand of $\mathfrak{A}$. Since $\mathfrak{L} \cap \{b_1\} = \{0\}$ and $nb_1 = 0$, this contradicts the maximality of $\mathfrak{L}$.

Q.E.D.

To work with the core sentences $Q_\alpha x(px = 0 \wedge p^n|x)$ we need

<u>Lemma 10.</u> Assume $n < m$ and

$$\mathfrak{A} \models \neg Q_\alpha x(px = 0 \wedge p^m|x) \wedge \bigwedge_{n<i\leq m} \neg \zeta_3(p,i,k_i)$$

for some $k_i$. Then

$$\mathfrak{A} \models \neg Q_\alpha x(px = 0 \wedge p^n|x).$$

Proof. Assume $\mathfrak{A} \models Q_\alpha x(px = 0 \wedge p^n | x)$. Then there exists some representation $\mathfrak{A} \cong \bigoplus_{n<j\leq m} (\bigoplus \mathfrak{Z}(p^j))^{\lambda_j} \oplus \mathfrak{B}$, where every $\lambda_j$ is chosen maximal. By the premise of the lemma, $\lambda_j < k_j$. Then $\mathfrak{B} \models Q_\alpha x(px = 0 \wedge p^n | x)$. You get some subgroup $\mathfrak{C} \subseteq \mathfrak{B}$ such that $\mathfrak{C} = \bigoplus_{\gamma<\omega_\alpha} \mathfrak{C}_\gamma$ and $\mathfrak{C}_\gamma \cong \mathfrak{Z}(p^{n+1})$. Now we can prove the existence of some $c \in \mathfrak{C}_\gamma$ with $c \neq 0$ and $p^m | c$ for every $\gamma < \omega_\alpha$. This is the desired contradiction. Otherwise there is some $\gamma$ and some $c \in \mathfrak{C}_\gamma$ such that $c \neq 0$, $pc = 0$, $p^n | c$, and $p^m \nmid c$. Let $k$ be maximal such that $p^k | c$. Assume $p^k c_1 = c$. Then $\{c_1\}$ is pure and by Theorem B, $\{c_1\}$ is a direct summand of $\mathfrak{B}$ isomorphic to $\mathfrak{Z}(p^{k+1})$, where $n < k + 1 \leq m$. This contradicts the maximality of $\lambda_{k+1}$.

Q.E.D.

## § 5. $Th_\alpha(AG)$ is decidable

Let $K_\alpha$ be the class of all groups

$$\bigoplus_{p|k} \Big( \bigoplus_{p^n|k} (\bigoplus \mathfrak{Z}(p^n))^{\lambda(p,n)} \oplus (\bigoplus \mathfrak{Z}(p^\infty))^{\lambda(p,\omega)} \oplus (\bigoplus \mathfrak{Z}_p)^{\varkappa(p)} \Big),$$

where $k$ is a natural number, $\varkappa(p) < \omega$, $\lambda(p,\mu) < \omega$ or $\lambda(p,\mu) = \omega_\alpha$, and for every $p | k$ there exists at most one $\mu$ with $\lambda(p,\mu) = \omega_\alpha$.

Theorem 11.

(i) $Th_\alpha(AG)$ is decidable.

(ii) If $\varphi \in L(Q_\alpha)$ and $\mathfrak{A} \models \varphi$ for some Abelian group $\mathfrak{A}$, then $\mathfrak{A}' \models \varphi$ for some $\mathfrak{A}' \in K_\alpha$.

(iii) $\text{'}Th_\alpha(AG)\text{'} = \text{'}Th_\beta(AG)\text{'} \neq \text{'}Th_0(AG)\text{'}$ for $\alpha, \beta > 0$.

Proof. By a model of a set $\Sigma$ of sentences of $L(Q_\alpha)$ an Abelian group $\mathfrak{A}$ with $\mathfrak{A} \models \Sigma$ is meant. According to Theorem 8 it is sufficient to decide for every finite set $\Sigma$ of negated and unnegated core sentences whether there is a model of $\Sigma$ or not. If we get a positive answer, we construct some model $\mathfrak{A} \in K_\alpha$. $\text{'}Th_\alpha(AG)\text{'} = \text{'}Th_\beta(AG)\text{'}$ for $\alpha, \beta > 0$ follows because our decision procedure depends on $\alpha > 0$ and $\alpha = 0$ only. Let be

$$\psi^\alpha = \neg Q_\alpha x(q|x) \wedge \exists x(px = 0 \wedge x \neq 0) \wedge \forall x(p|x),$$

where $p$ and $q$ are different primes. Lemma 9 implies $Th_0(AG) \models \neg\psi^0$. But if $\alpha > 0$, then $\mathfrak{Z}(p^\infty) \oplus (\bigoplus \mathfrak{Z}(q))^{\omega_\alpha}$ is a model of $\{\psi^\alpha\}$. Therefore $\text{'}Th_\alpha(AG)\text{'} \neq \text{'}Th_0(AG)\text{'}$ for $\alpha > 0$.

Now let $\Sigma$ be a finite set of core sentences and negated core sentences. Let $\Sigma_p$ be the subset of all $\varphi \in \Sigma$ and $\neg\varphi \in \Sigma$, where $\varphi$ is of the form $\zeta_i(p,n,k)$ for some i or $Q_\alpha x(px = 0 \wedge p^n|x)$. Furthermore define $R(\Sigma) = \{p: \Sigma_p \neq 0\}$. The following condition is necessary for the existence of a model of $\Sigma$:

(1) $\Sigma \cap L$ has a model.

You can decide (1) by Theorem 1(iii).

Define that $\varphi$ is of the form $\neg(p,m)$ iff $\varphi = (\neg Q_\alpha x(px = 0 \wedge p^m|x))$, or $\varphi = (\neg\zeta_1(p,m+1,k))$, or $\varphi = (\neg Q_\alpha x(d|x))$, or $\varphi = (\zeta_4(d))$, where $p^m|d$ and $p^{m+1}\nmid d$. Lemma 10 implies that the following condition is necessary:

(2) If $Q_\alpha x(px = 0 \wedge p^n|x) \in \Sigma$ and there is some $\varphi$ in $\Sigma$ of the form $\neg(p,m)$, then there exists some j such that $h < j$ for every $Q_\alpha x(px = 0 \wedge p^h|x) \in \Sigma$, $j \leq m$ for every sentence of the form $\neg(p,m)$ in $\Sigma$, and neither $\neg\zeta_3(p,j,k)$ nor $\neg\zeta_2(p,i,k)$ for any k and any $i \leq j$ are in $\Sigma$.

By Lemma 9 you get $\zeta_1(p,n,\mathfrak{A}) = \zeta_2(p,n,\mathfrak{A})$ for every $\mathfrak{A}$ with $\mathfrak{A} \models \neg Q_0 x(k|x)$. Therefore we can assume w.l.o.g.:

(3) If $\alpha = 0$ and $\neg Q_0 x(k|x) \in \Sigma$, then $\zeta_i(p,n,r) \in \Sigma$ implies $\zeta_{3-i}(p,n,r) \in \Sigma$ and $\neg\zeta_i(p,n,r) \in \Sigma$ implies $\neg\zeta_{3-i}(p,n,r)$ for $i \in \{1,2\}$.

From Lemma 9 it follows further that

$$Th_\alpha(AG) \models \neg Q_\alpha(k|x) \longrightarrow (Q_\alpha x(m|x) \longleftrightarrow \bigvee_{p^n|k,\ p^n\nmid m} Q_\alpha x(px=0 \wedge p^{n-1}|x)).$$

Considering a finite set of such sets of unnegated and negated core sentences instead of $\Sigma$ we can assume w.l.o.g.:

(4) If $\neg Q_\alpha x(k|x) \in \Sigma$, then there is no element $\neg Q_\alpha x(m|x)$ with $m \neq k$ or $Q_\alpha x(r|x)$ in $\Sigma$.

By Theorem A you can assume w.l.o.g.:

(5) If $\zeta_4(k) \in \Sigma$, then there is no element $\zeta_4(m)$ with $m \neq k$ or $\neg\zeta_4(r)$ or $Q_\alpha x(r|x)$ or $\neg Q_\alpha x(r|x)$ in $\Sigma$.

Now it is sufficient to show that every finite set $\Sigma$ of core sentences and negated core sentences fulfilling (1) - (5) has a model in $K_\alpha$. By (1) and Theorem 1(ii) there is some direct sum $\mathfrak{A}$ of finitely many $\mathfrak{Z}(p^n)$, $\mathfrak{Z}(p^\omega)$, and $\mathfrak{Z}_p$ such that $\mathfrak{A} \models \Sigma \cap L$.

If $\alpha = 0$ and $\neg Q_0 x(k|x) \in \Sigma$ for some k, you can choose $\mathfrak{A}$ finite by (3). If not $Q_\alpha x(px = 0 \wedge p^n|x) \in \Sigma$, then $\mathfrak{A} \models \Sigma_p$. Otherwise you have $\mathfrak{A} \oplus (\bigoplus \mathfrak{Z}(p^\omega))^{\omega_\alpha} \models \Sigma_p$ if there is not an element of the form $\neg(p,m)$ in $\Sigma$, and $\mathfrak{A} \oplus (\bigoplus \mathfrak{Z}(p^j))^{\omega_\alpha} \models \Sigma_p$ if there is an element of the form $\neg(p,m)$ in $\Sigma$ and j is chosen according to (2). Applying these considerations to every $p \in R(\Sigma)$ you get some $\mathfrak{A}' \in K_\alpha$ such that $\mathfrak{A}' \models (\Sigma \cap L) \cup \bigcup_{p \in R(\Sigma)} \Sigma_p$, where $\mathfrak{A}'$ is a direct sum of cyclic groups if $\alpha = 0$ and $\neg Q_0 x(k|x) \in \Sigma$ for some k. If $\zeta_4(k) \in \Sigma$ or $\neg Q_\alpha x(k|x)$, then $\mathfrak{A}' \models \Sigma$ by (4) and (5) and by construction of $\mathfrak{A}'$. (Do not forget the application of (3) above.) Otherwise the sentences of $\Sigma$ not considered up to now are of the form $Q_\alpha x(k|x)$. It is sufficient to choose some prime q such that $q \notin R(\Sigma)$ and $q \nmid k$ for every $Q_\alpha x(k|x) \in \Sigma$. Then $\mathfrak{A}' \oplus (\bigoplus \mathfrak{Z}(q^1))^{\omega_\alpha}$ is a model of $\Sigma$ from $K_\alpha$.

Q.E.D.

## § 6. Abelian p-groups and the quantifier $Q_\alpha$

Let AGp be the class of Abelian p-groups. Contrary to AG this is not an elementary class in the wider sense. Using Theorem 8 we want to prove the decidability of $Th_\alpha(AGp)$.

Theorem 12.
(i) $Th_\alpha(AGp)$ is decidable.
(ii) If $\varphi \in L(Q_\alpha)$ and $\mathfrak{A} \models \varphi$ for some p-group $\mathfrak{A}$, then there is some p-group $\mathfrak{A}'$ in $K_\alpha$ such that $\mathfrak{A}' \models \varphi$.
(iii) $`Th_\alpha(AGp)` = `Th_\beta(AGp)` \neq `Th_0(AGp)`$ for $\alpha, \beta > 0$.

Proof. We describe a procedure similar to that in the proof of Theorem 11. By a p-model of a set $\Sigma$ of sentences of $L(Q_\alpha)$ an Abelian p-group $\mathfrak{A}$ with $\mathfrak{A} \models \Sigma$ is meant. Using Theorem 8 we have to decide whether a finite set $\Sigma$ of negated and unnegated core sentences of $L(Q_\alpha)$ has a p-model or not. If such a p-model exist, we construct one of $K_\alpha$. (iii) is proved similarly as above.

W.l.o.g. you can suppose that the core sentences occuring in $\Sigma$ are of the form $\zeta_i(p,n,k)$ $(1 \leq i \leq 3)$, $Q_\alpha x(px = 0 \wedge p^n|x)$, and $Q_\alpha x(p^n|x)$ only. We call them the core sentences of $Th_\alpha(AGp)$. Furthermore, it is possible to assume that $\Sigma$ fulfils the conditions (1) - (4) in the proof of Theorem 11, where in (1) "model" is replaced by "p-model". If $\alpha > 0$, Lemma 6 implies

$$Th_\alpha(AGp) \models Q_\alpha x(p^n|x) \longleftrightarrow Q_\alpha x(px = 0 \wedge p^n|x).$$

In this case let us replace all subformulas $Q_\alpha x(p^n|x)$ by $Q_\alpha x(px = 0 \wedge p^n|x)$ and assume:

(5) If $\alpha > 0$, then $Q_\alpha x(p^n|x)$ and $\neg Q_\alpha x(p^n|x)$ are not in $\Sigma$.

Further we can confine us to sets $\Sigma$ such that

(6) If $\alpha = 0$, $Q_0 x(p^m|x) \in \Sigma$, and $\neg\zeta_1(p,n,k) \in \Sigma$, then $m \geq n$ and there exists some $j < k$ such that $\zeta_1(p,n,j) \in \Sigma$ and $\neg\zeta_2(p,n,j) \in \Sigma$.

Let $\mathfrak{A}$ be a p-model of $\Sigma$. By Theorems C and D, $\mathfrak{A} \models \neg\zeta_1(p,n,k)$ implies

(*) $\mathfrak{A} \xleftrightarrow{\simeq} \bigoplus_{i<\omega} (\bigoplus \mathfrak{Z}(p^i))^{\lambda(i)} \oplus (\bigoplus \mathfrak{Z}(p^\infty))^{\lambda(\omega)}$, where $\sum_{n \leq i < \infty} \lambda(i) + \lambda(\omega) < \omega$.

You can suppose $m \geq n$, because otherwise you can consider

$$\Sigma_1 = (\Sigma \setminus \{Q_0 x(p^m|x)\}) \cup \{Q_0 x(px = 0 \wedge p^m|x)$$

and

$$\Sigma_2 = (\Sigma \setminus \{Q_0 x(p^m|x)\}) \cup \{\neg Q_0 x(px = 0 \wedge p^m|x), Q_0 x(p^n|x)\}$$

instead of $\Sigma$. By $m \geq n$ and $\mathfrak{A} \models Q_0 x(p^m|x)$ you get $\lambda(\omega) > 0$ in (*) and therefore $\zeta_2(p,n,\mathfrak{A}) < \zeta_1(p,n,\mathfrak{A})$. You obtain (6) replacing by all

$$\Sigma_j = \Sigma \cup \{\neg\zeta_2(p,n,j), \zeta_1(p,n,j)\} \quad \text{for } j < k.$$

Using tautologies of $Th_\alpha(AG)$ we finally suppose

(7) For $Q_\alpha x(p^n|x)$, $\neg Q_\alpha x(p^n|x)$, $Q_\alpha x(px = 0 \wedge p^n|x)$, and $\neg Q_\alpha x(px = 0 \wedge p^n|x)$ there is in each case at most one formula in $\Sigma$.

As shown above it is sufficient to construct a p-model in $K_\alpha^\bullet$ for every $\Sigma$ fulfilling (1) - (7). By (1) there is some p-model of $\Sigma \cap L$. Using Theorem 1 it is not difficult to prove that there is a direct sum $\mathfrak{A}$ of finitely many groups $\mathfrak{Z}(p^n)$, $\mathfrak{Z}(p^\infty)$ such that $\mathfrak{A} \models \Sigma \cap L$.

<u>1. Case.</u> $\neg Q_\alpha x(p^m|x) \in \Sigma$. By (5), $\alpha = 0$. Assume $\mathfrak{A} \cong ((+)\,\mathfrak{Z}(p^\infty))^k \oplus \mathfrak{A}'$, where $\mathfrak{A}'$ does not contain any non-trivial divisible subgroup. Then, for sufficiently large n by (3), you have $((+)\,\mathfrak{Z}(p^n))^k \oplus \mathfrak{A}' \models \Sigma \cap L$. If $Q_0x(px = 0 \wedge p^n|x)$ is not in $\Sigma$ for every n, (4) implies $((+)\,(p^n))^k \oplus \mathfrak{A}' \models \Sigma$. Otherwise take the natural number j that exists by (2). By (2) and (4)

$$((+)\,\mathfrak{Z}(p^n))^k \oplus \mathfrak{A}' \oplus ((+)\,\mathfrak{Z}(p^j))^\omega \models \Sigma.$$

<u>2. Case.</u> Neither $\neg Q_\alpha x(p^m|x)$ nor $Q_\alpha x(p^m|x)$ is in $\Sigma$. If $Q_\alpha x(px = 0 \wedge p^n|x) \notin \Sigma$ for every n, then $\mathfrak{A} \models \Sigma$. In the case $Q_\alpha x(px = 0 \wedge p^n|x) \in \Sigma$ apply (2) as above if the premise of (2) is fulfilled. Otherwise $\mathfrak{A} \oplus ((+)\,\mathfrak{Z}(p^\infty))^{\omega_\alpha}$ is a p-model of $\Sigma$ in $K_\alpha$.

<u>3. Case.</u> $\neg Q_\alpha x(p^m|x) \notin \Sigma$ for every m, but $Q_\alpha x(p^k|x) \in \Sigma$ for some k. By (5) you have $\alpha = 0$. By (7) and the second case there is a p-model $\mathfrak{B}$ of $\Sigma \setminus \{Q_0x(p^k|x)\}$ in $K_\alpha$. If $\neg\zeta_1(p,n,r) \notin \Sigma$ for every n and r, $\mathfrak{B} \oplus \mathfrak{Z}(p^\omega)$ is a p-model of $\Sigma$ in $K_\alpha$. Otherwise (6) implies the existence of some n, r, and j such that $n \leq k$, $j < r$, $\neg\zeta_1(p,n,r) \in \Sigma$, $\zeta_1(p,n,j) \in \Sigma$, and $\neg\zeta_2(p,n,j) \in \Sigma$. Then there is some direct summand in $\mathfrak{B}$ isomorphic to $\mathfrak{Z}(p^\omega)$ (Theorems C and D). It follows that $\mathfrak{B} \models \Sigma$.

Q.E.D.

# CHAPTER 4

# GRAPHS AND FORESTS

As a graph we denote a relational structure with one binary, irreflexive and symmetric relation. In the first paragraph of this chapter we give a generalization of the so-called sphere-theorem of Hanf. This result is applied in the second paragraph to prove the decidability of the $Q_1$-theory of forests.

## § 1. Graphs, shrinking types and n-equivalence for $L(Q_1)$

In [Hanf, 1965] the notion of n-spheres was introduced and a sufficient condition for the elementary equivalence of two graphs was given. This result found many applications, especially in decidability theory. Hanf restricted his attention to graphs of finite valency. In [Korec, Rautenberg, 1976] a criterion for the n-equivalence of two graphs was given, omitting the restriction of finite valency. We shall give here a somewhat different criterion for the $(n,Q_1)$-equivalence of graphs. This result is a direct generalization from [Seese, Tuschik, 1977]. Before we can state this result we have to give some definitions.

Let $\mathfrak{A} = \langle A,R,\ldots\rangle$ be a relational structure with one binary relation R. We shall say that $\mathfrak{A}$ is a graph if there are no relations except R. In case that there are only unary relations beside R, $\mathfrak{A}$ is called a coloured graph. In the following definitions we shall restrict our attention to graphs. But it is obvious that they can also be given for any relational structure with a designated binary relation.

A finite sequence $\bar{a}$ of elements of $|\mathfrak{A}|$ is said to be a path in $\mathfrak{A}$ iff

(1) for all $i,j < lh(\bar{a})$, $\bar{a}(i) = \bar{a}(j)$ implies $i = j$,

and

(2) for all $i < lh(\bar{a}) - 1$, $\langle\bar{a}(i),\bar{a}(i+1)\rangle \in R$ or $\langle\bar{a}(i+1),\bar{a}(i)\rangle \in R$.

The length of the path $\bar{a}$ is $lh(\bar{a}) - 1$. The points $\bar{a}(0)$ and $\bar{a}(lh(\bar{a})-1)$ are the first and the last point of this path, respectively. Two points a and b are said to be connected in $\mathfrak{A}$ if there is a path in $\mathfrak{A}$ with first point a and last point b. $\mathfrak{A}$ is said to be connected iff any two points of $\mathfrak{A}$ are connected. A substructure $\mathfrak{B} \subseteq \mathfrak{A}$ is a component of $\mathfrak{A}$ iff $\mathfrak{B}$ is a maximal connected substructure of $\mathfrak{A}$. Suppose a and b are connected in $\mathfrak{A}$, then the distance of a and b, denoted by $d_{\mathfrak{A}}(a,b)$, is defined by

$$d_{\mathfrak{A}}(a,b) := \min(\{lh(\bar{a}): \bar{a} \text{ is a path in } \mathfrak{A} \text{ with first point a and last point b}\}).$$

If a and b are not connected in $\mathfrak{A}$, we define

$$d_{\mathfrak{A}}(a,b) := \infty.$$

From the definitions we immediately get

(1) $d_{\mathfrak{A}}(b,a) = d_{\mathfrak{A}}(a,b)$,

(2) $d_{\mathfrak{A}}(a,b) = 0$ iff $a = b$.

For nonempty sets $C, D \subseteq |\mathfrak{A}|$ we define

$$d_{\mathfrak{A}}(C,D) := \min(\{d_{\mathfrak{A}}(a,b): a \in C \text{ and } b \in D\}).$$

We define $d_{\mathfrak{A}}(C,D) := \infty$ if for every $a \in C$ and $b \in D$, $d_{\mathfrak{A}}(a,b) = \infty$. If C contains only one element a, then we write $d_{\mathfrak{A}}(a,D)$ instead of $d_{\mathfrak{A}}(C,D)$.

For $B \subseteq |\mathfrak{A}|$ we define

$$S^n_{\mathfrak{A}}(B) := \{a: a \in |\mathfrak{A}| \text{ and there is some } b \in B \text{ such that } d_{\mathfrak{A}}(a,b) \leq n\}$$

and

$$D^n_{\mathfrak{A}}(B) := S^n_{\mathfrak{A}}(B) \setminus S^{n-1}_{\mathfrak{A}}(B), \text{ where we set } S^{-1}_{\mathfrak{A}}(B) := \emptyset \text{ in case } n = 0.$$

Let $\underline{S}^n_{\mathfrak{A}}(B)$ be the restriction of $\mathfrak{A}$ to $S^n_{\mathfrak{A}}(B)$. If it is clear which structure we consider, then we shall omit the subscript $\mathfrak{A}$.

An illustration of the definitions we see in fig. 1. $S^2_{\mathfrak{A}}(\{a\})$

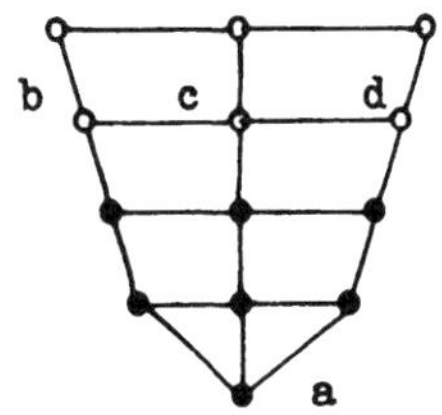

$\mathfrak{A}$
fig. 1

is the set of all full points (●). $D^3_{\mathfrak{A}}(\{a\}) = \{b,c,d\}$, while $S^1_{\mathfrak{A}}(\{a,b,c,d\})$ is the set $|\mathfrak{A}|$.

Let us now introduce a game which is slightly different from the game for $L(Q_1)$ introduced in Chapter 1, § 2. For this purpose let $\mathfrak{A}_0$ and $\mathfrak{A}_1$ be coloured graphs of the same type, and let $a_0^0 \in |\mathfrak{A}_0|$ and $a_0^1 \in |\mathfrak{A}_1|$. Assume that the game is played over $k < n$ rounds. As a result we have two sequences -

$$s_k^0 = \langle a_0^0,\dots,a_k^0\rangle \in |\mathfrak{A}_0|^{k+1} \quad\text{and}\quad s_k^1 = \langle a_0^1,\dots,a_k^1\rangle \in |\mathfrak{A}_1|^{k+1}.$$

Let $|s_k^i| := \{a_0^i, \dots, a_k^i\}$ for $i = 0,1$. The game will be continued as follows:

<u>Case 1.</u> Player I chooses an element $a_{k+1}^i$ from $S^{2^{n-k-1}}_{\mathfrak{A}_i}(|s_k^i|)$. Then player II is forced to choose an element $a_{k+1}^{1-i}$ from $S^{2^{n-k-1}}_{\mathfrak{A}_{1-i}}(|s_k^{1-i}|)$. The sequences $s_k^0$ and $s_k^1$ can be extended to $s_{k+1}^0 := \langle a_0^0,\dots,a_{k+1}^0\rangle$ and $s_{k+1}^1 := \langle a_0^1,\dots,a_{k+1}^1\rangle$, respectively.

<u>Case 2.</u> Player I chooses a subset $A_i \subseteq S^{2^{n-k-1}}_{\mathfrak{A}_i}(|s_k^i|)$ with $\mathrm{card}(A_i) \geq \omega_1$. Then player II has to choose a subset $A_{1-i} \subseteq S^{2^{n-k-1}}_{\mathfrak{A}_{1-i}}(|s_k^{1-i}|)$ with $\mathrm{card}(A_{1-i}) \geq \omega_1$. Now player I takes an element $a_{k+1}^{1-i}$ from $A_{1-i}$ and player II takes an element $a_{k+1}^i \in A_i$. Both sequences $s_k^0$ and $s_k^1$ can be extended to $s_{k+1}^0 := \langle a_0^0,\dots,a_{k+1}^0\rangle$ and $s_{k+1}^1 := \langle a_0^1,\dots,a_{k+1}^1\rangle$, respectively.

Player II wins the game after n rounds if the assignment $a_k^0 \longmapsto a_k^1$ for $0 \leq k \leq n$ is a partial isomorphism.

If player II has a winning strategy, then we say $\langle\mathfrak{A}_0,a_0^0\rangle$ and $\langle\mathfrak{A}_1,a_0^1\rangle$ are n-equivalent with respect to the shrinking $Q_1$-game: We write

$$\langle\mathfrak{A}_0,a_0^0\rangle \sim_n^{S1} \langle\mathfrak{A}_1,a_0^1\rangle.$$

Obviously we have

(i) $\langle\mathfrak{A}_0,a_0^0\rangle \sim_n^{S1} \langle\mathfrak{A}_1,a_0^1\rangle$ iff $\langle \underline{S}^{2^{n-1}}_{\mathfrak{A}_0}(\{a_0^0\}),a_0^0\rangle \sim_n^{S1} \langle \underline{S}^{2^{n-1}}_{\mathfrak{A}_1}(\{a_0^1\}),a_0^1\rangle$,

(ii) $\langle \mathfrak{A}_0, a_0^0 \rangle \sim_n^1 \langle \mathfrak{A}_1, a_0^1 \rangle$ implies $\langle \mathfrak{A}_0, a_0^0 \rangle \sim_n^{s1} \langle \mathfrak{A}_1, a_0^1 \rangle$,

(iii) $\langle S_{\mathfrak{A}_0}^{2^{n-1}}(\{a_0^0\}), a_0^0 \rangle \sim_n^1 \langle S_{\mathfrak{A}_1}^{2^{n-1}}(\{a_0^1\}), a_0^1 \rangle$ implies $\langle \mathfrak{A}_0, a_0^0 \rangle \sim_n^{s1} \langle \mathfrak{A}_1, a_0^1 \rangle$.

The inversions of (ii) and (iii) are not true in general. An example of two structures which are 4-equivalent with respect to the shrinking $Q_1$-game, but not 4-equivalent with respect to the usual $Q_1$-game can be found in fig. 2.

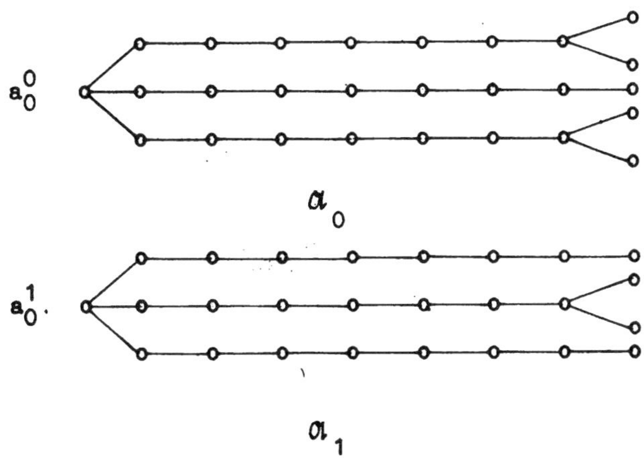


fig. 2

The idea of shrinking games improves the idea of "range of influence" (Einflußbereich), which was suggested by Hauschild. The choice of $2^{n-k-1}$-spheres in the definition of shrinking games is caused by the following lemma.

<u>Lemma 1.</u> Let $\mathfrak{A}$ and $\mathfrak{B}$ be two coloured graphs of the same type and let $a_0$, $a_1$ and $b_0$, $b_1$ be points of $\mathfrak{A}$ and $\mathfrak{B}$, respectively. Suppose that $\langle \mathfrak{A}, a_0, a_1 \rangle \sim_n \langle \mathfrak{B}, b_0, b_1 \rangle$ for a natural number n. Then we have:

(i) If $d_{\mathfrak{A}}(a_0, a_1) \leq 2^n$, then $d_{\mathfrak{A}}(a_0, a_1) = d_{\mathfrak{B}}(b_0, b_1)$,

(ii) $d_{\mathfrak{A}}(a_0, a_1) > 2^n$ implies $d_{\mathfrak{B}}(b_0, b_1) > 2^n$.

(ii) follows immediately from (i). The proof of (i) is an easy induction and is left to the reader (hint: choose a point a such that $|d_{\mathfrak{A}}(a_0,a) - d_{\mathfrak{A}}(a,a_1)| \leq 1$).

The following lemma shows that it is not possible to enlarge $2^n$ in Lemma 1.

<u>Lemma 2.</u> Let $\mathfrak{G}$ be the graph $\langle A,R\rangle$, where A is the set of all integers and R is the relation

$$\{(a,b)\colon a \in A \text{ and } b \in A \text{ and } (a - b = 1 \text{ or } b - a = 1)\}.$$

Let $\bar{a}_0,\ldots,\bar{a}_k$, $\bar{b}_0,\ldots,\bar{b}_k$ be finite sequences of integers such that

$$\bar{a}_i = \langle a_{i0},\ldots,a_{ik_i}\rangle \text{ for all } i \leq k;$$

$$d_{\mathfrak{G}}(a_{js},a_{ir}) > 2^n \quad \text{if } i < j \leq k,\ r \leq k_i \text{ and } s \leq k_j;$$

$$d_{\mathfrak{G}}(a_{is},a_{i,s+1}) \leq 2^n \text{ and } a_{is} \leq a_{i,s+1} \quad \text{if } i \leq k \text{ and } s < k_i;$$

$$\bar{b}_i = \langle b_{i0},\ldots,b_{ik_i}\rangle \text{ for all } i \leq k;$$

$$d_{\mathfrak{G}}(b_{js},b_{ir}) > 2^n \quad \text{if } i < j \leq k,\ r \leq k_i \text{ and } s \leq k_j;$$

$$d_{\mathfrak{G}}(b_{is},b_{ir}) = d_{\mathfrak{G}}(a_{is},a_{ir}) \text{ and } b_{ir} \leq b_{is} \quad \text{if } i \leq k \text{ and } r < s \leq k_i.$$

Then $\langle A,R,\bar{a}_0,\ldots,\bar{a}_k\rangle \sim_n \langle A,R,\bar{b}_0,\ldots,\bar{b}_k\rangle$ holds.

The proof of this lemma, which is an easy induction on n, is left as an exercise to the reader.

Let $\mathfrak{A}$ be a graph. Then $\sim_n^{s1}$ defines an equivalence relation $\sim$ on $|\mathfrak{A}|$:

$$a \sim b \quad \text{iff} \quad \langle \mathfrak{A},a\rangle \sim_n^{s1} \langle \mathfrak{A},b\rangle.$$

The equivalence classes will be denoted as shrinking n-types. Let $\dot{a}$ denote the shrinking n-type of a in $\mathfrak{A}$.

<u>Definition.</u> $\langle C,\ldots,a_{ij},\ldots,\ldots,B_{kj},\ldots\rangle$ with $1 \leq i \leq l$, $1 \leq k \leq m$ and $1 \leq j \leq n$ is called a <u>characteristic shrinking n-type system</u> (in the following shortly denoted by n-css) for $\mathfrak{A}$ iff

(i) $C \subseteq |\mathfrak{A}|$, $B_{kj} \subseteq |\mathfrak{A}|$, $a_{ij} \in |\mathfrak{A}|$ for $1 \leq i \leq l$, $1 \leq j \leq n$ and $1 \leq k \leq m$;

(ii) $\dot{a}_{ij} = \dot{a}_{i'j'}$ iff $i = i'$, for $1 \le i,i' \le l$ and $1 \le j,j' \le n$;
$\dot{b}_{kj} = \dot{b}_{k'j'}$ iff $k = k'$, for all $b_{kj} \in B_{kj}$, $b_{k'j'} \in B_{k'j'}$, $1 \le k,k' \le m$, $1 \le j,j' \le n$;

(iii) $\mathrm{card}(\dot{a}_{i1} \setminus C) < \omega_1$ for all $1 \le i \le l$;

(iv) for every $c \in (|\mathfrak{A}| \setminus C)$ there exists an element $d \in \{a_{i1}: 1 \le i \le l\} \cup \bigcup_{k=1}^{m} B_{k1}$ such that $\dot{c} = \dot{d}$;

(v) $\mathrm{card}(B_{kj}) \ge \omega_1$ for all $1 \le k \le m$ and $1 \le j \le n$;

(vi)
$$\left.\begin{array}{l} d_{\mathfrak{A}}(a_{ij}, a_{i'j'}) > 2^n \\ d_{\mathfrak{A}}(B_{kj}, B_{k'j'}) > 2^n \\ d_{\mathfrak{A}}(C, a_{ij}) > 2^n \\ d_{\mathfrak{A}}(C, B_{kj}) > 2^n \\ d_{\mathfrak{A}}(B_{kj}, a_{ij'}) > 2^n \end{array}\right\} \quad 1 \le i,i' \le l,\ 1 \le k,k' \le m,\ 1 \le j,j' \le n.$$

An illustration of this definition and the following construction can be found in fig. 3 (n = 2).

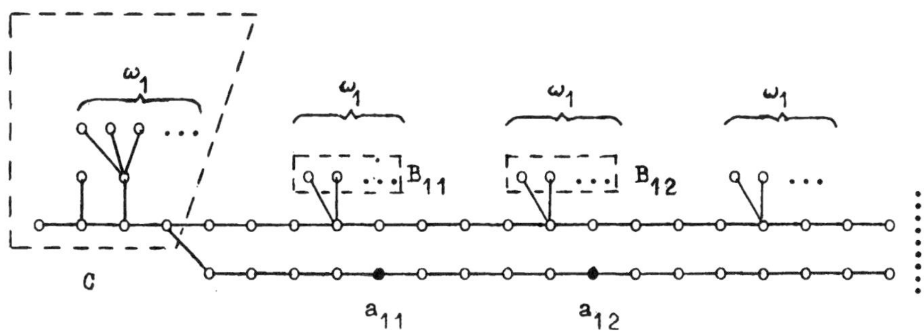


fig. 3

Let the n-css $\langle|\mathfrak{A}|\rangle$ (with $l = m = 0$) denote the trivial n-css for $\mathfrak{A}$.

There are structures which possess a non-trivial n-css, too. Let $\mathfrak{A}$ be a graph and $a_0$ an element of $|\mathfrak{A}|$. We suppose that $|\mathfrak{A}| = \bigcup_{r \in \omega} D^r(\{a_0\})$ and that $D^r_{\mathfrak{A}}(\{a_0\}) \ne \emptyset$ for every $r \in \omega$. Supposing this we get that $\mathfrak{A}$ has a non-trivial n-css for every natural number n.

To prove this let $\dot{d}_1, \dots, \dot{d}_s$ be all shrinking n-types of $\mathfrak{A}$. Obviously the number of these types is finite. A shrinking n-type $\dot{d}_i$ is said to be bounded ($\omega_1$-bounded) if there is a $t_i \in \omega$ such that $\dot{d}_i \subseteq S^{t_i}_{\mathfrak{A}}(\{a_0\})$ $(\mathrm{card}(\dot{d}_i \setminus S^{t_i}_{\mathfrak{A}}(\{a_0\})) < \omega_1$ respectively). We fix for every

$\omega_1$-bounded $\dot{d}_i$ a bound $t_i$ and define:

$$t_0 := \max(\{t_i : \dot{d}_i \text{ is } \omega_1\text{-bounded and } t_i \text{ is its fixed bound}\} \cup \{-1\}).$$

Let $\dot{a}_1,\dots,\dot{a}_l$ be all shrinking n-types which are not bounded, but which are $\omega_1$-bounded. Let $\dot{b}_1,\dots,\dot{b}_m$ be all shrinking n-types which are not $\omega_1$-bounded. Using our suppositions we get that $l + m \neq 0$ since not all shrinking n-types can be bounded. Now let C be the set $S_{\mathfrak{A}}^{t_0}(\{a_0\})$. Since the types $\dot{a}_1$ are not bounded, it is easy to choose elements $a_{i1},\dots,a_{in}$ from $\dot{a}_i$ for every i $(1 \le i \le l)$ such that $d_{\mathfrak{A}}(a_{ij},a_{i'j'}) > 2^n$ and $d_{\mathfrak{A}}(C,a_{ij}) > 2^n$ (for every $1 \le i,i' \le l$ and $1 \le j,j' \le n$). Let $r_0$ be defined by

$$r_0 := \max(\{t_0\} \cup \{r: a_{ij} \in D_{\mathfrak{A}}^r(\{a_0\}) \text{ for some } i,j\}).$$

For every $\dot{b}_k$ and every $r > r_0$ there exists a $t > r$ such that

$$\mathrm{card}(D_{\mathfrak{A}}^t(\{a_0\}) \cap \dot{b}_k) \geq \omega_1,$$

since $\dot{b}_k$ is not bounded and $\mathrm{cf}(\omega_1) > \omega$. Using that

$$d_{\mathfrak{A}}(D_{\mathfrak{A}}^{t_1}(\{a_0\}), D_{\mathfrak{A}}^{t_2}(\{a_0\})) = |t_1 - t_2|,$$

it is easy to find sets $B_{kj}$ $(1 \le k \le m,\ 1 \le j \le n)$ such that $\langle C,\dots,a_{ij},\dots,\dots,B_{kj},\dots\rangle$ is a non-trivial n-css for $\mathfrak{A}$.

From the above we get the following result.

<u>Lemma 3.</u> Let $\mathfrak{A}$ be a graph and let $a_0$ be an element of $|\mathfrak{A}|$ such that $|\mathfrak{A}| = \bigcup_{r\in\omega} D_{\mathfrak{A}}^r(\{a_0\})$. If $D_{\mathfrak{A}}^r(\{a_0\}) \neq 0$ for every $r \in \omega$, then $\mathfrak{A}$ has a non-trivial n-css, for every n.

Now we can give the main result of this paragraph.

<u>Theorem 1.</u> Let $\mathfrak{M}$ and $\mathfrak{N}$ be relational structures of the same signature with one designated binary relation. For a natural number n let $\langle C,\dots,a_{ij},\dots,\dots,B_{kj},\dots\rangle$ with $1 \le i \le l$, $1 \le k \le m$, $1 \le j \le n$ ($\langle C',\dots,a'_{ij},\dots,\dots,B'_{kj},\dots\rangle$ with $1 \le i \le l$, $1 \le k \le m$ and $1 \le j \le n$) be an n-css for $\mathfrak{M}$ ($\mathfrak{N}$, respectively). Suppose that the following

conditions are fulfilled:

(i) $\mathfrak{M} \upharpoonright S^{2^n}_{\mathfrak{M}}(C) \cong \mathfrak{N} \upharpoonright S^{2^n}_{\mathfrak{N}}(C')$;

(ii) $\langle \mathfrak{M}, a_{ij} \rangle \sim^{s1}_{n} \langle \mathfrak{N}, a'_{ij} \rangle$ for all $1 \leq i \leq l$ and $1 \leq j \leq n$;

(iii) $\langle \mathfrak{M}, b_{kj} \rangle \sim^{s1}_{n} \langle \mathfrak{N}, b'_{kj} \rangle$ for every $k$, $j$, $b_{kj}$, $b'_{kj}$ with $1 \leq k \leq m$, $1 \leq j \leq n$, $b_{kj} \in B_{kj}$ and $b'_{kj} \in B_{kj}$.

Then $\mathfrak{M} \sim^{1}_{n} \mathfrak{N}$ holds.

Proof. Without loss of generality we can assume that condition (i) is strengthened to $\mathfrak{M} \upharpoonright S^{2^n}_{\mathfrak{M}}(C) = \mathfrak{N} \upharpoonright S^{2^n}_{\mathfrak{N}}(C')$.

The winning strategy for player II runs as follows. Assume that the game is played over h rounds. As a result we have two sequences

$$s^h_{\mathfrak{M}} = \langle c_1, \ldots, c_h \rangle \in |\mathfrak{M}|^h \quad \text{and} \quad s^h_{\mathfrak{N}} = \langle d_1, \ldots, d_h \rangle \in |\mathfrak{N}|^h$$

and a partition of $\{1, \ldots, h\}$ into the sets $I_1 = \{i_{11}, \ldots, i_{1s_1}\}$, ..., $I_u = \{i_{u1}, \ldots, i_{us_u}\}$, where in each $I_v$ the numbers are in their natural ordering, such that

(i) $c_{i_{v,t+1}} \in S^{2^{n-1-t}}_{\mathfrak{M}}(\{c_{i_{v1}}, \ldots, c_{i_{vt}}\})$,
$d_{i_{v,t+1}} \in S^{2^{n-1-t}}_{\mathfrak{N}}(\{d_{i_{v1}}, \ldots, d_{i_{vt}}\})$
for all $1 \leq v \leq u$ and $1 \leq t \leq s_v$;

(ii) for all $1 < v \leq u$ and for all $1 \leq t < v$
$c_{i_{v1}} \notin S^{2^{n-1-w_{vt}}}_{\mathfrak{M}}(\{c_{i_{t1}}, \ldots, c_{i_{tw_{vt}}}\})$ and
$d_{i_{v1}} \notin S^{2^{n-1-w_{vt}}}_{\mathfrak{N}}(\{d_{i_{t1}}, \ldots, d_{i_{tw_{vt}}}\})$,
where $w_{vt} = \max(\{z : i_{tz} < i_{v1}\})$;

(iii) $\langle \mathfrak{M}, c_{i_{t1}} \rangle \sim^{s1}_{n-1} \langle \mathfrak{N}, d_{i_{t1}} \rangle$ and the sequences $\langle c_{i_{t1}}, \ldots, c_{i_{ts_t}} \rangle$ and $\langle d_{i_{t1}}, \ldots, d_{i_{ts_t}} \rangle$ are built according to the winning strategy for player II in the shrinking $Q_1$ game for $\langle \mathfrak{M}, c_{i_{t1}} \rangle \sim^{s1}_{n-1} \langle \mathfrak{N}, d_{i_{t1}} \rangle$ for all $1 \leq t \leq u$;

(iv) if $c_{i_{t1}} \in S^{2^n}_{\mathfrak{M}}(C)$ or $d_{i_{t1}} \in S^{2^n}_{\mathfrak{N}}(C')$ for some $1 \leq t \leq u$, then $c_{i_{tv}} = d_{i_{tv}}$ for all $1 \leq v \leq s_t$.

We extend both sequences such that the conditions (i) - (iv) are also fulfilled for the extended sequences.

1. Suppose player I chooses an element $c_{h+1}$ from $|\mathfrak{M}|$ (the case for $|\mathfrak{N}|$ is analogous).

1.1. $c_{h+1} \in S^{2^{n-1-s_t}}_{\mathfrak{M}}(\{c_{i_{t1}},\dots,c_{i_{ts_t}}\})$ for some $t$, $1 \leq t \leq u$. Then, by the winning strategy for player II in the shrinking $Q_1$ game, player II can choose $d_{h+1} \in S^{2^{n-1-s_t}}_{\mathfrak{N}}(\{d_{i_{t1}},\dots,d_{i_{ts_t}}\})$.
If $c_{i_{t1}}$ satisfies condition (iv), then we can choose $d_{h+1} = c_{h+1}$, because of condition (i) of the theorem.

1.2. The other case.

1.2.1. $c_{h+1} \in S^{2^{n-1-h}}_{\mathfrak{M}}(C)$, then set $d_{h+1} = c_{h+1}$.

1.2.2. $c_{h+1} \notin S^{2^{n-1-h}}_{\mathfrak{M}}(C)$. Since we have an n-css, there is some $\dot{a}_{ij}$ or some $b_{kj} \in B_{kj}$ such that $c_{h+1} = a_{ij}$ or $c_{h+1} = b_{kj}$. Assume $c_{h+1} = a_{ij}$ (the other case is similar). Let

$$j_0 := \min(\{j\colon 1 \leq j \leq h+1 \text{ and } \{d_1,\dots,d_h\} \cap S^{2^{n-1}}_{\mathfrak{N}}(\{a'_{ij}\}) = \emptyset\}).$$

Then player II chooses the element $a'_{ij_0}$.

2. Suppose player I chooses an uncountable subset $D \subseteq |\mathfrak{M}|$ (the case for $\mathfrak{N}$ is analogous). Then regard the associated subsets

$$D^h := S^{2^{n-1-h}}_{\mathfrak{M}}(C) \cap D,$$

$$D_t := S^{2^{n-1-s_t}}_{\mathfrak{M}}(\{c_{i_{t1}},\dots,c_{i_{ts_t}}\}) \cap D,$$

$$D_{a_{i1}} := [D \setminus (D^h \cup D_1 \cup \dots \cup D_u)] \cap \dot{a}_{i1},$$

$$D_{B_{k1}} := [D \setminus (D^h \cup D_1 \cup \dots \cup D_u)] \cap \dot{b}_{k1},$$

where $1 \leq t \leq u$, $1 \leq i \leq l$, $1 \leq k \leq m$ and $b_{k1} \in B_{k1}$. Let $J_1 := \{i: D_{a_{i1}} \neq \emptyset\}$ and let $J_2 := \{k: D_{B_{k1}}$ is uncountable$\}$. Player II then chooses the subset

$$E := D^h \cup E_1 \cup \dots \cup E_u \cup \bigcup_{i \in J_1} (\{a'_{ij'_i}\}) \cup \bigcup_{k \in J_2} (B'_{kj''_k}),$$

where

$$j'_i := \min(\{j: 1 \leq j \leq h+1 \text{ and } \{d_1,\dots,d_h\} \cap S^{2^{n-1}}_{\mathfrak{N}}(\{a'_{ij}\}) = \emptyset\}),$$

$$j''_k := \min(\{j: 1 \leq j \leq h+1 \text{ and } \{d_1,\dots,d_h\} \cap S^{2^{n-1}}_{\mathfrak{N}}(B'_{kj}) = \emptyset\})$$

and $E_t$ is chosen according to the winning strategy of player II in the shrinking $Q_1$ game for $\langle \mathfrak{M}, c_{i_{t1}} \rangle \sim^{s1}_{n-1} \langle \mathfrak{N}, d_{i_{t1}} \rangle$ in the round $(s_t + 1)$; and in case $c_{i_{t1}}$ satisfies condition (iv), we set $E_t := D_t$.

Now player I chooses an element $d_{h+1}$ from E. Then $d_{h+1} \in D^h$, or $d_{h+1} \in E_t$ for some t, or $d_{h+1} = a_{ij'_i}$ for some i, or $d_{h+1} \in B'_{kj''_k}$ for some k. In all cases player II can choose a corresponding element $c_{h+1}$ from D such that the extended sequences satisfy the conditions (i) - (iv). These conditions yield in the last round that our strategy is a winning strategy. So we have proved the theorem.

Q.E.D.

Remark. Theorem 1 can be strengthened by introducing shrinking $Q_1$ games not only for points, but also for sets of points. The condition (i) in the theorem can be weakened to

(i') $\langle \mathfrak{M}, C \rangle \sim^{s1}_{n} \langle \mathfrak{N}, C' \rangle$.

Since there are no applications to this extension at the moment, we shall not give the details here.

## § 2. Constructions of forests

**Definition.** A (coloured) graph $\mathfrak{M} = \langle A,R \rangle$ is a <u>forest</u> iff

(i) R is irreflexive and symmetric, and
(ii) any two connected points a and b are connected by at most one path with first point a and last point b.

A component of a forest is sometimes denoted as a <u>tree</u>.

**Remark.** Condition (i) is useful for technical reasons, but not essential.

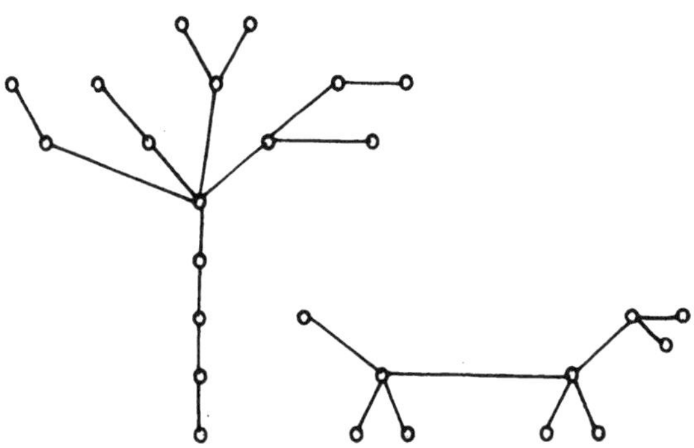

fig. 4

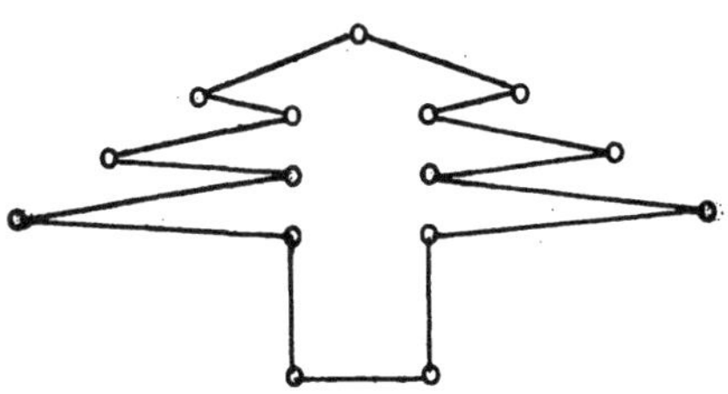

fig. 5

In fig. 4 a forest which contains two trees as components is to be seen, while the graph in fig. 5 is not a forest and hence not a tree.

Let $L_0$ be the first order language with equality and the binary predicate R. For some constructions we need special coloured forests. We introduce some helpful unary predicates W and for all $n \in \omega$, $P_n$. Let $L_1$ be the extension of $L_0$ which contains these additional predicates. Let $\mathfrak{M}$ be a model for the language $L_1$. Then $W(\mathfrak{M})$, $P_n(\mathfrak{M})$, and $R(\mathfrak{M})$ denote the corresponding interpretations of W, $P_n$, and R, respectively. The elements of $W(\mathfrak{M})$ will be denoted as roots while the elements of $\bigcup_{n\in\omega} P_n(\mathfrak{M})$ will be denoted as buds of the forest.

<u>**Definition.**</u> A model $\mathfrak{M}$ for $L_1$ is an <u>s-forest</u> (forest in spring or spring forest) iff

(i) $\mathfrak{M}$ is a forest,
(ii) any two different roots are not connected,
(iii) every bud has a unique color,
(iv) every bud a is connected with at most one element different from a.

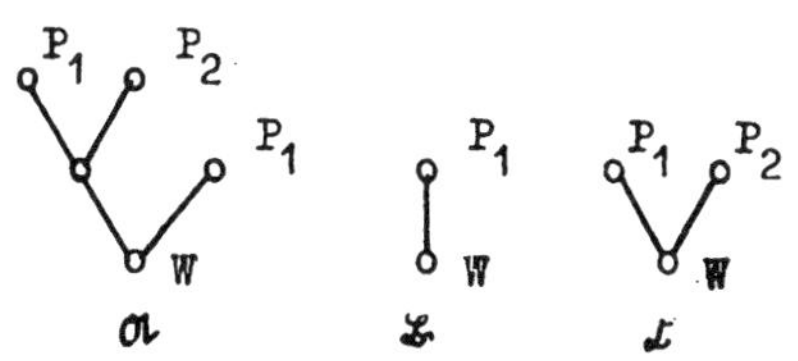


fig. 6

To illustrate the above definition we present an s-forest in fig. 6. The predicates $P_n$ are empty for all $n > 2$.

s-forests are useful to construct new forests from given forests. Thus we obtain more complicated forests from simpler ones by certain operations. Sometimes we need a copy of a given structure $\mathfrak{M}$ with different universe. For that reason we introduce the following notation:

Let P be any object, then $\mathfrak{M}(P)$ is the unique model with the universe $\{\langle P,a\rangle : a \in |\mathfrak{M}|\}$ such that the canonical map $a \longmapsto \langle P,a\rangle$ is an isomorphism.

Let $\mathfrak{M}$ and $\mathfrak{N}$ be two s-forests, then we define:

$$\mathfrak{M} + \mathfrak{N} := \mathfrak{M}(0) \cup \mathfrak{N}(1),$$

$$\Sigma\mathfrak{M} := \bigcup_{n<\omega} \mathfrak{M}(n),$$

$$\Sigma_1\mathfrak{M} := \bigcup_{\alpha<\omega_1} \mathfrak{M}(\alpha).$$

$\bigcup$ denotes the union of models defined as usually.

$S\mathfrak{M}$ is the s-forest described as follows:

$$|S\mathfrak{M}| := |\mathfrak{M}(0)| \cup W(\mathfrak{M}(1)),$$

$$R_1 := \{\langle\langle 0,a\rangle,\langle 1,a\rangle\rangle : a \in W(\mathfrak{M})\},$$

$$R(S\mathfrak{M}) := R(\mathfrak{M}(0)) \cup R_1 \cup R_1^{-1},$$

$$W(S\mathfrak{M}) := W(\mathfrak{M}(1))$$

and for all $n < \omega$

$$P_n(S\mathfrak{M}) := P_n(\mathfrak{M}(0)).$$

In fig. 7 we present an isomorphic copy of the s-forest $S\mathfrak{M}$ for the s-forest $\mathfrak{M}$ presented in fig. 6.

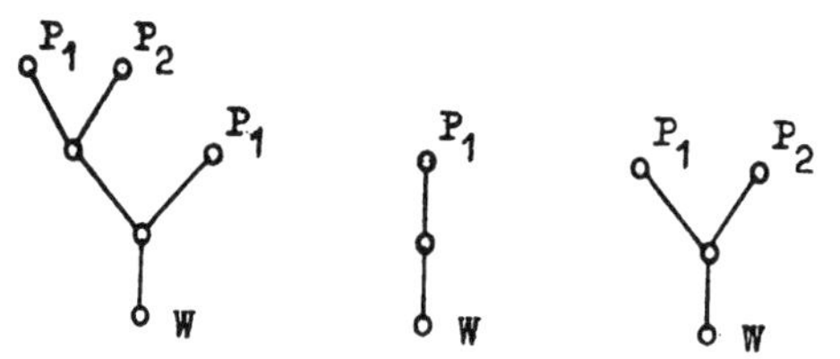


fig. 7

At last the s-forest $\Pi\mathfrak{M}$ arises from $\mathfrak{M}$ by identifying roots. We set $a \sim b$ iff a and b are both roots which are not coloured or $a = b$. Clearly $\sim$ is a congruence relation.

Set

$$\Pi\mathfrak{M} := \mathfrak{M}/\sim,$$

i.e. $\Pi\mathfrak{M}$ is the homomorphic image of the canonical homomorphism induced by $\sim$.

$\Pi\mathfrak{M}$ of the s-forest $\mathfrak{M}$ from fig. 6 is presented in fig. 8.

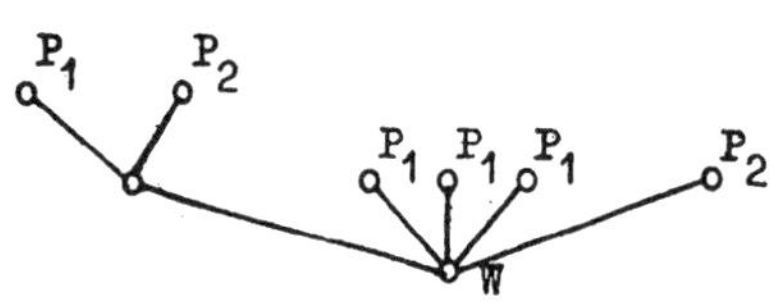


fig. 8

A connected s-forest which possesses a root is called an s-tree. A tree valuation f is a map into the class of s-trees. Let $\mathfrak{M}$ be an s-forest. Then $\mathfrak{M}^-$ is the s-forest which arises from $\mathfrak{M}$ by setting $W(\mathfrak{M}^-) := \emptyset$ and leave all the other relations unchanged. $\mathfrak{M}^+$ arises from $\mathfrak{M}$ by setting $P_n(\mathfrak{M}^+) := \emptyset$ for all n, and leave all the other relations unchanged. Suppose f is a tree valuation with $\mathrm{dom}(f) = \bigcup_{n\in\omega} P_n(\mathfrak{M})$ ($\mathrm{dom}(f) = \omega$). We define

$$\mathfrak{M}_f := \mathfrak{M}^+(\mathfrak{M}) \cup \bigcup_{n<\omega} \bigcup_{a\in P_n(\mathfrak{M})} (f(a))^-(a)$$

$$(\mathfrak{M}_f := \mathfrak{M}^+(\mathfrak{M}) \cup \bigcup_{n<\omega} \bigcup_{a\in P_n(\mathfrak{M})} (f(n))^-(a)).$$

In $\mathfrak{M}_f$ we identify the elements a and $\langle a,b\rangle$, where b is the root of f(a) (f(n), respectively) and $a \in \bigcup_{n\in\omega} P_n(\mathfrak{M})$ ($a \in P_n(\mathfrak{M})$, respectively). The resulting s-forest is denoted by $\sigma^1(\mathfrak{M},f)$ in both cases. $\sigma^1(\mathfrak{M},f)$ has a simple meaning: at $a \in \bigcup_{n\in\omega} P_n(\mathfrak{M})$ ($a \in P_n(\mathfrak{M})$, respectively) we substitute a copy of the s-forest f(a) (f(n), respectively). Let g be a tree valuation with dom(g) = $\omega$ and $\{\langle 1,\mathfrak{B}\rangle,\langle 2,\mathfrak{C}\rangle\} \subseteq g$, where $\mathfrak{A}$, $\mathfrak{B}$ and $\mathfrak{C}$ are as in fig. 6. Then $\sigma^1(\mathfrak{A},g)$ is presented in fig. 9.

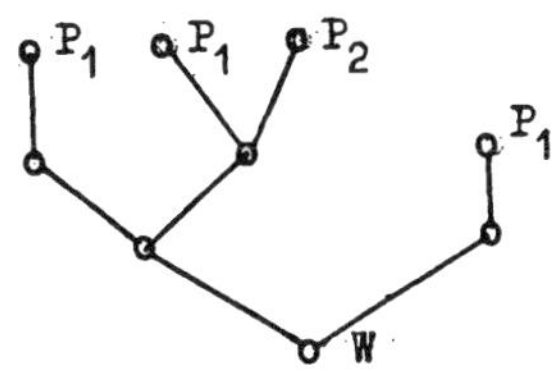


fig. 9

In the case with dom(f) = $\omega$ we can iterate the above process by defining:

$$\sigma^{m+1}(\mathfrak{M},f) := \sigma^1(\sigma^m(\mathfrak{M},f),f).$$

Obviously, the canonical map $h: |\mathfrak{M}| \longrightarrow |\sigma^1(\mathfrak{M},f)|$ defined by $h(a) := \langle\mathfrak{M},a\rangle$ is an embedding of $\mathfrak{M}$ into $\sigma^1(\mathfrak{M},f)$ with respect to the predicates R and W. Let $\sigma(\mathfrak{M},f)$ be the direct limit of $\{\sigma^i(\mathfrak{M},f)^+ : i \in \omega\}$. $\sigma(\mathfrak{A},g)$ (for $\mathfrak{A}$ and g as above) can be found in fig. 10.

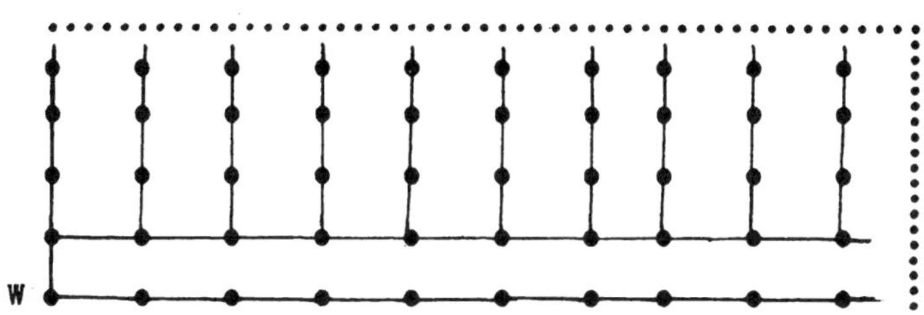


fig. 10

The following results are sometimes useful.

<u>Lemma 4.</u> Let n be a natural number and let C be a set. Moreover, let $\{\mathfrak{A}_c : c \in C\}$ and $\{\mathfrak{B}_c : c \in C\}$ be sets of structures with the following properties:

(i) $|\mathfrak{A}_c| \cap |\mathfrak{A}_d| = |\mathfrak{B}_c| \cap |\mathfrak{B}_d| = \emptyset$ for all $c,d \in C$ with $c \neq d$,

(ii) $\mathfrak{A}_c \sim_n^1 \mathfrak{B}_c$ for all $c \in C$.

Then the following holds:

$$\bigcup_{c\in C} \mathfrak{a}_c \sim_n^1 \bigcup_{c\in C} \mathfrak{b}_c .$$

The proof of this simple lemma we leave as an exercise to the reader (hint: play componentwise, but pay attention in case that player I chooses a set; use here cf $\omega_1 = \omega_1$)

The next lemma has a similar proof as the preceding one.

<u>Lemma 5.</u> Let $\mathfrak{M}$ be an s-forest and let f and g be tree valuations with

$$\mathrm{dom}(f) \cap \mathrm{dom}(g) \supseteq \bigcup_{n\in\omega} P_n(\mathfrak{M}).$$

Suppose that for all $a \in \bigcup_{n\in\omega} P_n(\mathfrak{M})$, $f(a) \sim_n^1 g(a)$. Then

$$\sigma^1(\mathfrak{M},f) \sim_n^1 \sigma^1(\mathfrak{M},g).$$

Let $\mathfrak{M}$ be an s-tree and let $a_0$ be the root of $\mathfrak{M}$. If $D^r_{\mathfrak{M}}(\{a_0\}) \neq \emptyset$ for every $r \in \omega$, then we get, as in the proof of Lemma 3, an n-css for every n. Let $\langle C,\dots,a_{ij},\dots,\dots,B_{kj},\dots\rangle$ $(1 \leq i \leq l,\ 1 \leq k \leq m,\ 1 \leq j \leq n)$ be an n-css, the construction of which is indicated in the proof of Lemma 3. Such an n-css has the property that there exists a natural number t such that

$$(*)\begin{cases} S^t_{\mathfrak{M}}(\{a_0\}) \supseteq C \cup \bigcup_{1\leq j\leq n} \Big( \bigcup_{1\leq i\leq l} \{a_{ij}\} \cup \bigcup_{1\leq k\leq m} B_{kj}\Big), \\ \text{there exists a } t_0 \in \omega \text{ such that } S^{t_0}_{\mathfrak{M}}(\{a_0\}) = C, \\ \text{for each } k,\ j \text{ (with } 1 \leq k \leq m,\ 1 \leq j \leq n) \text{ there exists a} \\ t_{kj} \in \omega \text{ such that } D^{t_{kj}}_{\mathfrak{M}}(\{a_0\}) = B_{kj}. \end{cases}$$

This leads to the following definition.

<u>Definition.</u> Let $\mathfrak{M}$ be a coloured graph and let n and t be natural numbers. t is said to be a <u>universal n-bound</u> for $\mathfrak{M}$ if there is an $a_0$ and an n-css $\langle C,\dots,a_{ij},\dots,\dots,B_{kj},\dots\rangle$ (with $1 \leq i \leq l,\ 1 \leq k \leq m,\ 1 \leq j \leq n$) such that (*) from above is fulfilled.

We get the following lemma.

**Lemma 6.** Let $\mathfrak{M}$ and $\mathfrak{N}$ be coloured graphs which are coloured with finitely many predicates and let t be a universal bound for $\mathfrak{M}$. Moreover, let $\langle C,\dots,a_{ij},\dots,\dots,B_{kj},\dots\rangle$ (with $1 \le i \le l$, $1 \le k \le m$, $1 \le j \le n$) be an n-css for $\mathfrak{M}$. Suppose $\mathfrak{M} \sim^{1}_{r} \mathfrak{N}$ for the natural number $r = t + n(l + m) + 2^n + 2$. Then t is a universal n-bound for $\mathfrak{N}$.

Using the definition of an n-css and the definition of an n-bound it is not difficult to construct a formula $\varphi$ with $qr(\varphi) = r$ such that:

> $\mathfrak{N} \models \varphi$ iff there exists an $a_0'$ and an n-css $\langle C',\dots,a'_{ij},\dots,\dots,B'_{kj},\dots\rangle$ (with $1 \le i \le l$, $1 \le k \le m$, $1 \le j \le n$) for $\mathfrak{N}$ for which t is a universal n-bound.

We use here claim (iii) from page 31 and Theorem 3 from Chapter 1. Using the same theorem again we get the conclusion of Lemma 6. The details we leave to the reader.

## § 3. A dense system for the $Q_1$-theory of trees

Let us consider the operations +, $\Sigma$, $\Sigma_1$, S, and $\Pi$ as function symbols of a first order language. Let 1 be a constant and $\{v_i : i \in \omega\}$ be the set of variables of this language. The set T of terms is built as usual. $T^{\Pi S}$ is the set of terms of the form $\Pi S\alpha$, where $\alpha$ is a term. A finite sequence $\langle\alpha_0,\dots,\alpha_n\rangle$ of terms is called an extended term iff

(i) $\alpha_1,\dots,\alpha_n$ are terms from $T^{\Pi S}$,

(ii) in each term $\alpha_i$ at most $v_1,\dots,v_n$ occur.

The set of extended terms is denoted by E. Let EC be the set

$$\{\langle\alpha_0,\dots,\alpha_n\rangle : \langle\alpha_0,\dots,\alpha_n\rangle \in E \text{ and } \alpha_0 \in T^{\Pi S}\}.$$

Now we shall see how every extended term represents a certain s-forest and that the class of represented s-forests is a nice subclass of all s-forests.

Suppose $\alpha$ is an extended term, then $\alpha^*$ denotes the s-forest represented by $\alpha$.

**Remark.** We identify terms with extended terms of length 1.

Firstly let us define which s-forests are represented by 1 and $v_i$ for $i \in \omega$.

$$1^* := \langle \{0\}, R, W, \langle P_j \rangle_{j \in \omega} \rangle \text{ with } R = \emptyset,\ W = \{0\} \text{ and for all } j \in \omega:\ P_j = \emptyset.$$

$$v_i^* := \langle \{0\}, R_1, W_1, \langle P_j^i \rangle_{j \in \omega} \rangle \text{ with } R_1 = \emptyset,\ W_1 = \{0\} \text{ and for all } j \neq i:\ P_j^i = \emptyset, \text{ while } P_i^i = \{0\}.$$

It is clear how to expand the representation to terms:

$$(\alpha + \beta)^* := \alpha^* + \beta^*, \quad (\Sigma\alpha)^* := \Sigma\alpha^*,$$

$$(\Sigma_1\alpha)^* := \Sigma_1\alpha^*, \quad (S\alpha)^* := S\alpha^* \quad \text{and} \quad (\Pi\alpha)^* := \Pi\alpha^*.$$

Thus we have defined the s-forest $\alpha^*$ for every term $\alpha$.

Let $\alpha$ be an extended term, i.e. $\alpha = \langle \alpha_0, \ldots, \alpha_n \rangle$, where $\alpha_0, \ldots, \alpha_n$ are terms. Let $f_\alpha$ be a tree valuation such that $\mathrm{dom}(f_\alpha) := \omega$ and for $1 \leq i \leq n$ we have $f_\alpha(i) := \alpha_i^*$. Then $\alpha^*$ is defined to be $\sigma(\alpha_0^*, f_\alpha)$.

For $\alpha \in E$ and $\varphi$ a sentence of $L_1(Q_1)$ we write $\alpha \models \varphi$ if $\alpha^* \models \varphi$.

<u>Theorem 2.</u> (i) The sets E and EC are recursive.
(ii) For every $\alpha \in E$ the set $\{\varphi : \alpha \models \varphi\}$ is recursive.

<u>Proof.</u> (i) follows immediately.

(ii) For any term $\alpha$ we see immediately that there is a recursive set $M_\alpha$ of axioms in the language $L_1(Q_1)$ such that all models of cardinality at most $\omega_1$ of this set $M_\alpha$ are isomorphic to $\alpha^*$. (Hint: induction using the definition of terms; the distance of every point of $\alpha^*$ from the root is universally bounded.)

Let $\beta = \langle \beta_0, \ldots, \beta_n \rangle$ be an extended term with $n > 0$, and let $\alpha = \langle \alpha_0, \ldots, \alpha_n \rangle$ be the extended term $\langle \Pi S\beta_0, \beta_1, \ldots, \beta_n \rangle$. Obviously $\beta^*$ is definable in $\alpha^*$. So it is sufficient to prove the decidability of $\{\varphi : \alpha \models \varphi\}$. To do this we recall that $\alpha^*$ is an s-forest without buds (the predicates $P_i$ are empty for all $i \in \omega$). By definition $\alpha^* = \sigma(\alpha_0^*, f_\alpha)$. $\alpha^*$ is the direct limit of $\sigma^n(\alpha_0^*, f_\alpha)^+$. $\alpha^*$ has no buds, but every $\sigma^n(\alpha_0^*, f_\alpha)$ has buds (in general). Now we expand $\alpha^*$ by defining the predicates $P_k(\alpha^*)$ (for $1 \leq k \leq n$) as follows:

$$a \in P_k(\alpha^*) \text{ iff } a \in |\sigma^m(\alpha_0^*, f_\alpha)| \text{ for some m and } a \in P_k(\sigma^m(\alpha_0^*, f_\alpha)).$$

Here we define $\sigma^0(\alpha_0^*, f_\alpha) := \alpha_0^*$. Let us call the resulting structure $\alpha'$ and let $a_0$ be the root of $\alpha'$. Let $\alpha''$ be the structure $\langle |\alpha'|, R'', \{a_0\}, P_1(\alpha^*), \dots, P_n(\alpha^*)\rangle$, where $R''$ is the set $\{\langle a,b\rangle : \langle a,b\rangle \in R(\alpha^*)$ and $d(a_0,a) < d(a_0,b)\}$. In this structure we can describe the process of constructing $\sigma(\alpha_0^*, f_\alpha)$ recursively:

At $a \in P_k(\alpha^*)$ we define $A_a$ to be the set of all points b such that a and b are connected by a path with length at most the greatest length of a path in $f_\alpha(k) = \alpha_k^*$ and between a and b there are no points which are in some predicate $P_j(\alpha^*)$ and if $\langle a, a_1, \dots, a_l, b\rangle$ is the path which connects a and b, then $\langle a, a_1\rangle \in R''$, $\langle a_l, b\rangle \in R''$ and $\langle a_i, a_{i+1}\rangle \in R''$ for $0 < i < l$.

$A_a$ is definable in $\alpha''$ in the language $L_1$. Now we describe the tree on $A_a$. $\alpha' \upharpoonright A_a$ has to be isomorphic to $\alpha_k^*$ (we can here omit the predicates $P_i$ with $i > n$, because they are empty in $\alpha_k^*$). But this is axiomatizable by $M_{\alpha_k}$. Following the above description we easily get a recursive set $M'_\alpha$ of axioms with the following properties:

> $\alpha''$ is a model of $M'_\alpha$.
> For any model $M'_\alpha$ with cardinality at most $\omega_1$ the component which contains the root is isomorphic to $\alpha''$.
> For all n all models of $M'_\alpha$ have the same n-css with respect to the isomorphic embedding of $\alpha''$ (hint: use that all models of $M'_\alpha$ are in a certain sense locally isomorphic).

Using Theorem 1 we get that $M'_\alpha$ is complete, hence $\{\varphi : \alpha'' \models \varphi\}$ is decidable. But obviously $\{\varphi : \alpha \models \varphi\}$ is interpretable in $\{\varphi : \alpha'' \models \varphi\}$ and consequently decidable.

Q.E.D.

<u>Definition.</u> Let $\mathfrak{M}$ be an s-forest and $a \in |\mathfrak{M}|$ such that a is connected with a root $a_0$ of $\mathfrak{M}$. Then set $T_a(\mathfrak{M}) := \{b : b \in |\mathfrak{M}|$ and b is connected with $a_0$ and every path which connects b and $a_0$ contains a as a point$\}$. $\mathcal{T}_a(\mathfrak{M})$ is defined to be the tree with universe $T_a(\mathfrak{M})$ such that $\mathcal{T}_a(\mathfrak{M})^-$ is a substructure of $\mathfrak{M}$ and $W(\mathcal{T}_a(\mathfrak{M})) := \{a\}$.

Now we prove that every tree $\mathfrak{M}$ is n-equivalent with respect to $L_0(Q_1)$ to some tree represented by an extended term.

Theorem 3. Let $\mathfrak{M}$ be a tree and n be a natural number. Then there is an extended term $\alpha$ such that $\mathfrak{M} \sim_n^1 \alpha^*$ with respect to $L_0(Q_1)$.

Proof. If $\mathfrak{M} \cong 1^*$, the statement is true (we have to omit here the unary predicates in $1^*$, since $\mathfrak{M}$ is not a coloured graph). Let $\mathfrak{M}$ not be isomorphic to $1^*$. $\sim_r^1$ defines an equivalence relation on the class of s-trees without buds (all $P_k$ are empty) which are not isomorphic to $1^*$. By Theorem 3 of Chapter 1 this relation has finitely many equivalence classes. Let $\mathfrak{M}_1^r, \dots \mathfrak{M}_{k(r)}^r$ be a complete system of representatives with respect to $\sim_r^1$ over the class of these s-trees. We define an equivalence on $D^1_{\mathfrak{M}_i^r}(W(\mathfrak{M}_i^r))$:

$$a \sim b \quad \text{iff} \quad \mathcal{T}_a(\mathfrak{M}_i^r) \sim_r^1 \mathcal{T}_b(\mathfrak{M}_i^r).$$

Let $\mathcal{D}(\mathfrak{M}_i^r)$ be the substructure with respect to R and W of $\mathfrak{M}_i^r$ with universe $S^1_{\mathfrak{M}_i^r}(W(\mathfrak{M}_i^r))$ and set

$$P_j(\mathcal{D}(\mathfrak{M}_i^r)) := \{a : a \in D^1_{\mathfrak{M}_i^r}(W(\mathfrak{M}_i^r)) \text{ and } \mathcal{T}_a(\mathfrak{M}_i^r) \sim_r^1 \mathfrak{M}_j^r\} \text{ for } 1 \le j \le k(r)$$

and

$$P_j := \emptyset \text{ otherwise.}$$

As an illustration regard fig. 11.

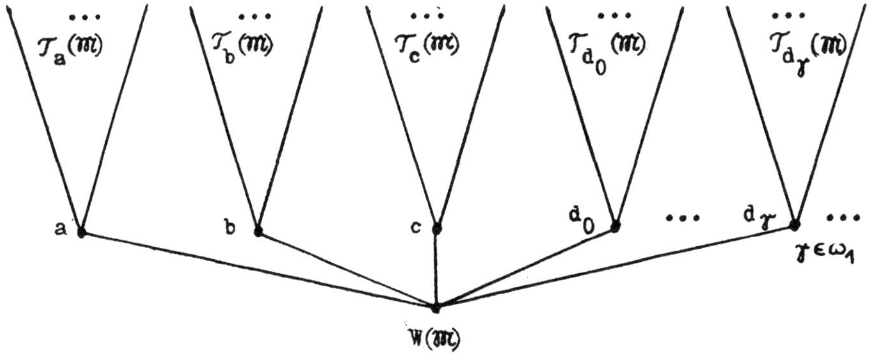


fig. 11

Obviously, $\mathcal{D}(\mathfrak{M}_i^r)$ is represented by a term $\alpha_{ri} \in T^{\Pi S}$. Moreover, it is possible to prove that there exists an $\alpha_{ri} \in T^{\Pi S}$ which does not contain the operation symbol $\Sigma$ and which represents $\mathcal{D}(\mathfrak{M}_i^r)$. Let $f_r$ be a tree valuation such that $f_r(i) := \mathfrak{M}_i^r$ for $1 \leq i \leq k(r)$, otherwise let $f_r$ be arbitrary. Then we have $\mathfrak{M}_i^r \sim_r^1 \sigma^1((\alpha_{ri})^*, f_r)$ by Lemma 5. We define by induction:

$$\alpha_{ri}^0 := \alpha_{ri}$$
$$\alpha_{ri}^{m+1} := \alpha_{ri}^m(\alpha_{r1}, \ldots, \alpha_{rk(r)}).$$

Here $\alpha_{ri}^m(\alpha_{r1}, \ldots, \alpha_{rk(r)})$ is obtained from $\alpha_{ri}^m$ by substituting the term $\alpha_{ri}$ for the variable $v_i$. Then we also have for all $m \in \omega$: $\mathfrak{M}_i^r \sim_r^1 \sigma^1((\alpha_{ri}^m)^*, f_r)$. Now let $\mathfrak{M}$ be an s-tree as above. Using the remarks following Lemma 5 we get the existence of a universal n-bound t for $\mathfrak{M}$. Moreover, we get the existence of a corresponding n-css $\langle C, \ldots, a_{ij}, \ldots, \ldots, B_{kj}, \ldots \rangle$ (with $1 \leq i \leq l$, $1 \leq k \leq m$, $1 \leq j \leq n$). Let $r := t + n(l + m) + 2^n + 2$. There is some i, $1 \leq i \leq k(r)$, such that $\mathfrak{M} \sim_r^1 \mathfrak{M}_i^r$. By Lemma 6 t is a universal n-bound for $\mathfrak{M}_i^r$ and for all models $\sigma^1((\alpha_{ri}^k)^*, f_r)$, $k \in \omega$. Let $\langle C', \ldots, a'_{ij}, \ldots, \ldots, B'_{kj}, \ldots \rangle$ $(1 \leq i \leq l, 1 \leq k \leq m, 1 \leq j \leq n)$ be a corresponding n-css for $\sigma^1((\alpha_{ri}^r)^*, f_r)$. We define $\alpha$ to be the extended term $\langle \alpha_{ri}, \alpha_{r1}, \ldots, \alpha_{rk(r)} \rangle$. Then we get that t is also a universal n-bound for $\alpha^*$ and that $\langle C', \ldots, a'_{ij}, \ldots, \ldots, B'_{kj}, \ldots \rangle$ $(1 \leq i \leq l, 1 \leq k \leq m, 1 \leq j \leq n)$ is a corresponding n-css for $\alpha^*$ up to isomorphic embedding. This follows from the proof of Lemma 6, the definition of $\alpha^*$ and the fact that for all $k \in \omega$:

$$\alpha^* \upharpoonright S_{\alpha^*}^k(W(\alpha^*)) \text{ is isomorphic to } \sigma^1((\alpha_{ri}^k)^*, f_r) \upharpoonright S_{(\alpha_{ri}^k)^*}^k(W((\alpha_{ri}^k)^*).$$

Hint: Use claim (i) from page 30.

By Theorem 1 we get $\sigma^1((\alpha_{ri}^r)^*, f_r) \sim_n^1 \alpha^*$. Thus we have proved:

$$\mathfrak{M} \sim_r^1 \mathfrak{M}_i^r \sim_r^1 \sigma^1((\alpha_{ri}^r)^*, f_r) \sim_n^1 \alpha^*.$$

Q.E.D.

Moreover, we can prove:

<u>Theorem 3'.</u> Let $\mathfrak{M}$ be a tree and let n be a natural number. Then there is an extended term $\alpha$ not containing the operation symbol $\Sigma$ such that $\mathfrak{M} \sim_n^1 \alpha^*$ (with respect to $L_0(Q_1)$).

Let CF (F) be the class of trees (forests) and let $CF(\omega_1)$ $(F(\omega_1))$ be the class of uncountable trees (uncountable forests).

<u>Corollary</u> ([Herre, 1976], [Seese, Tuschik, 1977]). The theories $Th_1(CF)$, $Th_1(F)$, $Th_1(CF(\omega_1))$, and $Th_1(F(\omega_1))$ are decidable.

<u>Proof.</u> The decidability of $Th_1(F)$ follows from Theorem 2, Theorem 3, and Theorem 7 of Chapter 1. The decidability of the other theories we get by interpretability.

Q.E.D.

# CHAPTER 5

# LINEAR ORDERINGS

In many branches of mathematics linear orderings occur, e.g. in the formulation of Zorn's Lemma, or the Well-ordering Principle, or forming chains of structures, or simply the ordering of the natural numbers. Thus it is very natural to study the general properties of linear orderings more precisely. At first let us note that the notion of a linear ordering is expressible in a first order language L. A. Ehrenfeucht proved in [Ehrenfeucht, 1959] that the elementary theory LO of linear orderings is decidable. This very remarkable result shed some light on the expressive power of LO. By adding the generalized quantifiers $Q_{\alpha_0},\dots,Q_{\alpha_m}$ to L we get the stronger language $L(Q_{\alpha_0},\dots,Q_{\alpha_m})$. E.g. assume that the extended language contains the quantifier $Q_0$, then the natural numbers with their usual ordering possess a categorical finite system of axioms in the extended language. Suppose $\alpha_0 = 0$ and $\alpha_1,\dots,\alpha_m$ are successor ordinals, then the more complex theory $LO_{\bar{\alpha}}$ remains decidable where $\bar{\alpha} = \langle\alpha_0,\dots,\alpha_m\rangle$. This was proved by [H. P. Tuschik, 1977] under various set-theoretical assumptions weaker than the generalized continuum hypothesis (GCH). In the following we will work in ZFC + GCH, i.e. Zermelo-Fraenkel set theory with the generalized continuum hypothesis. Our considerations are very similar to the one made by [H. Läuchli and J. Leonhard, 1966] and in fact we extend their methods to generalized quantifiers.

## § 1. Notations

Let L be the first order language with identity and one binary predicate $<$. $L(Q_{\alpha_0},\dots,Q_{\alpha_m})$ arises from L by adding the generalized quantifiers $Q_{\alpha_0},\dots,Q_{\alpha_m}$ to it. We suppose in the following that $\alpha_1,\dots,\alpha_m$ are successor ordinals in their natural ordering, i.e. $\alpha_1 < \alpha_2 < \dots < \alpha_m$. Since we want to have the important quantifiers $Q_0$

and $Q_1$ among $Q_{\alpha_0},\dots,Q_{\alpha_m}$, we set $\alpha_0 = 0$ and $\alpha_1 = 1$. For notational convenience we shorten $L(Q_{\alpha_0},\dots,Q_{\alpha_m})$ by $L(Q_{\bar{\alpha}})$. The sequence $\bar{\alpha} = \langle\alpha_0,\dots,\alpha_m\rangle$ is assumed to be fixed throughout the whole chapter. All properties and all notions introduced are regarded with respect to $\bar{\alpha}$, so we omit the subscript $\bar{\alpha}$. In case we have to consider the quantifiers $Q_{\alpha_0},\dots,Q_{\alpha_m}$ one by one, then $Q_{\bar{\alpha}}$ stands for any $Q_{\alpha_i}$.

In the language L we formulate the elementary theory LO consisting of the axioms

(1) $\neg x < x$,

(2) $x < y \wedge y < z \rightarrow x < z$,

(3) $x < y \vee y < x \vee x \doteq y$,

i.e. $<$ is irreflexive, transitive and linear (any two elements are comparable). $a < b$ is read as "a is smaller than b" or equivalently "b is greater than a". The models of LO are called linear orderings or sometimes ordered sets and are denoted by $\mathfrak{A}, \mathfrak{B},\dots$ or $\langle A, <\rangle$, $\langle B,<\rangle,\dots$ or simply $A,B,\dots$ if necessary, we write $x < y \pmod A$ to indicate that x is smaller than y in the ordering of A. To emphasize that A is regarded as a set rather than an ordering we write $|A|$ and call it as usual the field of A. In connection with $<$ we use the defined symbols $\leq$, $\geq$, and $>$ with their obvious meanings. To be complete we cite some common notions. Let A be a linear ordering, $X \subseteq A$ be a given subset, and $a,b \in A$ be elements of A. Then

(a) a is an upper bound for X iff for all $x \in X$, $x \leq a \pmod A$,
(b) a is a lower bound for X iff for all $x \in X$, $a \leq x \pmod A$,
(c) a is a greatest element of A iff a is an upper bound for A,
(d) a is a least element of A iff a is a lower bound for A,
(e) X is an initial segment of A (with bound a) denoted by $A^{<a}$ iff $X = \{x \in A: x < a \pmod A\}$,
(f) X is a final segment of A (with bound a) denoted by $A^{>a}$ iff $X = \{x \in A: x > a \pmod A\}$,
(g) X is the open interval $(a,b)_A$ iff $X = \{x \in A: a < x < b \pmod A\}$,
(h) X is the left-open and right-closed interval $(a,b]_A$ iff $X = \{x \in A: a < x \leq b \pmod A\}$,
(i) X is the left-closed and right-open interval $[a,b)_A$ iff $X = \{x \in A: a \leq x < b \pmod A\}$,

(j) X is the closed interval $[a,b]_A$ iff $X = \{x \in A: a \leq x \leq b(\mathrm{mod}\ A)\}$,
(k) X is cofinal in A iff for every $x \in A$ there is some $y \in X$ so that $x \leq y$.

If we use the notations $(a,b)_A$, $[a,b)_A$, $(a,b]_A$, or $[a,b]_A$, we always assume that $a \leq b(\mathrm{mod}\ A)$. We will omit the subscript A if A is clear from the context.

Now we give some important examples of linear orderings:

(1) $\underline{1}$ denotes the unique one-element linear ordering with field $\{0\}$.
(2) For every ordinal $\alpha$ let $\langle \omega_\alpha, < \rangle$, or simply $\omega_\alpha$, denote the linear ordering of the ordinals smaller than $\omega_\alpha$. Especially $\omega = \omega_0$ denotes the linear ordering of the natural numbers with their usual ordering.
(3) Let $\langle P, < \rangle$ be the set of rational numbers with their usual ordering.
(4) $\langle R, < \rangle$ is the set of real numbers with their usual ordering.

To get a feeling about the expressive power of the elementary theory of linear orderings and of generalized quantifiers, respectively, we list some properties expressible in the corresponding calculus. First remark that the structure $\underline{1}$ is finite, hence there is a sentence $\varphi$ in the elementary language so that every model of $\varphi$ is isomorphic to $\underline{1}$. For the ordered sets mentioned in (2) - (4) this is not the case.

The linear orderings cited in (1) and (2) possess a least element while P and R do not so. This is expressible in the elementary language:

$$A \models \forall y(a \leq y) \quad \text{iff} \quad a \text{ is the least element of } A.$$

Furthermore, it is expressible that an element x has an immediate predecessor and an immediate successor, respectively. In $\omega$ it holds that 0 is the least element and every element different from 0 has an immediate predecessor and an immediate successor, while in P and R no element will have an immediate predecessor and successor, respectively.

The structures P and R are dense linear orderings without endpoints, i.e. they are models of the theory DLO consisting of the axioms

(1) LO,

(2) $x < y \longrightarrow \exists z(x < z < y)$,

(3) $\forall x \exists y \exists z(y < x < z)$.

Many years ago G. Cantor showed that any two countable models of DLO are isomorphic. By Vaught's test DLO is complete, hence P and R can not be distinguished by elementary sentences.

Every initial segment of $\omega$ is finite. This fact can not be expressed by an elementary sentence. But with the additional quantifier $Q_0$ this property is formulated by the sentence $\forall x \neg Q_0 y(y < x)$. Furthermore, we even get a finite set of axioms for $\omega$ which is categorical.

Since $Q_0$ is eliminable from DLO (see [Tuschik, 1978]) P and R are also equivalent in the extended language. Yet with the additional quantifier $Q_1$, P and R can be distinguished. The theory DLO $\cup \{\neg Q_1 x(x = x)\}$ is a categorical set of sentences for P while R clearly is not a model of this theory.

Next we introduce some important dense linear orderings. Let $\varkappa$ be an uncountable cardinal number and $T_\varkappa$ be the set of all proper cofinal subsets of $\varkappa$,

$$T_\varkappa = \{x \subseteq \varkappa : x \neq \varkappa \text{ and } x \text{ is cofinal in } \varkappa\}.$$

For all natural numbers m and n let

$$(\varkappa)^{m,n} = \{\beta \in \varkappa : \beta = \omega^2 \cdot \alpha + \omega \cdot m + n \text{ for some } \alpha \in \varkappa\}$$

and $T_\varkappa^{m,n}$ be the set of all $x \in T_\varkappa$ which are "almost" a subset of $(\varkappa)^{m,n}$, where x is "almost" a subset of $(\varkappa)^{m,n}$ if x differs from a subset of $(\varkappa)^{m,n}$ by a bounded subset of $\varkappa$ only. Now $CT_\varkappa^{m,n}$ is the subset of $T_\varkappa^{m,n}$, which consists of all those x that are "almost" a final segment of $(\varkappa)^{m,n}$, i.e. x differs from a final segment of $(\varkappa)^{m,n}$ by a bounded subset of $\varkappa$ only. The order relation on $T_\varkappa$ is defined by

$x < y (\text{mod } T_\varkappa)$ iff there exists some ordinal $\alpha \in y - x$ so that for all $\beta < \alpha$ it holds that $\beta \in x \longleftrightarrow \beta \in y$.

The orders of $T_{\varkappa}^{m,n}$ and $CT_{\varkappa}^{m,n}$ are defined so that they become substructures of $T_{\varkappa}$.

For $\varkappa = \omega$ we have to modify the above construction. Instead of $\omega$ we use the ordinal $\omega^{\omega}$, e.g.

$$T_{\omega} = \{x \subseteq \omega^{\omega}: x \neq \omega^{\omega} \text{ and } x \text{ is cofinal in } \omega^{\omega}\}.$$

Furthermore, that x is "almost" a subset of $(\omega)^{m,n}$ means that x differs from a real subset of $(\omega)^{m,n}$ by a finite set. The details are left to the reader.

Lemma 1. For any infinite cardinal number $\varkappa$ and all natural numbers m and n the following holds:

(1) $T_{\varkappa}$, $T_{\varkappa}^{m,n}$, and $CT_{\varkappa}^{m,n}$ are dense linear orderings without endpoints,

(2) card $T_{\varkappa}^{m,n} = 2^{\varkappa}$ and card $CT_{\varkappa}^{m,n} = \varkappa$,

(3) for any different elements $x,y \in T_{\varkappa}$, $\text{card}((x,y)_{T_{\varkappa}} \cap T_{\varkappa}^{m,n}) = 2^{\varkappa}$ and $\text{card}((x,y)_{T_{\varkappa}} \cap CT_{\varkappa}^{m,n}) = \varkappa$.

Proof. Clearly, $T_{\varkappa}$ and $T_{\varkappa}^{m,n}$ have cardinality $2^{\varkappa}$ while $CT_{\varkappa}^{m,n}$ has cardinality $\varkappa$. Remark that $CT_{\varkappa}^{m,n}$ has the same cardinality as the set of all bounded subsets of $\varkappa$ which has cardinality $\sum_{\lambda<\varkappa} 2^{\lambda} = \varkappa$. Now let $x,y \in T_{\varkappa}$, $x < y (\text{mod } T_{\varkappa})$, and $\alpha$ be the least ordinal in $y - x$. Since y is cofinal in $\varkappa$, there is a least $\beta \in y$ with $\beta > \alpha$. Set

$$T_{\varkappa}^{m,n}(y \upharpoonright \beta) = \{z \in T_{\varkappa}^{m,n}: \forall \gamma < \beta(\gamma \in z \leftrightarrow \gamma \in y) \wedge \beta \notin z\}.$$

Clearly, we have card $T_{\varkappa}^{m,n}(y \upharpoonright \beta) = 2^{\varkappa}$. Now assume that $z \in T_{\varkappa}^{m,n}(y \upharpoonright \beta)$, then $\beta \in y - z$ and for all $\gamma < \beta$, $\gamma \in y \leftrightarrow \gamma \in z$, hence $z < y$. Furthermore, $x < z$ because of $z \cap (\alpha + 1) = y \cap (\alpha + 1)$ and by the definition of $\alpha$. Similarly we can define $CT_{\varkappa}^{m,n}(y \upharpoonright \beta)$ and obtain card $CT_{\varkappa}^{m,n}(y \upharpoonright \beta) = \varkappa$. But this implies (3) of the lemma.

Q.E.D.

From $T_{\varkappa}^{m,n}$ and $CT_{\varkappa}^{m,n}$ we get some derived linear orderings. Set

$$S_{\varkappa}^{m,n} = \bigcup_{0<i\leq m} CT^{i,0} \cup \bigcup_{0<j\leq n} T^{0,j}$$

and regard $S_{\varkappa}^{m,n}$ as a substructure of $T_{\varkappa}$. For later use we remark that there is a mapping $f : S_{\varkappa}^{m,n} \longrightarrow m + n + 1$ defined by

$$f(x) = \begin{cases} i & \text{if } x \in CT_{\varkappa}^{i,0}, \\ i + m & \text{if } x \in T_{\varkappa}^{0,i}. \end{cases}$$

Every subset $B \subseteq A$ of an ordered set A can be regarded as a substructure by restricting the order relation of A to B. Since LO consists of universal axioms only, every substructure of a linear ordering is again a linear ordering. Some important substructures are the convex substructures also called segments.

Definition 1. A substructure $B \subseteq A$ of the ordered set A is a segment or a convex substructure iff for all elements $x,y \in B$, $(x,y)_A \subseteq B$. B is a bounded segment iff B is a segment which is included in some closed interval of A.

Examples of segments are the initial segments, the final segments, and all kinds of intervals.

As for other classes of structures homomorphisms are an important tool for studying structures.

Definition 2. Let A and B be linear orderings and $f : A \longrightarrow B$ be a mapping. f is called monotone iff for all elements $x,y \in A$, $x \leq y$ implies $f(x) \leq f(y)$.

We remark that a monotone mapping is nothing else but a homomorphism with respect to the relation $\leq$. In the following we are mainly concerned with monotone mappings which are onto. Let $f : A \longrightarrow B$ be a monotone-onto mapping, then we define the factor ordering or quotient ordering, respectively, A/f as usual. The field of A/f consists of the equivalence classes of the canonical equivalence relation $\underset{f}{\sim}$ and the ordering of A/f is the induced ordering. To be precise: For $x,y \in A$ set

$$x \underset{f}{\sim} y \quad \text{iff} \quad f(x) = f(y).$$

Clearly, $\underset{f}{\sim}$ is an equivalence relation. Since f is monotone, $\underset{f}{\sim}$ is a congruence relation with respect to $\leq$. Thus A/f is well-defined and is really a linear ordering. As for other algebraic structures we have a Homomorphy Theorem.

Theorem 2 (Homomorphy Theorem). Let $f : A \longrightarrow B$ be a monotone-onto map of linear orderings, then
(1) $A/f$ and $B$ are isomorphic,
(2) the canonical map $\bar{f} : A \longrightarrow A/f$ is monotone and onto,
(3) for every $b \in B$, $f^{-1}(b)$ is a segment of $A$.

Proof. By routine calculations (1) - (3) are easily verified.

Q.E.D.

An ordered set $\mathfrak{A}$ is said to be a splitting of A if there is a monotone-onto map $f : A \longrightarrow \mathfrak{A}$ so that $\mathfrak{A} = A/f$. The elements of $\mathfrak{A}$ are called the parts of the splitting. By (3) above the parts of $\mathfrak{A}$ are segments of A.

Beside substructures and homomorphisms products play an important role. We introduce a general product. Let $(A_i : i \in I)$ be a family of ordered sets and $f : A \longrightarrow I$ be a mapping of the ordered set A onto the index set I. The general product $S_f(A_i : i \in I)$ consists of the field $\{(b,a) : a \in A \text{ and } b \in A_{f(a)}\}$ and the order relation defined by

$$(b,a) < (d,c)(\text{mod } S_f(A_i : i \in I)) \text{ iff } a < c(\text{mod } A) \text{ or } a = c \text{ and } b < d(\text{mod } A_{f(a)}).$$

For some important cases we use a special notation for the general product.

$\underline{2}$ is the ordered set with two elements 0 and 1 and with the usual ordering.

(1) $A = I = \underline{2}$ and $f : \underline{2} \longrightarrow \underline{2}$ is the identity, then $S_f(A_i : i \in I)$ is denoted by $A_0 + A_1$.

(2) $I = 1$, then $f : A \longrightarrow 1$ is uniquely determined and $S_f(A_i : i \in I)$ is denoted by $A_0 \cdot A$.

(3) Let $F = (A_i : 0 < i \leq m)$ and $G = (B_j : 0 < j \leq n)$ be two finite sequences of ordered sets. $(F,G)$ denotes the sequence obtained from F and G by concatenation. Let A be the linear ordering $S_\varkappa^{m,n}$ and $f : S_\varkappa^{m,n} \longrightarrow m + n + 1$ be the mapping defined above after the definition of $S_\varkappa^{m,n}$. Then $S_f(F,G)$ is denoted by $S_\varkappa^{m,n}(F,G)$, or simply $S_\varkappa(F,G)$.

Finally we introduce an operation $^*$ which gives the inverse relation, more precisely, $A^*$ has the same field as A and

$$x < y(\text{mod } A^*) \quad \text{iff} \quad y < x(\text{mod } A).$$

Mainly axiom (3) in the definition of LO implies that $A^*$ is again a linear ordering.

## § 2. Games for linear orderings

In the first chapter we introduced games for generalized quantifiers. If we have more than one generalized quantifier at the same time, the games are easily extended. Assume that the additional generalized quantifiers are $Q_{\alpha_0},\dots,Q_{\alpha_m}$, then player I decides at each step which quantifier round he wants to play. Suppose he decided for the quantifier $Q_{\alpha_i}$, then the game continues in the same way as for the quantifier $Q_{\alpha_i}$ alone. Games are a good tool for studying relational structures. In the case of linear orderings we get a simpler form of these games, namely the equivalence relations $\overset{n}{\sim}$ (we omit the subscript $\bar{\alpha} = (\alpha_0,\dots,\alpha_m)$ since $\bar{\alpha}$ is fixed). The main problem in this paragraph is to find satisfactory conditions on structures to be $\overset{n}{\sim}$-equivalent. Suppose A and B are ordered sets with $a \in A$ and $b \in B$. We say $X \subseteq A$ is the set of realizations of $(B,b)_n$ in A if
(1) every $a \in A$ so that $(A,a) \overset{n}{\sim} (B,b)$ belongs to X,
(2) for all $x \in X$, $(A,x) \overset{n}{\sim} (B,b)$.

We call the equivalence classes $(B,b)_n$ n-types and say that some type $(B,b)_n$ occurs in A iff there is some $a \in A$ so that $(A,a) \overset{n}{\sim} (B,b)$, and $(B,b)_n$ occurs $\omega_\alpha$-many in A iff $(B,b)_n$ has $\omega_\alpha$-many realizations in A.

Definition 3. For any two ordered sets A, B (possibly empty) and any elements $a \in A$, $b \in B$

(1) $A \overset{0}{\sim} B$ and $(A,a) \overset{0}{\sim} (B,b)$,
(2) $A \overset{n+1}{\sim} B$ iff every n-type $(X,x)_n$
  (i) occurs in A iff it occurs in B,
  (ii) occurs $\omega_{\bar{\alpha}}$-many in A iff it occurs $\omega_{\bar{\alpha}}$-many in B,
(3) $(A,a) \overset{n+1}{\sim} (B,b)$ iff $A^{<a} \overset{n}{\sim} B^{<b}$ and $A^{>a} \overset{n}{\sim} B^{>b}$.

Remark. We recall our convention on $^-$. Thus (ii) above reads: For every $j \leq m$, $(X,x)_n$ occurs $\omega_{\alpha_j}$-many in A iff it occurs $\omega_{\alpha_j}$-many in B.

Lemma 3. For each n there are only finitely many equivalence classes of $\overset{n}{\sim}$.

Proof. This easily follows from the definition.

Q.E.D.

Lemma 4. Let A, B, C be linear orderings so that there are monotone mappings $f : A \longrightarrow C$ and $g : B \longrightarrow C$. Suppose that for all $c \in C$ the equivalence $f^{-1}(c) \overset{n}{\sim} g^{-1}(c)$ is satisfied, then $A \overset{n}{\sim} B$.

Proof. The lemma is trivial for $n = 0$. Assume then that it is true for j and that we are given monotone mappings $f : A \longrightarrow C$ and $g : B \longrightarrow C$ so that for all $c \in C$, $f^{-1}(c) \overset{j+1}{\sim} g^{-1}(c)$. Let a A, then there is some $b \in g^{-1}(f(a))$ so that

$$(f^{-1}(f(a)),a) \overset{j}{\sim} (g^{-1}(f(a)),b).$$

Hence $f \upharpoonright A^{<a}$ and $g \upharpoonright B^{<b}$ satisfy the hypothesis of the lemma for j. By the induction hypothesis $A^{<a} \overset{j}{\sim} B^{<b}$. For the same reason we also obtain $A^{>a} \overset{j}{\sim} B^{>b}$. Then by definition $(A,a) \overset{j}{\sim} (B,b)$. Similarly we get for every $b \in B$ some $a \in A$ so that $(A,a) \overset{j}{\sim} (B,b)$. Thus any type $(X,x)_j$ occurs in A iff it occurs in B. Suppose some type $(X,x)_j$ occurs $\omega_{\bar{\alpha}}$-many in A, then by the regularity of $\omega_{\bar{\alpha}}$ there is some $c \in C$ so that $\omega_{\bar{\alpha}}$-many realizations belong to $f^{-1}(c)$ or there are $\omega_{\bar{\alpha}}$-many elements $c \in C$ so that $f^{-1}(c)$ contains at least one realization of $(X,x)_j$. In both cases the same holds for B, thus $A \overset{j+1}{\sim} B$ by definition.

Q.E.D.

Corollary 5. The relations $\overset{n}{\sim}$ are compatible with the operations $A + B$, $A \cdot \omega_\alpha$, and $A \cdot \omega_\alpha^*$, for any ordinal $\alpha$.

Proof. Assume we are given linear orderings A, B, C, and D so that $A \overset{n}{\sim} C$ and $B \overset{n}{\sim} D$. Define monotone mappings $f : A + B \longrightarrow \underline{2}$ and $g : C + D \longrightarrow \underline{2}$

$$f(x) = \begin{cases} 0 \text{ if } x \in A, \\ 1 \text{ if } x \in B, \end{cases} \quad \text{and} \quad g(x) = \begin{cases} 0 \text{ if } x \in C, \\ 1 \text{ if } x \in D. \end{cases}$$

Then the hypothesis of the lemma is satisfied, hence $A + B \overset{n}{\sim} C + D$.

By the definition of $A\cdot\omega_\alpha$ its field consists of pairs $(a,b)$, where $a \in A$ and $b \in \omega_\alpha$. It is easily checked that the projection on the second coordinate is a monotone mapping $f : A\cdot\omega_\alpha \longrightarrow \omega_\alpha$ onto $\omega_\alpha$. Now assume that there is another structure $B \overset{n}{\sim} A$, then let $g : B\cdot\omega_\alpha \longrightarrow \omega_\alpha$ be the projection on the second coordinate. Obviously the hypothesis of the lemma is satisfied, hence $A\cdot\omega_\alpha \overset{n}{\sim} B\cdot\omega_\alpha$.

Q.E.D.

Lemma 6. Let A, B, C, and D be linear orderings so that there are monotone mappings $f : A \longrightarrow C$ and $g : B \longrightarrow D$. Suppose $C \overset{n}{\sim} D$ and for all $c \in C$ and $d \in D$ it holds that $f^{-1}(c) \overset{n}{\sim} g^{-1}(d)$, then $A \overset{n}{\sim} B$.

Proof. The lemma is proved by induction on n. For $n = 0$ the lemma is obviously true. Assume then that it is true for j and we are given monotone mappings $f : A \longrightarrow C$ and $g : B \longrightarrow D$ so that for all $c \in C$ and all $d \in D$

$$f^{-1}(c) \overset{j+1}{\sim} g^{-1}(d) \quad \text{and} \quad C \overset{j+1}{\sim} D.$$

Let $a \in A$, then choose some $b' \in D$, $b \in B$ so that $g(b) = b'$ and

$$(C,f(a)) \overset{j}{\sim} (D,b'), \quad (g^{-1}(b'),b) \overset{j}{\sim} (f^{-1}(f(a)),a).$$

By the hypothesis b and b' really exist. Let $f{\restriction}$ and $g{\restriction}$ be the restrictions of f and g to $f^{-1}(C^{<f(a)})$ and $g^{-1}(D^{<g(b)})$, respectively. $f{\restriction}$ and $g{\restriction}$ satisfy the hypothesis of the lemma for j, thus by the induction hypothesis

$$f^{-1}(C^{<f(a)}) \overset{j}{\sim} g^{-1}(D^{<g(b)}).$$

Clearly,

$$A^{<a} = f^{-1}(C^{<f(a)}) + f^{-1}(a)^{<a}$$

and

$$B^{<b} = g^{-1}(D^{<g(b)}) + g^{-1}(b)^{<b},$$

then $A^{>a} \overset{j}{\sim} B^{>b}$ by Corollary 5. In the same way we obtain $A^{>a} \overset{j}{\sim} B^{>b}$, hence $(A,a) \overset{j}{\sim} (B,b)$. Thus some type $(X,x)_j$ occurs in A iff it occurs in B. If some type $(X,x)_j$ occurs $\omega_\alpha$-many in A, then for some $c \in C$ there are $\omega_\alpha$-many realizations of $(X,x)_j$ in $f^{-1}(c)$ or there are

$\omega_{\bar{\alpha}}$-many elements $c \in C$ so that $f^{-1}(c)$ contains at least one realization of $(X,x)_j$. In both cases the same holds for B. Interchanging A with B we get similar results, thus $A \overset{j+1}{\sim} B$ by definition.

Q.E.D.

Let A be an ordered set and $\mathfrak{A}$ be a splitting of A, i.e. $\mathfrak{A} = A/f$ for some monotone onto mapping f. We say that the n-type of an ordered set B occurs $\omega_\alpha$-dense in $\mathfrak{A}$ if for all $Y,Z \in \mathfrak{A}$, $Y < Z$, there are at least $\omega_\alpha$-many $X \in (Y,Z)_{\mathfrak{A}}$ with $X \overset{n}{\sim} B$.

<u>Lemma 7.</u> If ordered sets A, B admit splittings $\mathfrak{A}$, $\mathfrak{B}$, respectively, so that

(1) $\mathfrak{A}$ and $\mathfrak{B}$ have no endpoints,

(2) if some n-type X occurs in $\mathfrak{A}$ or $\mathfrak{B}$, then the n-type occurs $\omega$-dense in $\mathfrak{A}$ and $\mathfrak{B}$,

(3) any n-type X occuring at least $\omega_{\bar{\alpha}}$-many in $\mathfrak{A}$ or $\mathfrak{B}$ occurs $\omega_{\bar{\alpha}}$-dense in $\mathfrak{A}$ and $\mathfrak{B}$,

then $A \overset{n}{\sim} B$ is valid.

<u>Proof.</u> By induction on n. Clearly, the lemma holds for n = 0. Assume then that it is true for j and we are given two ordered sets A,B which have splittings $\mathfrak{A}$, $\mathfrak{B}$, respectively, so that the conditions (1) - (3) above are satisfied for j + 1. For $a \in A$ let X be the part of $\mathfrak{A}$ which contains the element a. Choose $Y \in \mathfrak{B}$ so that $X \overset{j+1}{\sim} Y$ and some $b \in Y$ with

(i) $(X,a) \overset{j}{\sim} (Y,b)$.

$\mathfrak{A}^{<X}$ and $\mathfrak{B}^{<Y}$ form splittings of

$$A^{<X} = \bigcup_{C \in \mathfrak{A}^{<X}} C \quad \text{and} \quad B^{<Y} = \bigcup_{C \in \mathfrak{B}^{<Y}} C,$$

respectively, which satisfy the conditions (1) - (3) of the lemma for j. By the induction hypothesis

(ii) $A^{<X} \overset{j}{\sim} B^{<Y}$.

$A^{<X}$ and $X^{<a}$ partition $A^{<a}$ while similarly $B^{<Y}$ and $Y^{<b}$ partition $B^{<b}$. By (i), (ii), and Corollary 5 we get $A^{<a} \overset{j}{\sim} B^{<b}$ and for the same reason $A^{>a} \overset{j}{\sim} B^{>b}$, thus $(A,a) \overset{j}{\sim} (B,b)$.

Suppose there are $\omega_{\bar{\alpha}}$-many realizations of the j-type $(X,x)_j$ in A, then by the regularity $\omega_{\bar{\alpha}}$-many realizations belong to one part of the splitting or at least $\omega_{\bar{\alpha}}$-many parts contain at least one realization of $(X,x)_j$. By the assumptions on $\mathfrak{A}$ and $\mathfrak{B}$ made in the lemma we easily conclude that at least $\omega_{\bar{\alpha}}$-many realizations of $(X,x)_j$ are in B. Interchanging A with B we get the other direction, thus $A \overset{j+1}{\sim} B$.

Q.E.D.

Let $X = \{x_\alpha : \alpha < \varkappa\}$ be a given set, then we can define a canonical order relation on X, namely

$$x_\alpha < x_\beta \quad \text{iff } \alpha < \beta.$$

Suppose $X = \{x_\alpha : \alpha < \varkappa\}$ is a subset of an ordered set A, $X \subseteq A$, and X with its canonical ordering is a substructure of A, then X is called a $\varkappa$-sequence in A.

<u>Definition 4.</u> A subset $X \subseteq A$ of an ordered set A is called <u>n-homogeneous</u> if for all elements $a,b,c,d \in X$, $a < b$ and $c < d$ imply $(a,b) \overset{n}{\sim} (c,d)$.

<u>Lemma 8.</u> Suppose A, B are ordered sets which possess n-homogeneous $\varkappa$-sequences $X \subseteq A$ and $Y \subseteq B$, respectively, so that

(1) X and Y are cofinal in A and B, respectively,

(2) $x_0$ and $y_0$ are the least elements of A and B, respectively,

(3) $(x_0,x_1)_A \overset{n}{\sim} (y_0,y_1)_B$,

then $A \overset{n}{\sim} B$ is valid.

<u>Proof.</u> By induction on n. Trivially the lemma is true for n = 0. Assume then that it is true for j and that we are given two ordered sets A and B which possess $\varkappa$-sequences $X \subseteq A$ and $Y \subseteq B$, respectively, so that the hypotheses of the lemma are satisfied for j + 1. For any $a \in A$ we will show that $(A,a)_j$ is realized in B. Given $a \in A$ there is by cofinality of X some $x_\beta$ with $a \leq x_\beta$. Suppose $x_\alpha < a$, then clearly there is some $b \in (y_\alpha, y_\beta]_B$ so that

$$(x_\alpha, a)_A \overset{j}{\sim} (y_\alpha, b)_B \quad \text{and} \quad (a, x_\beta)_A \overset{j}{\sim} (b, y_\beta)_B,$$

since the conditions of the lemma imply

$$(x_\alpha, x_\beta)_A \overset{j+1}{\sim} (y_\alpha, y_\beta)_B$$

for any ordinals $\alpha,\beta < \varkappa$. By the induction hypothesis it follows that $A^{\geq x_\beta} \overset{j}{\approx} B^{\geq y_\beta}$, hence

$$(A,a) \overset{j}{\approx} (B,b).$$

If $(A,a)_j$ occurs at least $\omega_{\bar{\alpha}}$-many in A, then there is either some $x_\alpha$ so that $\omega_{\bar{\alpha}}$-many realizations of $(A,a)_j$ are smaller than $x_\alpha$ or the set of realizations is cofinal. In both cases we get that $(A,a)_j$ has also $\omega_{\bar{\alpha}}$-many realizations in B. Thus $A \overset{j+1}{\approx} B$.

Q.E.D.

Let B be a linear ordering and $f : B \longrightarrow \omega + (\omega^* + \omega)\cdot\varkappa$ be a monotone onto mapping. By the definition of the general product in the first paragraph the elements of $\omega + (\omega^* + \omega)\cdot\varkappa$ have the form $(m,0)$ if they belong to the first $\omega$ and $(((n,k),\alpha),1)$, respectively, where n is an arbitrary natural number, $k = 0$ or $k = 1$, and $\alpha$ is an ordinal smaller than $\varkappa$. We hope the reader has a clear picture of the "simple" ordering $\omega + (\omega^* + \omega)\cdot\varkappa$, and we only use the above complicated notation of its elements to be precise. Suppose now that $f^{-1}(z)$ always contains a greatest element for any $z \in \omega + (\omega^* + \omega)\cdot\varkappa$. Let $\alpha$ be an ordinal with $\omega \leq \alpha$ and $\alpha < \varkappa$. Then there is a unique ordinal $\beta$ and a finite ordinal n so that $\alpha = \omega + \omega\cdot\beta + n$.

Now assume $\alpha < \varkappa$ is given.

1. $\alpha$ is finite. Then let $y_\alpha$ be the greatest element of B in the part $f^{-1}((\alpha,0))$.

2. $\omega \leq \alpha$. Then let $\beta$ and n be so that we have a decomposition $\alpha = \omega + \omega\cdot\beta + n$. Define the element $y_\alpha$ to be the greatest element in the part $f^{-1}((((n,1),\beta),1))$.

Set $Y_f = \{y_\alpha : \alpha < \varkappa\}$, where $y_\alpha$ are the elements of B as defined above. It is easily checked that $Y_f$ is a cofinal $\varkappa$-sequence.

Lemma 9. Suppose A and B are linear orderings so that
(1) A possesses a cofinal, n-homogeneous $\varkappa$-sequence X,
(2) there is a monotone onto mapping $f : B \longrightarrow \omega + (\omega^* + \omega)\cdot\varkappa$ so that

$$f^{-1}((0,0)) \overset{n}{\approx} A^{<x_0}$$

and for all $z \in \omega + (\omega^* + \omega)\cdot\varkappa$, $z \neq (0,0)$, $f^{-1}(z)$ has a greatest element and

$$f^{-1}(z)^{<} \overset{n}{\sim} (x_0,x_1)_A.$$

Then
(i) the cofinal $\varkappa$-sequence $Y_f$ of B is n-homogeneous, and
(ii) $A \overset{n}{\sim} B$.

<u>Proof.</u> By induction on n. The lemma is trivial for n = 0. Assume then that it is true for n and that there are ordered sets A and B, and a monotone onto mapping $f : B \twoheadrightarrow \omega + (\omega^{*} + \omega)\cdot\varkappa$, and a cofinal $\varkappa$-sequence X in A so that the hypotheses of the lemma are satisfied for n + 1. We prove (i) by transfinite induction. Precisely, we prove for all $\alpha$ and $\beta$ that $(y_\alpha,y_\beta)_B \overset{n+1}{\sim} (x_\alpha,x_\beta)_A$ by induction on $\delta$, where $\beta = \alpha + \delta$. In the following, intervals with $x_\alpha$, $y_\alpha$ are always intervals in A and B, respectively, therefore we will avoid the subscript A and B, respectively.

<u>1. Case.</u> $\delta = \gamma + 1$. Then the interval $(y_\alpha,y_{\alpha+\delta})$ can be partitioned in the following way

$$(y_\alpha,y_{\alpha+\delta}) = (y_\alpha,y_{\alpha+\gamma}) + \{y_{\alpha+\gamma}\} + (y_{\alpha+\gamma},y_{\alpha+\gamma+1})$$

and by Lemma 4 $(y_\alpha,y_{\alpha+\delta}) \overset{n+1}{\sim} (x_\alpha,x_{\alpha+\delta})$ since by the induction hypothesis $(y_\alpha,y_{\alpha+\gamma}) \overset{n+1}{\sim} (x_\alpha,x_{\alpha+\gamma})$ and by the hypothesis of the lemma $(y_{\alpha+\gamma},y_{\alpha+\gamma+1}) \overset{n+1}{\sim} (x_0,x_1)$ and X is n-homogeneous.

<u>2. Case.</u> $\delta$ is a limit ordinal. Let be $b \in (y_\alpha,y_{\alpha+\delta})$.

2.1. $b \in (y_\alpha,y_{\alpha+\gamma})$ for some $\gamma < \delta$. By the induction hypothesis on $\gamma$

$$(y_\alpha,y_{\alpha+\gamma}) \overset{n+1}{\sim} (x_\alpha,x_{\alpha+\gamma}).$$

Then choose some $a \in (x_\alpha,x_{\alpha+\gamma})$ so that $(y_\alpha,b) \overset{n}{\sim} (x_\alpha,a)$ and $(b,y_{\alpha+\gamma}) \overset{n}{\sim} (a,x_{\alpha+\gamma})$. By the induction hypothesis on n we have

$$(y_{\alpha+\gamma},y_{\alpha+\delta}) \overset{n}{\sim} (x_{\alpha+\gamma},x_{\alpha+\delta}),$$

hence $(b,y_{\alpha+\delta}) \overset{n}{\sim} (a,x_{\alpha+\delta})$, i.e. a realizes the n-type $((y_\alpha,y_{\alpha+\delta}),b)_n$ in $(x_\alpha,x_{\alpha+\delta})$.

2.2. $f(b) = (((i,0),\gamma),1)$ for some ordinal $\gamma$ by the definition of $y_{\alpha+\delta}$. By the induction hypothesis and Lemma 4

$$f^{-1}((((i,0),\gamma),1)) + \ldots + f^{-1}((((1,0),\gamma),1)) + f^{-1}((((0,0),\gamma),1))$$

is (n+1)-equivalent to

$$(x_{\alpha+\delta}, x_{\alpha+\delta+i+1})$$

since X is (n+1)-homogeneous, i.e. all occuring parts are (n+1)-equivalent to $(x_0,x_1)$. Then choose some $c \in (x_{\alpha+\delta}, x_{\alpha+\delta+1})$ so that

$$(x_{\alpha+\delta}, c) \overset{n}{\sim} f^{-1}((((i,0),\gamma),1))^{<b}$$

and

$$(c, x_{\alpha+\delta+1}) \overset{n}{\sim} f^{-1}((((i,0),\gamma),1))^{>b}.$$

From the induction hypothesis on n it follows that

$$\bigcup_{\lambda<\delta} (y_\alpha, y_{\alpha+\lambda}) \cup \bigcup_{k>i} f^{-1}((((k,0),\gamma),1)) \overset{n}{\sim} (x_\alpha, x_{\alpha+\delta}).$$

Now, as above by Lemma 4, b and c realize the same n-type in $(y_\alpha, y_{\alpha+\delta})$ and $(x_\alpha, x_{\alpha+\delta+i+1})$, respectively. Since $(x_\alpha, x_{\alpha+\delta})$ and $(x_\alpha, x_{\alpha+\delta+i+1})$ are (n+1)-equivalent, we can find some element $a \in (x_\alpha, x_{\alpha+\delta})$ which realizes the common n-type of c and b. Thus we have proved that every n-type which is realized in $(y_\alpha, y_{\alpha+\delta})$ is realized in $(x_\alpha, x_{\alpha+\delta})$, too.

In the same way we prove that every n-type which is realized in $(x_\alpha, x_{\alpha+\delta})$ is realized in $(y_\alpha, y_{\alpha+\delta})$. Firstly note that $(x_\alpha, x_{\alpha+\delta})$ and $(x_0, x_2)$ realize exactly the same n-types since X is (n+1)-homogeneous. Let $a \in (x_0, x_2)$.

<u>1. Case.</u> $a \le x_1$. By the induction hypothesis $(x_0,x_1) \overset{n+1}{\sim} (y_\alpha, y_{\alpha+1})$ and let b realize the same n-type in $(y_\alpha, y_{\alpha+1})$ as a does in $(x_0,x_1)$. By the induction hypothesis on n, $(x_1,x_2) \overset{n}{\sim} (y_{\alpha+1}, y_{\alpha+\delta})$. By Lemma 4, a and b realize the same n-type in $(x_0,x_2)$ and $(y_\alpha, y_{\alpha+\delta})$, respectively. Remark: In case $a = x_1$, choose $b = y_{\alpha+1}$.

<u>2. Case.</u> $x_1 < a$. Let $\gamma$ be so that $\alpha + \delta = \omega + \omega\cdot\gamma$, then by the assump-

tions of the lemma

$$f^{-1}((((0,0),\gamma),1)) \overset{n+1}{\sim} (x_1,x_2)$$

and by the induction hypothesis on n

$$\bigcup_{\lambda<\delta}(y_\alpha,y_{\alpha+\lambda}) \cup \bigcup_{k>0} f^{-1}((((k,0),\gamma),1)) \overset{n}{\sim} (x_0,x_1).$$

Then choose some element b so that a and b satisfy the same n-type in $(x_1,x_2)$ and $f^{-1}((((0,0),\gamma),1))$, respectively. By Lemma 4 and the properties above a and b will realize the same n-types in $(x_0,x_2)$ and $(y_\alpha,y_{\alpha+\delta})$, respectively.

From the proof above and the regularity of $\omega_{\bar{\alpha}}$ it easily follows that some n-type $(X,x)_n$ is realized at least $\omega_{\bar{\alpha}}$-many in $(x_\alpha,x_{\alpha+\delta})$ iff it is realized at least $\omega_{\bar{\alpha}}$-many in $(y_\alpha,y_{\alpha+\delta})$, hence $(x_\alpha,x_{\alpha+\delta})$ and $(y_\alpha,y_{\alpha+\delta})$ are (n+1)-equivalent. This completes the proof of (i).

It is easily verified that $A^{\geq x_0}$ and $B^{\geq y_0}$ with the cofinal (n+1)-homogeneous $\varkappa$-sequences X and $Y_f$, respectively, satisfy the hypotheses of Lemma 8, hence $A^{\geq x_0} \overset{n+1}{\sim} B^{\geq y_0}$, thus $A \overset{n+1}{\sim} B$.

Q.E.D.

Now we will determine the precise connection between the relations $\overset{n}{\sim}$ and the game theoretical n-equivalence. At the beginning of the paragraph we introduced games for many generalized quantifiers. If two ordered sets are game-theoretical n-equivalent, i.e. player II has a winning strategy in the games over n rounds, we will indicate this by $\sim_n^{\bar{\alpha}}$. Since the generalized quantifiers remain fixed, we avoid the subscript $\bar{\alpha}$. We hope the reader has no trouble in distinguishing the relations $\sim_n$ and $\overset{n}{\sim}$. But if someone has trouble, then the following lemma shows that it does not matter.

<u>Lemma 10.</u> For every natural number n the relations $\sim_n$ and $\overset{n}{\sim}$ are the same.

<u>Proof.</u> By induction on n. For n = 0 the lemma is true by definition. (We set any structures as game-theoretical 0-equivalent.) Assume the lemma is true for n and we are given two ordered sets A and B.

<u>1. Case.</u> $A \overset{n+1}{\sim} B$. We describe a winning strategy for player II in games over n + 1 rounds.

1.1. Suppose player I has chosen structure A and some element $a \in A$. By the (n+1)-equivalence player II finds an element $b \in B$ so that

$$A^{<a} \overset{n}{\approx} B^{<b} \quad \text{and} \quad A^{>a} \overset{n}{\approx} B^{>b}.$$

By the induction hypothesis

$$A^{<a} \underset{n}{\simeq} B^{<b} \quad \text{and} \quad A^{>a} \underset{n}{\simeq} B^{>b}.$$

Now player II continues the game according to his winning strategies cited above.

1.2. Suppose player I chooses a subset $X \subseteq A$ which has cardinality at least $\omega_{\bar{\alpha}}$. Define an equivalence relation on X: Two elements are equivalent if they realize the same n-type in A. Then let X be partitioned $X = X_1 \cup \ldots \cup X_k$ so that the $X_i$ are equivalence classes of the equivalence relation defined above. Clearly, at least one of the classes $X_i$, say $X_1$, has cardinality at least $\omega_{\bar{\alpha}}$. By the (n+1)-equivalence of A and B player II finds a subset $Y \subseteq B$ so that $Y = Y_1 \cup \ldots \cup Y_k$ with

(1) all $Y_i$ are non-empty,
(2) the elements of $Y_i$ realize the same n-type as the elements of $X_i$,
(3) $Y_1$ has cardinality at least $\omega_{\bar{\alpha}}$.

If player I continues the game by choosing an element $a \in X_i$, then player II takes an element $b \in Y_i$. By the construction of Y we obtain

$$A^{<a} \overset{n}{\approx} B^{<b} \quad \text{and} \quad A^{>a} \overset{n}{\approx} B^{>b}.$$

By the induction hypothesis

$$A^{<a} \underset{n}{\simeq} B^{<b} \quad \text{and} \quad A^{>a} \underset{n}{\simeq} B^{>b}.$$

As in the case 1.1 player II now continues the game.

2. Case. $A \underset{n+1}{\simeq} B$. Let $a \in A$. Then we can assume that player I has chosen the element a at the first step of an (n+1)-round game. According to the winning strategy of player II we choose an element $b \in B$. Restricting the continuation of the game to elements smaller or greater than a and b, respectively, player II possesses a winning

strategy for the modified games over n rounds, thus

$$A^{<a} \underset{n}{\sim} B^{<b} \quad \text{and} \quad A^{>a} \underset{n}{\sim} B^{>b}.$$

By the induction hypothesis

$$A^{<a} \overset{n}{\sim} B^{<b} \quad \text{and} \quad A^{>a} \overset{n}{\sim} B^{>b}.$$

Thus b realizes the same n-type as a.

Suppose some n-type $(X,x)_n$ has $\omega_{\bar{\alpha}}$-many realizations in A, then let C be the set of elements of A which realize the type $(X,x)_n$. We can assume that player I chooses the subset C. By the winning strategy of player II we find a subset $D \subseteq B$ of cardinality at least $\omega_{\bar{\alpha}}$. Let $b \in D$ and $a \in C$, then as above

$$A^{<a} \underset{n}{\sim} B^{<b} \quad \text{and} \quad A^{>a} \underset{n}{\sim} B^{>b}.$$

By the induction hypothesis

$$A^{<a} \overset{n}{\sim} B^{<b} \quad \text{and} \quad A^{>a} \overset{n}{\sim} B^{>b},$$

thus a and b realize the same n-type in A and B, respectively. Since all $a \in C$ realize the n-type $(X,x)_n$, all elements $b \in D$ realize the n-type $(X,x)_n$, too. Hence $A \overset{n+1}{\sim} B$.

Q.E.D.

<u>Corollary 11.</u> Suppose A and B are linear orderings so that $A \overset{n}{\sim} B$. Then for every sentence $\varphi$ of $L(Q_{\bar{\alpha}})$ with at most n quantifiers it holds that

$$A \models \varphi \quad \text{iff} \quad B \models \varphi.$$

<u>Proof.</u> By the preceding lemma and Theorem 3 of the first chapter.

Q.E.D.

<u>Lemma 12.</u> For every natural number n and every linear ordering A there is a sentence $\varphi_{A,n}$ of $L(Q_{\bar{\alpha}})$ so that for all linear orderings B it holds:

$$B \overset{n}{\sim} A \quad \text{iff } B \models \varphi_{A,n}.$$

<u>Proof.</u> This is a general result for game-theoretical n-equivalence for

structures of finite similarity type. The proof is by induction on n. To get $\varphi_{A,n+1}$ we have to describe which n-types occur in A and to prove that no other n-types occur. But the n-type of an element $a \in A$ is expressible by $\varphi_{A^{<a},n} \wedge \varphi_{A^{>a},n}$ according to the induction hypothesis. If some n-type occurs $\omega_{\bar{\alpha}}$-many, then this is expressible with the generalized quantifier $Q_{\bar{\alpha}}$. We leave the details as an exercise to the reader.

Q.E.D.

Corollary 13. For any two ordered sets A and B

$$A \equiv B(\text{mod } L(Q_{\bar{\alpha}})) \quad \text{iff} \quad \text{for all } n,\ A \overset{n}{\sim} B.$$

Proof. By a routine calculation using Corollary 11 and Lemma 12.

Q.E.D.

## § 3. Combinatorics

To get deeper results we need a combinatorial lemma which is a substitute for the Ramsey-Theorem, which is not applicable to structures of higher cardinality. The following lemma is implied by the Ramsey-Theorem for additive colourings. We give [Shelah's, 1975] proof here.

Lemma 14. Let A be an ordered set with a least element. Suppose A possesses a cofinal $\varkappa$-sequence X. Then for every natural number n there exists a cofinal, n-homogeneous $\varkappa$-sequence $X_n$.

Proof. Define: For $x,y \in X$

$$x \sim y \quad \text{iff} \quad \text{there is a } z \in X,\ x,y < z, \text{ so that } (x,z) \overset{n}{\sim} (y,z);$$

clearly this implies that for any $z' \in X$, $z < z'$,

$$(x,z') \overset{n}{\sim} (y,z')$$

by Corollary 5.

It is easily verified that $\sim$ is an equivalence relation with finitely many equivalence classes only. So there is at least one equivalence class I which is an unbounded subset of X. Let $y_0$ be the first element of I. Furthermore, define an equivalence relation $\approx$ on I:

$$x \approx y \quad \text{iff} \quad (y_0,x) \overset{n}{\sim} (y_0,y).$$

There are only finitely many equivalence classes of $\approx$ on I. So there is at least one équivalence class $J \subseteq I$ which is cofinal. Then we define $X_n \subseteq J$, $X_n = \{y_\alpha : \alpha < \varkappa\}$, by induction:

(1) $y_0$ is already defined,
(2) $y_\alpha$ is the least element y of J so that for all $\beta < \alpha$, $(y_0,y) \overset{n}{\sim} (y_\beta,y)$.

Since I and J are well-orderings of type $\varkappa$, $X_n$ is well defined and has order type $\varkappa$. Now $X_n$ is the desired set because for every $y_\alpha, y_\beta \in X_n$, $\alpha < \beta$,

$$(y_\alpha,y_\beta) \overset{n}{\sim} (y_0,y_\beta)$$

by the definition of $y_\beta$ and

$$(y_0,y_1) \overset{n}{\sim} (y_0,y_\beta)$$

since $y_1 \approx y$, hence $X_n$ is n-homogeneous.

Q.E.D.

The use of the above lemma instead of the Ramsey-Theorem is the main difference between the elementary case and the case with generalized quantifiers.

## § 4. Well-orderings

The class of well-orderings is an important non-elementary subclass of the class of linear orderings. Especially the Axiom of Choice increases the importance of well-orderings since it implies that every set can be well-ordered. And we are able to prove theorems by transfinite induction. That a linear ordering A is a well-ordering can be expressed by: any substructure of A has a least element. Clearly this notion is not elementary, it is a second-order notion. We are mainly interested in the isomorphism types of well-orderings rather than in well-ordered sets. To avoid set-theoretical difficulties we identify the isomorphism type of a well-ordering with the ordinal contained in it. Now we can consider any operations on well-orderings as being operations on ordinals.

<u>Lemma 15.</u> For any ordinals $\alpha$, $\beta$, and $\gamma$ it holds that
(i) $\beta < \gamma$ implies $\alpha + \beta < \alpha + \gamma$,
(ii) $\alpha < \beta$ iff there is an ordinal $\delta > 0$ so that $\alpha + \delta = \beta$,
(iii) for every natural number n there are unique ordinals $\beta_0, \ldots, \beta_n$ with $\beta_i < \beta$ for $i < n$ so that

$$\alpha = \beta^n \cdot \beta_n + \ldots + \beta_0.$$

<u>Proof.</u> Let $\alpha$ and $\beta$ be ordinals, then $\alpha < \beta$ iff there is a proper embedding of $\alpha$ into $\beta$. For this reason (i) follows immediately. (ii) and (iii) are proved by transfinite induction on $\beta$.

Q.E.D.

Next we want to classify well-orderings in the elementary theory, i.e. in the following we assume that $\bar{\alpha}$ is the empty sequence. First we consider finite linear orderings more precisely. Clearly any finite linear ordering is a well-ordering. The following lemma shows that for finite linear orderings the equivalence relations $\overset{n}{\sim}$ can be characterized by cardinal numbers of the orderings.

<u>Lemma 16.</u> Let A and B be any two finite linear orderings with cardinality at least $2^n - 1$. Then $A \overset{n}{\sim} B$ $(\bar{\alpha} = \emptyset)$.

<u>Proof.</u> By induction on n. For $n = 0$ the lemma is trivially true. Assume now the lemma is true for n and that we are given two finite linear orderings A and B with cardinality at least $2^{n+1} - 1$. Let $a \in A$.

<u>1. Case.</u> card $A^{<a} < 2^n - 1$. Then choose the unique element b so that $A^{<a}$ and $B^{<b}$ are isomorphic. By the cardinality assumption on A and B

$$\text{card } A^{>a} \geq 2^n - 1 \quad \text{and} \quad \text{card } B^{>b} \geq 2^n - 1,$$

hence $A^{>a} \overset{n}{\sim} B^{>b}$ by the induction hypothesis. Thus a and b realize the same n-type in A and B, respectively.

<u>2. Case.</u> card $A^{>a} < 2^n - 1$. Then choose the unique element b so that $A^{>a}$ and $B^{>b}$ are isomorphic. As in the 1. case, a and b realize the same n-type in A and B, respectively.

<u>3. Case.</u> card $A^{<a} \geq 2^n - 1$ and card $A^{>a} \geq 2^n - 1$. Let b be the unique element of b so that card $B^{<b} = 2^n - 1$. Then card $B^{>b} \geq 2^n - 1$

because card $B \geq 2^{n+1} - 1$. By the induction hypothesis

$$A^{<a} \overset{n}{\approx} B^{<b} \quad \text{and} \quad A^{>a} \overset{n}{\approx} B^{>b}.$$

By all three cases above we can conclude that A and B realize exactly the same n-types, hence $A \overset{n+1}{\approx} B$.

Q.E.D.

Lemma 17. For any two ordinals $\alpha$ and $\beta$ and all natural numbers n:
(i) $\omega^n \cdot \alpha \overset{n}{\approx} \omega^n \cdot \beta$ and
(ii) $\omega^n \cdot \alpha \overset{n}{\approx} \omega^n$.

Proof. Clearly (i) and (ii) are equivalent. We prove (ii) by induction on n. For $n = 0$ the lemma becomes trivial as usual. Assume then that it is true for n and that we are given an ordinal $\alpha$.

1. Case. Let $a \in \omega^{n+1}$. Since we can regard $\omega^{n+1}$ as an initial segment of $\omega^{n+1} \cdot \alpha$, we may assume $a \in \omega^{n+1} \cdot \alpha$. Then $(\omega^{n+1})^{>a} \cong \omega^{n+1}$ and for some ordinal $\beta$, $(\omega^{n+1} \cdot \alpha)^{>a} \cong \omega^{n+1} \cdot \beta$. Since $\omega^{n+1} = \omega^n \cdot \omega$ and $\omega^{n+1} \cdot \beta = \omega^n \cdot \omega \cdot \beta$, $\omega^{n+1} \overset{n}{\approx} \omega^{n+1} \cdot \beta$ by the induction hypothesis, hence a realizes the same n-type in $\omega^{n+1} \cdot \alpha$ as in $\omega^{n+1}$.

2. Case. Suppose $a \in \omega^{n+1} \cdot \alpha$. a itself is again an ordinal. By Lemma 15 a can be represented as $a = \omega^{n+1} \cdot \beta + \omega^n \cdot p + q$ for some $\beta$, p, and q with p finite and $q < \omega^n$. By the induction hypothesis and Lemma 4:

$$\omega^n \cdot p + q \overset{n}{\approx} \omega^n \cdot (\omega \cdot \beta + p) + q.$$

But $b = \omega^n \cdot p + q$ is an element of $\omega^{n+1}$ which realizes the same n-type in $\omega^{n+1}$ as in $\omega^{n+1} \cdot \alpha$ since, as in the 1. case,

$$(\omega^{n+1})^{>b} \overset{n}{\approx} (\omega^{n+1} \cdot \alpha)^{>a}.$$

Thus $\omega^{n+1}$ and $\omega^{n+1} \cdot \alpha$ realize the same n-types.

Q.E.D.

Next we define the n-reduced $R_n^{\emptyset}(\alpha)$ for every ordinal $\alpha$. By Lemma 15 $\alpha$ can be represented as $\alpha = \omega^n \cdot \alpha_n + \ldots + \alpha_0$, where $\alpha_0, \ldots, \alpha_{n-1}$ are finite ordinals. Then set

$$R_n^{\emptyset}(\alpha) = \omega^n \cdot \varepsilon + \omega^{n-1} \cdot p_{n-1} + \ldots + p_0$$

where $\varepsilon = 0$ iff $\alpha_n = 0$ and $\varepsilon = 1$ iff $\alpha_n > 0$ and for $i < n$

$$p_i = \begin{cases} \alpha_i & \text{iff } \alpha_i < 2^n - 1, \\ 2^n - 1 & \text{iff } \alpha_i \geq 2^n - 1. \end{cases}$$

Thus by Lemma 16 for $i < n$, $p_i \overset{n}{\sim} \alpha_i$ and by Lemma 6

$$\omega^i \cdot p_i \overset{n}{\sim} \omega^i \cdot \alpha_i .$$

By Lemma 17

$$\omega^n \cdot \alpha_n \overset{n}{\sim} \omega^n \cdot \varepsilon$$

and by Lemma 4

$$\alpha \overset{n}{\sim} R_n^{\emptyset}(\alpha).$$

Thus we have proved:

Lemma 18. For any ordinal $\alpha$ and all natural numbers n

$$\alpha \overset{n}{\sim} R_n^{\emptyset}(\alpha).$$

At the same time $R_n^{\emptyset}$ denotes the set of n-reduced ordinals, i.e. all ordinals $\alpha$ so that $R_n^{\emptyset}(\alpha) = \alpha$. We have the explicite description of $R_n^{\emptyset}$:

$$R_n^{\emptyset} = \{\alpha : \alpha = \omega^n \cdot \varepsilon + \omega^{n-1} \cdot p_{n-1} + \dots + p_0, \text{ where } \varepsilon = 0 \text{ or } \varepsilon = 1 \text{ and } p_0, \dots, p_{n-1} \text{ are smaller than } 2^n\}.$$

Corollary 19. Every ordinal number is n-equivalent to some ordinal in $R_n^{\emptyset}$.

Thus $R_n^{\emptyset}$ is a finite set of representatives of the equivalence classes of the relation $\overset{n}{\sim}$. Let $P_n^{\emptyset}$ be the subset of $R_n^{\emptyset}$ consisting of those elements for which there is at least one $p_i$ with $p_i = 2^n - 1$.

Next we want to extend our results about the elementary language to languages with generalized quantifiers. Just now we work again in $L(Q_{\bar{\alpha}})$ for arbitrary possible sequence $\bar{\alpha} = \langle \alpha_0, \dots, \alpha_m \rangle$.

Lemma 20. For any two ordinals $\alpha$ and $\beta$ and every natural number n
(i) $\omega_{\alpha_m}^n \cdot \alpha \overset{n}{\approx} \omega_{\alpha_m}^n \cdot \beta$ and
(ii) $\omega_{\alpha_m}^n \cdot \alpha \overset{n}{\approx} \omega_{\alpha_m}^n$.

Proof. Obviously (i) and (ii) are equivalent. We prove (ii) by induction on n. For n = 0 the lemma is trivially true. Assume then that the lemma is true for n and that we are given an ordinal $\alpha > 0$.

1. Case. Let $a \in \omega_{\alpha_m}^{n+1}$. Then, as in Lemma 17, we can consider $\omega_{\alpha_m}^{n+1}$ as an initial segment of $\omega_{\alpha_m}^{n+1} \cdot \alpha$. Now $(\omega_{\alpha_m}^{n+1})^{>a} \cong \omega_{\alpha_m}^{n+1}$ and for some ordinal $\beta$

$$(\omega_{\alpha_m}^{n+1} \cdot \alpha)^{>a} \cong \omega_{\alpha_m}^{n+1} \cdot \beta,$$

hence by the induction hypothesis

$$(\omega_{\alpha_m}^{n+1} \cdot \alpha)^{>a} \overset{n}{\approx} (\omega_{\alpha_m}^{n+1})^{>a}$$

and a realizes the same n-type in $\omega_{\alpha_m}^{n+1} \cdot \alpha$ as in $\omega_{\alpha_m}^{n+1}$. Thus every n-type occuring in $\omega_{\alpha_m}^{n+1}$ occurs in $\omega_{\alpha_m}^{n+1} \cdot \alpha$ and if some n-type occurs $\omega_{\bar{\alpha}}$-many in $\omega_{\alpha_m}^{n+1}$, then it occurs $\omega_{\bar{\alpha}}$-many in $\omega_{\alpha_m}^{n+1} \cdot \alpha$.

2. Case. Suppose $a \in \omega_{\alpha_m}^{n+1} \cdot \alpha$. We can consider a as an ordinal which has by Lemma 15 a representation

$$a = \omega_{\alpha_m}^{n+1} \cdot \beta_2 + \omega_{\alpha_m}^{n} \cdot \beta_1 + \beta$$

for some ordinals $\beta$, $\beta_1$, and $\beta_2$ with $\beta_1 < \omega_{\alpha_m}$, $\beta < \omega_{\alpha_m}^n$ and for some $\gamma$

$$(\omega_{\alpha_m}^{n+1} \cdot \alpha)^{>a} \cong \omega_{\alpha_m}^{n+1} \cdot \gamma.$$

Suppose $\beta_1$ and $\beta_2$ are not both 0, then

$$\omega_{\alpha_m}^{n} \cdot (\omega_{\alpha_m} \cdot \beta_2 + \beta_1) \overset{n}{\approx} \omega_{\alpha_m}^{n} \cdot \gamma$$

by the induction hypothesis for any ordinal $\gamma$. For every $\gamma < \omega_{\alpha_m}$, $b_\gamma = \omega_{\alpha_m}^n \cdot \gamma + \beta$ is an element of $\omega_{\alpha_m}^{n+1}$ which realizes the same n-type as a does. Thus the n-type realized by a is also realized in $\omega_{\alpha_m}^{n+1}$.

If some n-type occurs $\omega_{\bar{\alpha}}$-many in $\omega_{\alpha_m}^{n+1}\cdot\alpha$, then

(i) $\omega_{\bar{\alpha}}$-many realizations are contained in $\omega_{\alpha_m}^{n+1}$, or
(ii) some realization a has the form

$$a = \omega_{\alpha_m}^{n}\cdot\beta_1 + \beta \quad \text{with } \beta_1 > 0 \text{ and } \beta < \omega_{\alpha_m}^{n},$$

then for every $\gamma < \omega_{\alpha_m}$, $b_\gamma = \omega_{\alpha_m}^{n}\cdot\gamma + \beta$ belongs to $\omega_{\alpha_m}^{n+1}$ and realizes the same n-type as a does, hence there are $\omega_{\bar{\alpha}}$-many relizations in $\omega_{\alpha_m}^{n+1}$.

Thus the lemma is proved.

Q.E.D.

Only in the following we now allow to vary the sequence $\bar{\alpha} = (\alpha_0,\dots,\alpha_m)$. Firstly we define the n-$\bar{\alpha}$-reduced $R_n^{\bar{\alpha}}(\beta)$ of an ordinal $\beta$. We do this by induction on the length of $\bar{\alpha}$. $R_n^{\emptyset}$ is already defined. Suppose $R_n^{\bar{\alpha}}(\beta)$ is defined for $\bar{\alpha} = (\alpha_0,\dots,\alpha_m)$. Let $\omega_{\alpha_{m+1}} > \omega_{\alpha_m}$ and $\overline{\alpha'} = (\alpha_0,\dots,\alpha_{m+1})$. By Lemma 15 $\beta$ can be represented as

$$\beta = \omega_{\alpha_{m+1}}^{n}\cdot\beta_n + \dots + \beta_0,$$

where $\beta_0,\dots,\beta_{n-1}$ are ordinals smaller than $\omega_{\alpha_{m+1}}$. Set

$$R_n^{\overline{\alpha'}}(\beta) = \omega_{\alpha_{m+1}}^{n}\cdot\varepsilon + \omega_{\alpha_{m+1}}^{n-1}\cdot R_n^{\bar{\alpha}}(\beta_{n-1}) + \dots + R_n^{\bar{\alpha}}(\beta_0),$$

where $\varepsilon = 0$ iff $\beta_n = 0$, and $\varepsilon = 1$ iff $\beta_n > 0$.

<u>Lemma 21.</u> For every ordinal $\beta$, any sequence $\bar{\alpha}$, and all natural numbers n: $R_n^{\bar{\alpha}}(\beta) \overset{n}{\sim} \beta$.

<u>Proof.</u> By induction on $\bar{\alpha}$. For $\bar{\alpha} = \emptyset$ this is just Lemma 18. Assume the lemma is true for $\bar{\alpha} = (\alpha_0,\dots,\alpha_m)$. Let $\overline{\alpha'} = (\alpha_0,\dots,\alpha_{m+1})$ and $\beta$ be a given ordinal. Suppose $\beta$ is represented as

$$\beta = \omega_{\alpha_{m+1}}^{n}\cdot\beta_n + \omega_{\alpha_{m+1}}^{n-1}\cdot\beta_{n-1} + \dots + \beta_0$$

with $\beta_i < \omega_{\alpha_{m+1}}$ for $i < n$. By definition

$$R_n^{\overline{\alpha'}}(\beta) = \omega_{\alpha_{m+1}}^{n}\cdot\varepsilon + \omega_{\alpha_{m+1}}^{n-1}\cdot R_n^{\bar{\alpha}}(\beta_{n-1}) + \dots + R_n^{\bar{\alpha}}(\beta_0)$$

and by the induction hypothesis $\beta_i \overset{n}{\sim} R_n^{\bar{\alpha}}(\beta_i)$, where $\overset{n}{\sim}$ is the n-equivalence with respect to $\bar{\alpha}$. However, since $\beta_i < \omega_{\alpha_{m+1}}$, this extends to n-equivalence with respect to the sequence $\overline{\alpha'}$. By Lemma 20, Lemma 6, and Lemma 4

$$R_n^{\overline{\alpha'}}(\beta) \overset{n}{\sim} \beta.$$

Q.E.D.

Now we define the important subsets $P_n^{\bar{\alpha}}$ and $CP_n^{\bar{\alpha}}$ of $R_n^{\bar{\alpha}}$. $P_n^{\bar{\alpha}}$ will contain all those ordinals of $R_n^{\bar{\alpha}}$ which are n-equivalent to at least $\omega_{\alpha_m}$-many ordinals while $CP_n^{\bar{\alpha}}$ contains all those ordinals of $R_n^{\bar{\alpha}}$ which represent some n-type which occurs cofinally in the class of all ordinals.

$$P_n^{\emptyset} = \{\beta \in R_n^{\emptyset}: \text{for some } p_i \text{ in the representation of } \beta: p_i = 2^n - 1\}$$

and

$$CP_n^{\emptyset} = \{\beta \in P_n^{\emptyset}: \varepsilon = 1 \text{ in the representation of } \beta\}.$$

Assume $P_n^{\bar{\alpha}}$ and $CP_n^{\bar{\alpha}}$ are defined for $\bar{\alpha} = (\alpha_0,\ldots,\alpha_m)$. Let $\alpha_{m+1} > \alpha_m$ and $\overline{\alpha'} = (\alpha_0,\ldots,\alpha_{m+1})$. Then $R_n^{\overline{\alpha'}}$ is already defined. Define

$$P_n^{\overline{\alpha'}} = \{\beta \in R_n^{\overline{\alpha'}}: p_i \in CP_n^{\bar{\alpha}} \text{ for some } p_i \text{ in the representation of } \beta\}$$

and

$$CP_n^{\overline{\alpha'}} = \{\beta \in P_n^{\overline{\alpha'}}: \varepsilon = 1 \text{ in the representation of } \beta\}.$$

The sets $R_n^{\bar{\alpha}}$ are important for deciding whether some n-type occurs in some given ordinal while $CP_n^{\bar{\alpha}}$ is important for deciding whether some n-type occurs cofinally in some given ordinal. We express this in the following lemma, which will later be useful in the decision procedure.

We will make use of the following fact without mentioning it every time: if two ordered sets A and B of cardinality smaller than $\omega_{\alpha_{m+1}}$ are n-equivalent with respect to the sequence $\bar{\alpha} = (\alpha_0,\ldots,\alpha_m)$, i.e. $A \sim_n^{\bar{\alpha}} B$, then they are n-equivalent with respect to the sequence $\overline{\alpha'} = (\alpha_0,\ldots,\alpha_{m+1})$, i.e. $A \sim_n^{\overline{\alpha'}} B$. This easily follows from the definition.

<u>Lemma 22.</u> Let $\bar{\alpha} = (\alpha_0,\dots,\alpha_m)$ be a given possible sequence and $\varkappa$ be an infinite cardinal number. Then

(i) some n-type occurs $\omega_{\alpha_m}$-many in $\varkappa$ iff it has some realization in $P_n^{\bar{\alpha}}$,

(ii) some n-type is realized cofinally in the class of ordinals iff it has a realization in $CP_n^{\bar{\alpha}}$.

<u>Proof.</u> The possible sequences were defined at the beginning of this chapter. We prove the lemma by induction on $\bar{\alpha}$. Suppose $\bar{\alpha} = \emptyset$, in case (i) there is nothing to prove. (ii) follows from Lemmata 17 and 18 and the definition of $P_n^{\emptyset}$.

Assume then the lemma is true for all sequences shorter than $\bar{\alpha}' = (\alpha_0,\dots,\alpha_{m+1})$ and we are given a cardinal $\varkappa$.

Proof of (i): I (left to right). Suppose the n-type $(X,x)_n$ occurs $\omega_{\alpha_{m+1}}$-many in $\varkappa$. Then $\varkappa \geq \omega_{\alpha_{m+1}}$, and there are $\omega_{\alpha_{m+1}}$-many realizations smaller than $\omega_{\alpha_{m+1}}$, or between $\omega^n_{\alpha_{m+1}}$ and $\varkappa$, or between $\omega^j_{\alpha_{m+1}}$ and $\omega^{j+1}_{\alpha_{m+1}}$ for some $1 \leq j < n$.

I.1. $(X,x)_n$ has $\omega_{\alpha_{m+1}}$ realizations in $(\omega^n_{\alpha_{m+1}},\varkappa)$. Let a be a realization of $(X,x)_n$ in the cited interval, then $\varepsilon = 1$ in $R_n^{\bar{\alpha}'}(a)$, thus by Lemma 21 $(X,x)_n$ has a realization in $P_n^{\bar{\alpha}'}$.

I.2. $(X,x)_n$ has $\omega_{\alpha_{m+1}}$-many realizations in $(\omega^j_{\alpha_{m+1}},\omega^{j+1}_{\alpha_{m+1}})$. Let a be a realization, then a can be represented as

$$a = \omega^j_{\alpha_{m+1}} \cdot a_j + \dots + a_0.$$

Since there are $\omega_{\alpha_{m+1}}$-many realizations, at least one coordinate has a cofinal sequence, i.e. for some realization a we have $\varepsilon = 1$ in the representation of $R_n^{\bar{\alpha}}(a_i)$ for some coefficient $a_i$, thus $(X,x)_n$ has a realization in $P_n^{\bar{\alpha}'}$.

I.3. $(X,x)_n$ has $\omega_{\alpha_{m+1}}$-many realizations smaller than $\omega_{\alpha_{m+1}}$. This is proved as for I.2.

II (right to left). Suppose the n-type $(X,x)_n$ has a realization in $\varkappa \cap P_n^{\bar{\alpha}'}$, then $\varkappa \geq \omega_{\alpha_{m+1}}$.

II.1. $\varkappa = \omega_{\alpha_{m+1}}$. Then the realization of $(X,x)_n$ in $\varkappa \cap P_n^{\overline{\alpha'}}$ belongs to $CP_n^{\overline{\alpha}}$, hence by the induction hypothesis for (ii), $(X,x)_n$ is cofinally realized. Thus $(X,x)_n$ is realized $\omega_{\alpha_{m+1}}$-many in $\varkappa$.

II.2. $\varkappa > \omega_{\alpha_{m+1}}$. Then the ordinals from $\varkappa \cap P_n^{\overline{\alpha'}}$ belong to $CP_n^{\overline{\alpha}}$ or have the form $\omega^n_{\alpha_{m+1}} + \beta$. In the first case we continue as for II.1. In the second case we obtain $\omega_{\alpha_{m+1}}$-many realizations of $(X,x)_n$ in $\varkappa$ by Lemma 20.

Part (ii) follows from Lemmata 20 and 21.

Q.E.D.

To get more information about well-orderings we introduce further subsets of $R_n^{\overline{\alpha}}$. Let $\overline{\alpha} = (\alpha_0,\dots,\alpha_m)$ be a possible sequence. Then define

$$(\overline{\alpha})_k = \begin{cases} (\alpha_0,\dots,\alpha_k) & \text{for } k \leq m, \\ \overline{\alpha} & \text{for } k > m. \end{cases}$$

Now set $P_n^{(\overline{\alpha})_k,0} = P_n^{(\overline{\alpha})_k}$ and

$$P_n^{(\overline{\alpha})_k,i+1} = \{\beta \in R_n^{(\overline{\alpha})_{k+i+1}} : \varepsilon = 1 \text{ or } p_j \in P_n^{(\overline{\alpha})_k,i} \text{ for some } j \text{ in the representation of } \beta\}.$$

<u>Lemma 23.</u> Let $\overline{\alpha} = (\alpha_0,\dots,\alpha_m)$ be a possible sequence and $\varkappa$ be an infinite cardinal. Then some n-type $(X,x)_n$ occurs $\omega_{\alpha_k}$-many in $\varkappa$ iff $(X,x)_n$ is realized by some element of $\varkappa \cap P_n^{(\overline{\alpha})_k,m}$.

<u>Proof.</u> I (right to left). Assume $(X,x)_n$ is realized by some element of $\varkappa \cap P_n^{(\overline{\alpha})_k,m}$ then by Lemma 6, Lemma 4, and by the definition of $P_n^{(\overline{\alpha})_k,m}$, $(X,x)_n$ has $\omega_{\alpha_k}$-many realizations in $\varkappa$.

II (left to right). By induction on $m - k$. $m - k = 0$, i.e. $m = k$. Then the lemma is just part one of the preceding lemma, hence already proved. Assume the lemma is true for $m - k = j$, and we are given some n-type $(X,x)_n$ and we have to prove the lemma for $m - k = j + 1$, i.e. $k = m - j - 1$.

If $(X,x)_n$ has a realization in $\varkappa$ which is greater than $\omega^n_{\alpha_m}$, then as in case I.1 of the preceding lemma we can conclude that $(X,x)_n$ has

$\omega_{\alpha_m}$-many realizations in $\varkappa$ and then it has a realization in $\varkappa \cap P_n^{(\bar{\alpha})_{k,m}}$ by the induction hypothesis.

Suppose $(X,x)_n$ has $\omega_{\alpha_k}$-many realizations in some interval $(\omega^j_{\alpha_m}, \omega^{j+1}_{\alpha_m})$, then there are $\omega_{\alpha_k}$-many realizations a of $(X,x)_n$ with representations

$$a = \omega^j_{\alpha_m} \cdot a_j + \ldots + a_0$$

so that corresponding coefficients are n-equivalent. Thus for some $i \leq j$ there is some n-type $(Y,y)_n$ which is realized $\omega_{\alpha_k}$-many in $\omega_{\alpha_m}$ by coefficients $a_i$ of the above kind. Set $\overline{\alpha'} = (\alpha_0, \ldots, \alpha_{m-1})$. Then, by the induction hypothesis on $m - 1 - k = j$, $(Y,y)_n$ is realized by some element $b_i$ of $P_n^{(\overline{\alpha'})_{k,m-1}}$. For $p \neq i$ let $b_p$ be any element of $R_n^{\overline{\alpha'}}$ which is n-equivalent to $a_p$, i.e. $b_p \overset{n}{\sim} a_p$. By the definition of $P_n^{(\bar{\alpha})_{k,m}}$ it is then easily checked

$$b = \omega^j_{\alpha_m} \cdot b_j + \ldots + b_0$$

belongs to $P_n^{(\bar{\alpha})_{k,m}}$. By Lemmata 4 and 6 it follows that b realizes the n-type $(X,x)_n$.

The case that $\omega_{\alpha_k}$-many realizations are smaller than $\omega_{\alpha_m}$ is completely similar to the above case. Thus the lemma is proved.

Q.E.D.

## § 5. Terms

We have already introduced a lot of special linear orderings like $\underline{1}$, $\omega$, $\omega_1, \ldots$ and operations on the class of orderings like + and $\cdot$. Now we want to define in a constructive manner a set of linear orderings each of which is highly constructive. However, first we define a set of terms as in the first order model theory. We do this completely independent of linear orderings. We look at $\underline{1}$ as being only a constant. In addition we have certain function symbols which we describe now. There are the unary function symbols $w, w_1, w_2, \ldots, w_m$, $w^*, w^*_1, \ldots, w^*_m$ and the binary function symbol +. For every $j \leq m$ and all natural numbers k, n there is a (k+n)-ary function symbol $S^{k,n}_j$. We will avoid the subscripts k and n since the arity is uniquely determined by the number of arguments following the function symbol.

Thus we simply write $S_j$ instead of $S_j^{k,n}$. So $S_j$ can be regarded as operating on finite sequences of terms rather than as a function symbol. The set $M_m$ of terms is defined inductively, i.e. the set of terms $M_m$ is the least set satisfying

(1) $\underline{1}$ is a term,
(2) if t is a term, then for all $j \leq m$, $t \cdot w_j$ and $t \cdot w_j^*$ are terms,
(3) if $t_1$ and $t_2$ are terms, then $t_1 + t_2$ is a term,
(4) if F and G are finite sequences of terms, then $S_j(F,G)$ is a term for all $j \leq m$.

By the remark on $S_j$ made above we can really regard $M_m$ as a set of terms of a first order language. Next we want to introduce an interpretation of terms as linear orderings. The linear ordering which corresponds to the term t will be denoted by $(t)^o$. Sometimes we identify t and $(t)^o$ in case that there is no misunderstanding; and we use mainly t rather than $(t)^o$ to denote the corresponding linear ordering. We do this for notational convenience.

We consider $(\ )^o$ as a function of $M_m$ into the class of linear orderings which is defined inductively.

(1) $(\underline{1})^o := \underline{1}$,

i.e. we make no notational distinction between the constant $\underline{1}$ and the ordered set $\underline{1}$. We think that no difficulties will arise by this identification.

(2) If $t = t_1 \cdot w_j$, then $(t)^o := (t_1)^o \cdot \omega_{\alpha_j}$,

(3) if $t = t_1 \cdot w_j^*$, then $(t)^o := (t_1)^o \cdot \omega_{\alpha_j}^*$,

(4) if $t = t_1 + t_2$, then $(t)^o := (t_1)^o + (t_2)^o$,

(5) if F and G are finite sequences of terms, then
$(S_j(F,G))^o := S_{\omega_{\alpha_j - 1}}(((F)^o,(G)^o))$.

Remark that + denotes a function symbol and at the same time it denotes an operation between ordered sets.

The terms carry a lot of information, especially about decompositions of the corresponding linear ordering. For example let $t = w + w$, then there is a monotone onto map

$$f : (t)^{o} \rightarrow \underline{2}$$

so that each part $f^{-1}(0)$, $f^{-1}(1)$ is isomorphic to the ordering of the natural numbers.

Here we are not interested in the word problem of terms, i.e. in the various representations of linear orderings as terms.

Definition 5. A linear ordering A is said to be n-term-like iff there is a term $t \in M_m$ so that $A \overset{n}{\sim} t$.

Lemma 24. The class of n-term-like orderings is closed under the operations $w, w_1, \ldots, w_m, w^*, w_1^*, \ldots, w_m^*, +, S_1, \ldots, S_m$.

**We recall our convention to avoid the subscripts k, n of** $S_j$, i.e. $S_j$ represents a class of operations indexed by k, n.

Proof. Assume A is an n-term-like linear ordering and t is a term which represents A, i.e. $A \overset{n}{\sim} t$. Then by Corrollary 5 $A \cdot w, \ldots, A \cdot w_m$, $A \cdot w^*, \ldots, A \cdot w_m^*$ are n-equivalent to the terms $t \cdot w, \ldots, t \cdot w_m$, $t \cdot w^*, \ldots, t \cdot w_m^*$, respectively.

Suppose $A \overset{n}{\sim} t_1$ and $B \overset{n}{\sim} t_2$, then by Corrollary 5

$$A + B \overset{n}{\sim} t_1 + t_2.$$

Given finite sequences $F = (A_i : i < k)$ and $G = (B_j : j < l)$ of ordered sets so that there are finite sequences $F_t = (t_0, \ldots, t_{k-1})$ and $G_t = (s_0, \ldots, s_{l-1})$ of terms so that corresponding orderings are n-equivalent, then by Lemma 4

$$S_{\omega_{\alpha_j - 1}}(F, G) \overset{n}{\sim} S_j(F_t, G_t).$$

Q.E.D.

Furthermore we get

Lemma 25. $\overset{n}{\sim}$ is a congruence relation on $M_m$ with respect to +, $w, w_1, \ldots, w_m$, $w^*, \ldots, w_m^*$ and $S_1, \ldots, S_m$.

Proof. The same proof as for the preceding lemma. Q.E.D.

Corollary 26. $M_m/\overset{n}{\sim}$ is a finite algebraic structure with respect to the operations +, $w, w_1, \ldots, w_m$, $w^*, \ldots, w_m^*$ and $S_1, \ldots, S_m$.

Proof. By Lemma 3 $M_m/\overset{n}{\sim}$ is finite and by the preceding lemma $\overset{n}{\sim}$ is a congruence relation with respect to the cited operations.

Q.E.D.

Next we prove two key lemmata about the representation of n-term-like linear orderings.

Lemma 27. Let A be an ordered set of cardinality $\lambda$. If every bounded segment of A is n-term-like, then A itself is n-term-like.

Proof. We may assume that A has a least element. The proof is similar in the other cases: if A has a greatest element, then we use an analogue of Lemma 14 for the inverse ordering; in case A has no least and no greatest element, then we partition A = B + C + D, where B has a greatest element, D has a least element, and C is bounded.

Clearly, there is a cofinal sequence $X \subseteq A$. If X is finite, then A is n-term-like by Lemma 24. We may assume that X is a $\varkappa$-sequence for some $\varkappa \leq \lambda$. By Lemma 14 there is a cofinal, n-homogeneous $\varkappa$-sequence, i.e. we may suppose X is n-homogeneous.

Set $B_\alpha := A_0 + A_1 \cdot w + A_1 \cdot (w^* + w) \cdot \alpha$, and let $f : B_\varkappa \longrightarrow w + (w^* + w) \cdot \varkappa$ be the canonical, monotone, onto map which results from the representation of $B_\varkappa$ as indicated above. Instead of giving an exact definition of f we demonstrate how f is defined in a simpler case. Assume $B = A \cdot w$, then $g : B \longrightarrow w$ is defined by $g((a,n)) = n$.

Clearly, by the definition $f^{-1}((0,0)) \cong A_0$ and for all $z \in w + (w^*+w) \cdot \varkappa$, $z \neq (0,0)$ it holds that $f^{-1}(z) \cong A_1$. Note that $A_0 = A^{\leq x_0}$ and $A_1 = (x_0, x_1]_A$ and therefore $A_0$ and $A_1$ have both a greatest element. Thus all hypotheses of Lemma 9 are satisfied, hence we can conclude $B_\varkappa \overset{n}{\sim} A$. Suppose $\varkappa$ is n-term-like, then there are terms t, $t_0$, and $t_2$ so that

$$A_0 \overset{n}{\sim} t_0, \quad A_1 \overset{n}{\sim} t_1, \quad \text{and} \quad \varkappa \overset{n}{\sim} t,$$

hence by Lemma 6 and Lemma 4

$$B_\varkappa \overset{n}{\sim} t_0 + t_1\cdot(w^* + w)\cdot t,$$

thus A is n-term-like.

It remains to show that $\varkappa$ is n-term-like. We prove the stronger result that every ordinal $\alpha$ is n-term-like. We do this by transfinite induction on $\alpha$.

1. Case. $\alpha = 1$. $\underline{1}$ itself is a term.

2. Case. $\alpha = \beta + 1$. By the induction hypothesis there is a term t with $t \overset{n}{\sim} \beta$, then by Corollary 5: $t + \underline{1} \overset{n}{\sim} \alpha$.

3. Case. $\alpha$ is a limit ordinal. Let $j \leq m$ be the greatest natural number so that card $|\alpha| \geq \omega_{\alpha_j}$.

3.1. card $|\alpha| > \omega_{\alpha_j}$. By the Löwenheim-Skolem-Theorem $\alpha$ possesses a $L(Q_{\bar{\alpha}})$-elementary substructure of cardinality $\omega_{\alpha_j}$. Since $\alpha$ is a well-ordering, this substructure itself is a well-ordering of smaller cardinality, hence by the induction hypothesis it is n-term-like. Thus $\alpha$ is also n-term-like by Corollary 13. (Remark: In the above proof we used to Löwenheim-Skolem-Theorem not completely correctly. We can argue as above with respect to the language $L(Q_{\alpha_0},\dots,Q_{\alpha_j})$, but since by our assumption on j, card $|\alpha| \leq \omega_{\alpha_{j+1}}$ the argumentation extends to $L(Q_{\bar{\alpha}})$.)

3.2. card $|\alpha| = \omega_{\alpha_j}$, and $\alpha \neq \omega_{\alpha_j}$. Now there is a cofinal $\varkappa$-sequence $X \subseteq \alpha$ for some $\varkappa \leq \omega_{\alpha_j}$, and by Lemma 14 we may assume X is n-homogeneous. Set $C_0 = \alpha^{\leq x_0}$ and $C_1 = (x_0, x_1]_\alpha$, then

$$\alpha \overset{n}{\sim} C_0 + C_1\cdot(w^* + w)\cdot\varkappa.$$

as above by Lemma 9. By our assumptions $\varkappa < \alpha$, thus by the induction hypothesis, $C_0$, $C_1$, and $\varkappa$ are n-term-like, hence by Lemma 6 and Lemma 4 $\alpha$ is also n-term-like.

3.3. $\alpha = \omega_{\alpha_j}$. Then $\alpha$ is isomorphic to the term $\underline{1}\cdot w_j$, hence n-term-like. This completes the proof.

Q.E.D.

Now we are ready to characterize the n-term-like linear orderings completely.

Theorem 28. Every linear ordering is n-term-like.

Proof. We prove the theorem in a series of steps. We start by defining an equivalence relation $\approx$ on A, where **A** is a given linear ordering:

$x \approx y$ iff every segment of the closed interval $[x,y]$ is n-term like.

We want to prove that all elements of A are equivalent to each other.

1. $\approx$ is an equivalence relation with the splitting property, i.e. the equivalence classes are convex.

Proof. By the definition and Lemma 24.

2. Every equivalence class C has the property that each segment of C is n-term-like.

Proof. By the definition of $\approx$ each bounded segment is n-term-like, then by Lemma 27 any segment itself is n-term-like.

3. Let $\mathfrak{A} = A/\approx$ be the splitting determined by $\approx$. Then $\mathfrak{A}$ has order type $\underline{1}$, i.e. $|\mathfrak{A}| = \{A\}$.

Proof. Assume that $\mathfrak{A}$ has at least two elements C and D with $C < D$. Suppose there is no element between C and D. Let $a \in C$ and $b \in D$ and E be a segment of $[a,b]_A$. If E is contained in C or D, then E is n-term-like by 2 above. In the other case C and D induce a splitting on E, the parts $C \cap E$ and $D \cap E$ are n-term-like, hence by Lemma 15 E is n-term-like. Thus $a \approx b$, which contradicts $a \in C$, and $b \in D$, and $C < D$. Conclusion: $\mathfrak{A}$ is a dense linear ordering.

Let $F(C,D) = (F_0,\dots,F_p)$ be a sequence of ordered sets so that every n-type $F_i$ occurs at least $\varkappa$-many between C and D, and every n-type occuring at least $\varkappa$-many between C and D is n-equivalent to some $F_i$.

Furthermore, let $G(C,D) = (G_0,\dots,G_j)$ be a sequence of ordered sets so that every n-type $G_k$ occurs at least $\varkappa^+$-many between C and D, and every n-type occuring at least $\varkappa^+$-many between C and D is n-equivalent to some $G_k$.

We may assume that the interval $(C,D)_{\mathfrak{A}}$ has the smallest cardinality that is possible in $\mathfrak{A}$. Say this is $\lambda$.

1. Case. $\lambda$ is a successor cardinal, then set $\lambda = \varkappa^+$. Now choose C and D so that $G_\varkappa(C,D)$ is minimal. From the set of all C and D which satisfy the above conditions take some so that $F_\varkappa(C,D)$ and $F_1(C,D)$ are minimal, too.

Let $a \in C$ and $b \in D$. Consider a segment B of the closed interval $[a,b]_A$ of A. We prove that B is n-term-like. This is obviously true if B is a segment of some $E \in \mathfrak{A}$. Otherwise B is of the form $B_1 + \bigcup\mathfrak{B} + B_2$, where $\mathfrak{B}$ is some segment of $(C,D)$ mod $\mathfrak{A}$ and $B_1$, $B_2$ are (possibly empty) segments of some classes $C_1, D_1 \in \mathfrak{B}$. $\bigcup\mathfrak{B}$ denotes the corresponding segment of A. By the minimality of $F_\varkappa(C,D)$ and $G_\varkappa(C,D)$ we have

$$F_\varkappa(C',D') = F_\varkappa(C,D) \quad \text{and} \quad G_\varkappa(C',D') = G_\varkappa(C,D)$$

for all $C',D' \in \mathfrak{B}$ with $C' < D'$ mod $\mathfrak{A}$. By 2 every $F_i$ and every $G_j$ are n-term-like, thus we can assume that $F_i$ and $G_j$ themselves are terms. By the minimality of $F_1(C,D)$ each n-type occuring between C and D occurs $\omega$-dense. If some n-type X occurs $\omega_{\bar\alpha}$-many, then X occurs dense in $(C,D)_{\mathfrak{A}}$. By the continuum hypothesis $\omega_{\bar\alpha} = \varkappa$ or $\omega_{\bar\alpha} = \varkappa^+$. Then by the minimality of $F_\varkappa(C,D)$ and $G_\varkappa(C,D)$ X occurs $\omega_{\bar\alpha}$-dense in $(C,D)_{\mathfrak{A}}$. Then by Lemma 7

$$\bigcup\mathfrak{B} \overset{n}{\approx} S_\varkappa(F_\varkappa(C,D), G_\varkappa(C,D)).$$

Let $j \leq m$ be the greatest natural number so that $\omega_{\alpha_j} \leq \varkappa^+$.

1.1. $\omega_{\alpha_j} = \varkappa^+$. Then

$$S_\varkappa(F_\varkappa(C,D), G_\varkappa(C,D)) = S_j(F_\varkappa(C,D), G_\varkappa(C,D)),$$

hence B is n-term like.

1.2. $\omega_{\alpha_j} < \varkappa^+$. Then it is easily checked that

$$S_\varkappa(F_\varkappa(C,D),G_\varkappa(C,D)) \overset{n}{\sim} S_j(\emptyset,(F_\varkappa(C,D), G_\varkappa(C,D))),$$

where $\emptyset$ denotes the empty sequence and $(F_\varkappa(C,D),G_\varkappa(C,D))$ is the sequence obtained by concatenating $F_\varkappa(C,D)$ and $G_\varkappa(C,D)$. Thus B is also in this case n-term-like. (Remark: To apply Lemma 7 we must be sure that $\mathfrak{L}$ has no endpoints. If $\mathfrak{L}$ has a least or greatest element, then omit the endpoints from $\mathfrak{L}$ and argue as before. Since $\mathfrak{A}$ is a dense ordered set, the segment without endpoints has no least or greatest element.)

2. Case. $\lambda$ is a limit cardinal, then set $\lambda = \varkappa$. Let $j \leq m$ be the greatest natural number so that $\omega_{\alpha_j} < \varkappa$. Then as in the first case we may assume that C and D are chosen so that $F_1(C,D)$ and $F_\varkappa(C,D)$ are minimal. By the continuum hypothesis every n-type X occuring dense in $(C,D)_{\mathfrak{A}}$ occurs $\varkappa$-dense in the interval. Thus as in the first case all elements of $F_\varkappa(C,D)$ can be assumed to be terms. Then by Lemma 7

$$S_j(\emptyset,F_\varkappa(C,D)) \overset{n}{\sim} S_\varkappa(F_\varkappa(C,D),\emptyset).$$

Hence B is n-term-like as in the first case.

We have just proved that every segment of the closed interval $[a,b]_A$ is n-term-like. Hence $a \approx b$, which contradicts the assumption $C < D$. Thus there is only one equivalence class A which is n-equivalent to some term by property 2.

Q.E.D.

Next we apply the results obtained till now to the language $L(Q_{\bar{\alpha}})$.

<u>Theorem 29.</u> Every sentence $\varphi \in L(Q_{\bar{\alpha}})$ which has an ordered set as a model has a model in $M_m$.

<u>Proof.</u> Let A be an ordered set which is a model of $\varphi$, and n be the number of quantifiers occuring in $\varphi$. By Theorem 28 there is a term t with $A \overset{n}{\sim} t$, and by Corollary 11, $t \models \varphi$.

Q.E.D.

By the property expressed in the above theorem $M_m$ is called a dense system of linear orderings.

## § 6. Decidability

When proving the decidability of certain classes of linear orderings in $L(Q_{\bar{\alpha}})$ some additional difficulties arise since the logic of $L(Q_{\bar{\alpha}})$ is not axiomatizable. Thus we will prove decidability results by methods which differ from those usually used for elementary theories. In the first paragraph we introduced the product operation $S_{\varkappa}$. It is clear that we can assume that certain parts of the ordering $S_{\varkappa}^{m,n}$ are effectively calculable. However, we do not want to specify these parts of $S_{\varkappa}^{m,n}$, it will be clear from the proof how this has to be done.

r, s, t, ... denote terms from $M_m$. $s \lhd t$ denotes that t is one of the terms $r + s$, $s + r$, $s \cdot w_j$, $s \cdot w_j^*$ for some $j \leq m$, or $S_j(F,G)$ and s belongs to F or G. Clearly the set $\{s\colon s \lhd t\}$ is finite for every $t \in M_m$. An element $a \in t$ is called effectively calculable if it is given in a constructive way, i. e. there is an effective procedure which generates a. All constants used in the following should be effectively calculable. By Lemma 12 the equivalence relations $\overset{n}{\sim}$ are expressible in the language $L(Q_{\bar{\alpha}})$. We get the following lemma.

Lemma 30. There is a unary recursive function g(n) so that for all linear orderings A, all elements $a \in A$, and for all n:
(i) there is some sentence $\varphi_{A,n} \in L(Q_{\alpha})$ so that for all B

$$B \overset{n}{\sim} A \quad \text{iff} \quad B \models \varphi_{A,n},$$

and $\varphi_{A,n}$ has at most g(n) quantifiers,
(ii) there is some formula $\psi(x) \in L(Q_{\bar{\alpha}})$ with at most g(n)-many quantifiers so that for all B and all $b \in B$:

$$(A,a) \overset{n}{\sim} (B,b) \quad \text{iff} \quad B \models \psi(b).$$

Proof. By Lemma 12. We only have to count the number of quantifiers to get the function value g(n). Since the proof of Lemma 12 is constructive, g(n) is indeed recursive. Thus (i) follows. (ii) is an easy consequence of (i).

Q.E.D.

Theorem 31.
(i) For every term $t \in M_m$ and every sentence $\varphi(a_1,\ldots,a_i)$, where $\varphi(x_1,\ldots,x_i)$ is a formula of $L(Q_{\bar{\alpha}})$, there is an effective decision

procedure which decides "$t \models \varphi(a_1,\dots,a_i)$" where $a_1,\dots,a_i$ are effectively given constants.

(ii) If $t \models \exists x\varphi(x,a_1,\dots,a_i)$, then there is an effective procedure which generates some $b \in t$ so that $t \models \varphi(b,a_1,\dots,a_i)$.

Proof. By induction on t. If $t = \underline{1}$, then the theorem follows at once. Now let t be given. Then we prove the theorem by induction on the complexity of $\varphi$. If $\varphi$ is atomic, then the theorem follows immediately. In case $\varphi$ has the form $\psi \wedge \chi$, or $\psi \vee \chi$, or $\neg\psi$, then the theorem (i) is a consequence of the induction hypothesis on $\psi$ and $\chi$.

We continue to prove (i) by cases.

1. Case. $\varphi(a_1,\dots,a_i) = \exists x\psi(x,a_1,\dots,a_i)$.

1.1. $t = r + s$. Suppose $\psi$ has n quantifiers. We assume that $a_1,\dots,a_i$ are ordered as follows: $a_1 < a_2 < \dots < a_i$. Let A be the interval $(a_j,a_{j+1})_r$ in r. By Lemma 30 (ii) there are finitely many $\varphi_1(x),\dots,\varphi_p(x)$ so that for every $a \in A$

$$A \models \varphi_k(a) \quad \text{for some } \varphi_k(x),\ 1 \le k \le p,$$

and for every $\varphi_k(x)$

$$A \models \exists x\varphi_k(x),$$

moreover, condition (ii) of Lemma 30 is satisfied and $\varphi_1(x),\dots,\varphi_p(x)$ are effectively obtained. If $a,b \in A$ and

$$A \models \varphi_k(a) \quad \text{and} \quad A \models \varphi_k(b),$$

then by Lemma 30 (ii), Lemma 4, and Corollary 11, a and b satisfy exactly the same formulae with at most n quantifiers and the constants $a_1,\dots,a_i$ in t. By the induction hypothesis on r there are effective procedures which generate $b_k$, $1 \le k \le p$, so that

$$A \models \varphi_k(b_k).$$

In this way we get effectively calculable elements $b_k^j$ for every interval $(a_j,a_{j+1})$ and similarly for $f^{<a_1}$ and $s^{>a_i}$ there are effectively calculable elements $b_k^0$ and $b_k^{j+1}$, respectively.

Now the set $\{b_k^j: 0 \leq j \leq i+1, 1 \leq k \leq p_j\}$ is finite and for every $b_k^j$ there is by the induction hypothesis a decision procedure for

$$"t \models \varphi_k^j(b_k^j, a_1, \ldots, a_i)".$$

From the construction it follows that: If $t = x\ (x, a_1, \ldots, a_i)$, then for some $b_k^j$

$$t \models \psi(b_k^j, a_1, \ldots, a_i),$$

which is decidable for every $b_k^j$ by the induction hypothesis on $\psi$. Thus we have a decision procedure for

$$"t \models \exists x \psi(x, a_1, \ldots, a_i)".$$

1.2. $t = s \cdot w$. Obviously, $s \cdot w$ is isomorphic to $s \cdot k + s \cdot w$. Then we may assume that all constants $a_1, \ldots, a_i$ occur in $a \cdot k$. Now, by the definition of t, t admits a splitting of type $\omega$ and each part of the splitting is isomorphic to s. For the same reason as in case 1.1 there are effective procedures which generate elements $b_q^i$, $1 \leq i \leq k + 2^n$ and $0 \leq q \leq p_i$, and $b_q^i$ belongs to the i-th part of the splitting. From Lemma 6 it follows that for every element $b \in t$, which belongs to some part $k_1$ with $k + 2^n < k_1$, there is some $b_q^i$ with

$$(t, b) \overset{n}{\sim} (t, b_q^i)$$

since finite linear orderings with more than $2^n$ elements are n-equivalent by Lemma 16. Thus we get a decision procedure as in case 1.1.

1.3. $t = s \cdot w_j$ for some $j \leq m$. Now by the definition of t, t admits a splitting of type $\omega_{\alpha_j}$ and each part of the splitting is isomorphic to s. Set

$$R_n^{\bar{\alpha}}(|\omega_{\alpha_j}|) = \{R_n^{\bar{\alpha}}(\beta): \beta < \omega_{\alpha_j}\}.$$

By the definition of $R_n^{\bar{\alpha}}$, $R_n^{\bar{\alpha}}(|\omega_{\alpha_j}|)$ is finite, and by Lemma 21, $\beta$ and $R_n^{\bar{\alpha}}(\beta)$ realize the same n-type in $\omega_{\alpha_j}$. $R_n^{\bar{\alpha}}(|\omega_{\alpha_j}|)$ is effectively calculable and there are effective procedures which generate $b_q^i$, $i \in R_n^{\bar{\alpha}}(|\omega_{\alpha_j}|)$, and $b_q^i$ belongs to the i-th part of the splitting.

From the definition of $R_n^{\bar{\alpha}}(|\omega_{\alpha_j}|)$ and by Lemma 6 it follows that for every element $b \in t$ there is some $b_q^i$ with

$$(t,b) \overset{n}{\approx} (t,b_q^i).$$

Then as in the first case we get a decision procedure.

1.4. $t = S_k(F,G)$. By the definition of t, t admits a splitting of type $S_k$. Consider those parts which contain some constants of $a_1,\dots,a_i$. Since these constants are effectively calculable, these parts are effectively given. As before (case 1.1) there are effective procedures which produce elements $b_q^i$. Now we choose effectively finitely many parts of the splitting for every $s \in |F| \cup |G|$, all of type s so that for any two constants of $a_1,\dots,a_i$, say $a_j$ and $a_{j+1}$, there is one of the effectively chosen parts between them (in case $a_j$ and $a_{j+1}$ do not belong to the same part). Moreover, we effectively choose two parts of type s which are greater and smaller, respectively, than all parts which contain some of the constants $a_1,\dots,a_i$. For all these parts we also have procedures which generate elements $b_q^j$. From Lemma 7 it follows that: For every $b \in S_k(F,G)$ there is some $b_q^j$ so that

$$(S_k(F,G),b) \overset{n}{\approx} (S_k(F,G),b_q^j).$$

As in case 1.1 we get a decision procedure.

2. Case. $(a_1,\dots,a_i) = Q_{\bar{\alpha}}x\psi(x,a_1,\dots,a_i)$. According to the definition of t there is a splitting on t. For

$$t \models Q_{\bar{\alpha}}x\psi(x,a_1,\dots,a_i)$$

to be valid there are two possibilities:

(i) There are $\omega_{\bar{\alpha}}$-many parts of the splitting which contain an element b with $t \models \psi(b,a_1,\dots,a_i)$, or
(ii) some part contains $\omega_{\bar{\alpha}}$-many elements b with $t \models \psi(b,a_1,\dots,a_i)$.

ad (i). This is without sense if the splitting of t has less than $\omega_{\bar{\alpha}}$ many parts. Thus t must have the form $s \cdot w_k$ or $S_k(F,G)$, where $\omega_{\alpha_k}$ is greater than or equal to $\omega_{\bar{\alpha}}$. In the decision procedure for "$t \models \exists x\psi(x,a_1,\dots,a_i)$" (1. case) we had to decide $t \models \psi(b_q^j,a_1,\dots,a_i)$. By Lemma 6 and Lemma 23 it follows that $\omega_{\alpha_1}$-many parts contain some b with

$$t \models \psi(b,a_1,\dots,a_i) \quad \text{iff} \quad t \models \psi(b_q^j,a_1,\dots,a_i)$$

for some $b_q^j$ with $j \in P_n^{(\bar{\alpha})1,m}$. However, the right hand side of the above equivalence is decidable by the induction hypothesis.

The other cases for t are similar.

ad (ii). In the first case we decided $t \models \exists x\psi(x,a_1,\dots,a_i)$ by deciding $t \models \psi(b_q^j,a_1,\dots,a_i)$ for effectively calculable elements $b_q^j$. Every $b_q^j$ was determined by an effective procedure generating an element so that $s \models \varphi_q^j(b_q^j)$. Then it follows from case 1.1 that in one part s there are $\omega_{\bar{\alpha}}$-many elements b so that

$$t \models \psi(b,a_1,\dots,a_i) \quad \text{iff} \quad \text{for some } \varphi_q^j(x),\ s \models Q_{\bar{\alpha}}x\varphi_q^j(x).$$

However, the latter proposition is decidable.

Thus we have proved (i) of the theorem. Now (ii) automatically follows from the case 1.

Q.E.D.

Theorem 32. The set of all sentences $\varphi \in L(Q_{\bar{\alpha}})$, so that for all terms $t \in M_m$: $t \models \varphi$, is recursive.

Proof. This follows from Theorem 31 and the following fact. By Corollary 26 $M_m/\overset{n}{\sim}$ is a finite algebraic structure which is generated by the equivalence class of $\underline{1}$. Now let $\varphi$ be a given sentence with at most n quantifiers. To decide whether $t \models \varphi$ for all $t \in M_m$, we only have to decide if for all $A \in M_m/\overset{n}{\sim}$ there is some $t \in M_m$ so that

$$A = \{s \in M_m : s \overset{n}{\sim} t\} \quad \text{and} \quad t \models \varphi.$$

Since $M_m/\overset{n}{\sim}$ is finite and generated by $\underline{1}$,we find effectively finitely many terms $t_1,\dots,t_q$ which form a complete system of representatives for the set of equivalence classes. Hence we only have to decide for finitely many terms $t_1,\dots,t_q$ whether $t_j \models \varphi$ or not. If for all j, $1 \leq j \leq q$, $t_j \models \varphi$, then $\varphi$ is valid in all terms $t \in M_m$. Now we have a decision method which answers the question: "Is $\varphi$ valid in every term $t \in M_m$?" Since the decision method is uniform, this is the same as if the set of all sentences, which are valid in all terms $t \in M_m$, were recursive.

Q.E.D.

Now we get the decidability of many classes of orderings in the language $L(Q_{\bar{\alpha}})$.

Let $LO_j$ be the set of all sentences of $L(Q_{\bar{\alpha}})$ which are true in all linear orderings with cardinality greater than or equal to $\omega_{\alpha_j}$, and $LO_{\bar{\alpha}}$ be the set of all sentences of $L(Q_{\bar{\alpha}})$ which are valid in all linear orderings. $DLO_j$ and $DLO_{\bar{\alpha}}$ are similaryl definea with the restriction that we only regard dense linear orderings.

Theorem 33. All the theories $LO_{\bar{\alpha}}$, $DLO_{\bar{\alpha}}$, and for all $j \leq m$, $LO_j$ and $DLO_j$ are decidable.

Proof. Theorem 29 implies that the set of sentences which are valid in all linear orderings is exactly the same as the set of those sentences which are valid in all terms $t \in M_m$. Then the decidability of $LO_{\bar{\alpha}}$ is an immediate consequence of Theorem 32. The other theories mentioned in the above theorem are finite extensions of $LO_{\bar{\alpha}}$, hence also decidable.

Q.E.D.

# CHAPTER 6

# BOOLEAN ALGEBRAS

Let Ba be the class of Boolean algebras. In this chapter we will show the decidability of $Th_\alpha(Ba)$ for any $\alpha \in On$. We will show that there are exactly three possibilities for $Th_\alpha(Ba)$ depending on some cardinality properties on $\omega_\alpha$.

## § 1. General remarks

Let L be a language, $\alpha \in On$. We assume that Q and $Q_\alpha$ are not symbols of L. We use $Form(L(Q_\alpha))$ and $Sent(L(Q_\alpha))$ to denote the set of formulas and sentences of $L(Q_\alpha)$, respectively. The language L(Q), its formulas and sentences are defined by the same rules as $L(Q_\alpha)$, but without any satisfaction relation.

Let $\varphi \in Form(L(Q))$. Then $\varphi^\alpha$ and $fv(\varphi)$ denote the formula obtained from $\varphi$ by replacing each occurrence of Q by $Q_\alpha$ and the set of free variables of $\varphi$, respectively.

The theory of Boolean algebras is formulated in the language $L_{Ba}$ with the following nonlogical symbols:

two constant symbols, 0 and 1;
one unary function symbol, $\overline{\phantom{x}}$;
two binary function symbols, $\cap$ and $\cup$.

An axiomatization of Ba is given by the following axioms:

(i) $\forall xy(x \cup y = y \cup x \wedge x \cap y = y \cap x)$;

(ii) $\forall xyz((x \cup y) \cup z = x \cup (y \cup z) \wedge (x \cap y) \cap z = x \cap (y \cap z))$;

(iii) $\forall xy((x \cap y) \cup y = y \wedge (x \cup y) \cap y = y)$;

(iv) $\forall xyz(x \cap (y \cup z) = (x \cap y) \cup (x \cap z) \wedge x \cup (y \cap z) = (x \cup y) \cap (x \cup z))$;

(v) $\forall x(x \cup \bar{x}) = 1 \wedge x \cap \bar{x} = 0)$;

(vi) $0 \neq 1$;

(vii) $\forall x(x \cup 0 = x \wedge x \cap 1 = x)$.

By definition we introduce the following two binary relations $\leq$ and $<$:

$$x \leq y \;:=\; x \cap y = x; \quad x < y \;:=\; x \leq y \wedge x \neq y.$$

Let $X \in \mathrm{Form}(L_{Ba}(Q_\alpha))$, P be a unary predicate symbol. The relativization of X on P, $X^P$, is defined as follows:

if $\varphi = (t_1 = t_2)$, then $\varphi^z = (t_1 \cap z = t_2 \cap z)$;

if $\varphi = \neg \psi$, then $\varphi^z = \neg \psi^z$;

if $\varphi = \psi_1 \wedge \psi_2$, then $\varphi^z = \psi_1^z \wedge \psi_2^z$;

if $\varphi = \exists x \psi$, then $\varphi^z = \exists x(x \leq z \wedge \psi^z)$;

if $\varphi = Q_\alpha x \psi$, then $\varphi^z = Q_\alpha x(x \leq z \wedge \psi^z)$.

Let $X \in \mathrm{Form}\,(L_{Ba}(Q_\alpha))$, P be a unary predicate symbol. The relativization of X on P, $X^P$, is defined as follows:

$$X^P := \{\forall z(P(z) \longrightarrow \varphi^z) : \varphi \in X,\ z \text{ does not occur in } \varphi\}.$$

By definition we introduce the following unary relations:

$$\mathrm{At}\ x \;:=\; x \neq 0 \wedge \forall y(y < x \longrightarrow y = 0);$$

$$\mathrm{at}\ x \;:=\; \forall y(y \leq x \wedge y \neq 0 \longrightarrow \exists z(z \leq y \wedge \mathrm{At}\ z));$$

$$\eta x \;:=\; x \neq 0 \wedge \neg \exists y(y \leq x \wedge \mathrm{At}\ y);$$

$$\eta_1 x \;:=\; \eta x \wedge \forall y(y \leq x \longrightarrow Q_\alpha z(z \leq y)).$$

(The special use of $\alpha$ is clear from the context.)

Let $\mathfrak{A} \in \mathrm{Ba}$, $A \subseteq |\mathfrak{A}|$. We write $\bigcup[a: a \in A]$ and $\bigcap[a: a \in A]$ for the least upper bound and the greatest lower bound, respectively. Let $\mathfrak{A} \in \mathrm{Ba}$, $a \in |\mathfrak{A}|$. $\mathfrak{A}[a]$ is the structure

$$\mathfrak{L} = \langle |\mathfrak{L}|, 0^{\mathfrak{A}}, a, \cup^{*}, \cap^{*}, {}^{-*} \rangle$$

with

$$|\mathfrak{L}| := \{b \in |\mathfrak{A}| : b \leq a\};$$

$$b \cup^{*} c := b \cup c; \quad b \cap^{*} c := b \cap c; \quad \bar{b}^{*} := \bar{b} \cap a.$$

If $a \neq 0$, $\mathfrak{A}[a]$ is a Boolean algebra; $\mathfrak{A}[0]$ has just one element and is called the trivial algebra. For any $a \in |\mathfrak{A}|$ we have $\mathfrak{A} \cong \mathfrak{A}[a] \times \mathfrak{A}[\bar{a}]$. Let $\mathfrak{A} \in \mathrm{Ba}$, $a, a_0, \ldots, a_{n-1} \in |\mathfrak{A}|$. $\{a_0, \ldots, a_{n-1}\}$ is a <u>partition</u> of a if $\bigcup[a_i : i < n] = a$ and the $a_i$'s are pairwise disjoint, that means, if $i < j < n$, then $a_i \cap a_j = 0$. If $a = 1$, then $\{a_0, \ldots, a_{n-1}\}$ is called a partition of $\mathfrak{A}$. If $\{a_0, \ldots, a_{n-1}\}$ is a partition of a, then $\mathfrak{A}[a] = \mathfrak{A}[a_0] \times \ldots \times \mathfrak{A}[a_{n-1}]$. If $\{a_i : i < n\}$ is a partition of a and $b \in |\mathfrak{A}|$, then $\{a_i \cap b : i < n\} \cup \{a_i \cap \bar{b} : i < n\}$ is also a partition of a.

Let $\mathfrak{A} \in \mathrm{Ba}$, $A \subseteq |\mathfrak{A}|$. Then $\mathfrak{A}[A]$ denotes the least subalgebra of $\mathfrak{A}$ containing A.

<u>Remark.</u> We have to distinguish between $\mathfrak{A}[a]$ and $\mathfrak{A}[\{a\}]$. While $\mathfrak{A}[a]$ is the Boolean algebra consisting of all elements $\leq a$, $\mathfrak{A}[\{a\}]$ is the four-element subalgebra of $\mathfrak{A}$ with the domain $\{0, a, \bar{a}, 1\}$.

If A is finite, then $\mathfrak{A}[A]$ is finite, too.

A sequence $\bar{a} = \langle a_0, \ldots, a_{n-1} \rangle \in |\mathfrak{A}|^{<\omega}$ is a <u>partition sequence for A</u> if $\bar{a} \in \mathfrak{A}[A]^{<\omega}$ and for $i < j < n$, $a_i \cap a_j = 0$.

Let $\bar{a} \in |\mathfrak{A}|^{<\omega}$, $f \in 2^{<\omega}$ with $\mathrm{lh}(\bar{a}) = \mathrm{lh}(f) = n$. We set

$$\langle f, \bar{a} \rangle := \bigcap[b_i : i < n]$$

with

$$b_i = \begin{cases} a_i & \text{if } f(i) = 0, \\ \bar{a}_i & \text{if } f(i) = 1. \end{cases}$$

Let $\{f_i : i < 2^n\}$ be the set of all $f \in 2^{<\omega}$ with $lh(f) = n$. Then the sequence $\bar{c} = \langle c_i : i < 2^n \rangle$ with $c_i = \langle f_i, \bar{a} \rangle$ is a partition sequence for $\langle a_i : i < n \rangle$. (We also simply say that $\bar{c}$ is a partition sequence for $\bar{a}$.)

Let n be any natural number; then there are exactly $2^{2^n}$ different terms (with respect to Th(Ba)) in the variables $v_i$ with $i < n$. Now each atomic formula in $L_{Ba}$ is of the form $t_1 = t_2$; thus there are at most $m = 2 \cdot 2^{2^n}$ atomic formulas in the variables $v_i$ $(i < n)$, which are different with respect to Th(Ba). Thus there are at most $2^{2^m}$ quantifier-free formulas in the variables $v_i$ $(i < n)$ which are different with respect to Th(Ba).

Let k be any natural number. Assume that there are $\leq m_1$ formulas $\varphi \in Form(L_{Ba}(Q_\alpha))$ with $qr(\varphi) = k$, $fv(\varphi) \subseteq \{v_i : i < n\}$ which are different with respect to $Th_\alpha(Ba)$. Then there are $\leq 2^{3^{m_1}}$ formulas $\varphi \in Form(L_{Ba}(Q_\alpha))$ with $qr(\varphi) \leq k + 1$ and $fv(\varphi) \subseteq \{v_i : i < n - 1\}$ which are different with respect to $Th_\alpha(Ba)$. (From each $\varphi$ with $qr(\varphi) = k$ there arise the three formulas $\exists v_{n-1}\varphi$, $\exists v_{n-1}\varphi \wedge \neg Q_\alpha v_{n-1}\varphi$, $Q_\alpha v_{n-1}\varphi$ with quantifier rank k + 1.) Now it follows that there is a recursive function $g : \omega \longrightarrow \omega$ so that for each k there are at most g(n) sentences $\varphi \in Sent(L_{Ba}(Q_\alpha))$ with $qr(\varphi) \leq k$ which are different with respect to $Th_\alpha(Ba)$.

Let $\mathfrak{A} \in Ba$, $a \in |\mathfrak{A}|$. cd(a), the <u>cardinality of a</u>, is defined as follows:

$$cd(a) := card\{b \in |\mathfrak{A}| : b \leq a\}.$$

a is <u>$\varkappa$-atomless</u> if $a \neq 0$ and for each $b \leq a$ with $0 \neq b$ we have $cd(b) \geq \varkappa$. a is <u>$\varkappa$-weakly-atomless</u> if $a \neq 0$ and for each $b \leq a$ with $b \neq 0$ there is $c \leq b$ with $c \neq 0$ and $cd(c) < \varkappa$. A subset D of $\mathfrak{A}$ is <u>dense</u> in $\mathfrak{A}$ if for each $a \in |\mathfrak{A}|$ with $a \neq 0$ there is $b \in D$ with $0 < b \leq a$.

<u>Lemma 0.</u> Let $\mathfrak{A} \in Ba$, $D \subseteq |\mathfrak{A}|$ be a dense subset, and for all $\lambda < card(|\mathfrak{A}|)$, $2^\lambda < card(|\mathfrak{A}|)$. Then $card(D) = card(|\mathfrak{A}|)$.

<u>Proof.</u> We define a function $f : |\mathfrak{A}| \longrightarrow pow(D)$ by $f(a) := \{d \in D : d \leq a\}$. We show that f is injective. Assume that this is not so. Then there are a,b with $f(a) = f(b)$ and $a \neq b$. Let be $c = (a \cap \bar{b}) \cup (\bar{a} \cap b)$. Then $c \neq 0$ and

thus there is a $d \in D$ with $d \leq c$. Thus $d \in f(a)$ iff $d \notin f(b)$, which contradicts the assumption. Thus $\mathrm{card}(|\mathfrak{A}|) = 2^{\mathrm{card}(D)}$ and this implies $\mathrm{card}(|\mathfrak{A}|) = \mathrm{card}(D)$.

Q.E.D.

## § 2. Special Boolean algebras and products of Boolean algebras

Let I be a set. Pow(I) denotes the Boolean algebra with $|\mathrm{Pow}(I)| = \mathrm{pow}(I)$, whose Boolean operations are the usual set-theoretical operations join, meet, and complement. Pow(I) is the power-set algebra on I. Fin(I) denotes the subalgebra of Pow(I) which is generated by the finite subsets of I.

Let $A \in \mathrm{pow}(I)$. Then we set

$$\mathrm{Pow}(I;A) = \mathrm{Pow}(I)[|\mathrm{Fin}(I)| \cup A].$$

Of course, we have for any $A \subseteq \mathrm{pow}(I)$,

$$\mathrm{Fin}(I) \subseteq \mathrm{Pow}(I;A) \subseteq \mathrm{Pow}(I).$$

Let $\mathfrak{M} = \langle |\mathfrak{M}|, < \rangle$ be a linear ordering with a least element. $\mathcal{J}(\mathfrak{M})$ denotes the subalgebra of $\mathrm{Pow}(|\mathfrak{M}|)$ which is generated by the left-closed right-open intervals of $\mathfrak{M}$, that means, by the sets of the form $\{x \in |\mathfrak{M}|: a \leq x < b\}$ for some $a,b \in |\mathfrak{M}|$. $\mathcal{J}(\mathfrak{M})$ is the interval algebra of $\mathfrak{M}$. We set

$$\mathcal{J}_1(\mathfrak{M}) = \mathrm{Pow}(|\mathfrak{M}|;|\mathcal{J}(\mathfrak{M})|),$$

that means, we get $\mathcal{J}_1(\mathfrak{M})$ from $\mathcal{J}(\mathfrak{M})$ by adjoining the finite subsets of $|\mathfrak{M}|$.

A segment of $\mathfrak{M}$ is a set X so that for $a,b \in X$, $c \in |\mathfrak{M}|$, if $a \leq c \leq b$, then $c \in X$. $\mathcal{J}_2(\mathfrak{M})$ denotes the subalgebra of $\mathrm{Pow}(|\mathfrak{M}|)$ generated by all segments of $\mathfrak{M}$. Then we have

$$\left.\begin{array}{ll}\mathrm{Fin}(|\mathfrak{M}|) & \subseteq \\ \mathcal{J}(\mathfrak{M}) & \subseteq\end{array}\right\} \mathcal{J}_1(\mathfrak{M}) \subseteq \mathcal{J}_2(\mathfrak{M}) \subseteq \mathrm{Pow}(|\mathfrak{M}|).$$

Let us give some examples:

If $\mathfrak{M}$ is a one-element linear ordering, then $\mathcal{J}(\mathfrak{M})$ is the two-element Boolean algebra. We denote this Boolean algebra by B(2).

Let $\mathcal{R}^+ = \langle|\mathcal{R}^+|,<\rangle$ be the set of non-negative rational numbers in their usual ordering. Then $\mathcal{J}(\mathcal{R}^+)$ is the countable atomless Boolean algebra. We denote this Boolean algebra by $B(\eta_0)$. $\mathcal{J}_1(\mathcal{R}^+)$ is a countable atomic Boolean algebra with the following property:

> if $a \in |\mathcal{J}_1(\mathcal{R}^+)|$ with $cd(a) = \omega_0$, then there are disjoint elements $b,c$ with $b \cup c = a$ and $cd(b) = cd(c) = \omega_0$.

Let $\alpha$ be an ordinal and let $|\mathcal{H}_\alpha|$ be the set of all sequences s with $lh(s) = \omega_\alpha$, $card(\{i < \omega_\alpha: s(i) = 0\}) < \omega_0$. Then $card(|\mathcal{H}_\alpha|) = \omega_\alpha$. Let $\langle|\mathcal{H}_\alpha|,<\rangle$ be $|\mathcal{H}_\alpha|$ with the usual lexicographic ordering, that means, if $s_1, s_2 \in |\mathcal{H}_\alpha|$, then $s_1 < s_2$ iff there is a $\beta < \omega_\alpha$ with $s_1(\beta) = 0$, $s_2(\beta) = 1$ and for all $\gamma < \beta$, $s_1(\gamma) = s_2(\gamma)$. Then $\mathcal{J}(\mathcal{H}_\alpha)$ is an $\omega_\alpha$-atomless Boolean algebra.

Let $\mathcal{N} = \langle|\mathcal{N}|,<\rangle$ be the set of natural numbers with their usual ordering. Then

$$\mathcal{J}(\mathcal{N}) \cong \mathcal{J}_1(\mathcal{N}) \cong \mathcal{J}_2(\mathcal{N}) \cong \mathrm{Fin}(|\mathcal{N}|).$$

Let I be a set and for each $x \in I$ let $\mathcal{B}_x \in Ba$. We use $\prod\{\mathcal{B}_x: x \in I\}$ for the direct product of the $\mathcal{B}_x$'s. $\prod^w\{\mathcal{B}_x: x \in I\}$ denotes the weak direct product of the $\mathcal{B}_x$'s (also denoted as the direct sum). That means $\prod^w\{\mathcal{B}_x: x \in I\}$ is the subalgebra of $\prod\{\mathcal{B}_x: x \in I\}$ generated by all $a \in |\prod\{\mathcal{B}_x: x \in I\}|$ with $card(\{x \in I: a(x) \neq 0\}) < \omega_0$. Let D be an ideal on Pow(I) so that each finite subset of I is in D. Let $\mathfrak{A} \subseteq Pow(I)$ with $D \subseteq |\mathfrak{A}|$. $\prod^{D,\mathfrak{A}}\{\mathcal{B}_x: x \in I\}$ denotes the subalgebra of $\prod\{\mathcal{B}_x: x \in I\}$ generated by all $a \in |\prod\{\mathcal{B}_x: x \in I\}|$ with $\{x \in I: a(x) \neq 0\} \in |\mathfrak{A}|$ and $\{x \in I: a(x) \neq 0 \text{ and } a(x) \neq 1\} \in D$. If D is the ideal of finite subsets of I, we write $\prod^{\mathfrak{A}}\{\mathcal{B}_x: x \in I\}$ instead of $\prod^{D,\mathfrak{A}}\{\mathcal{B}_x: x \in I\}$. We call the products described just above <u>generalized direct products.</u>

By definition it immediately follows that

$$\prod^w\{\mathcal{B}_x: x \in I\} \cong \prod^{Fin(I)}\{\mathcal{B}_x: x \in I\} \subseteq \prod^{Pow(I)}\{\mathcal{B}_x: x \in I\}$$
$$\subseteq \prod^{Pow(I),Pow(I)}\{\mathcal{B}_x: x \in I\} \cong \prod\{\mathcal{B}_x: x \in I\}.$$

Let us give an example:

Let $\mathcal{B} = \prod^{\mathcal{I}_2(\aleph_0)} B(2) \times B(\eta_0)$. Then $card(|\mathcal{B}|) = 2^{\omega_0}$, $\mathcal{B}$ contains countably many atoms and countably many atomless elements.

Let $\mathfrak{A}, \mathcal{B} \in Ba$ with $\mathfrak{A} \subseteq \mathcal{B}$, D be an ideal on $\mathcal{B}$. Instead of $\mathfrak{A}/D \cap |\mathfrak{A}|$ we simply write $\mathfrak{A}/D$.

If $\mathcal{B}_x = \mathcal{B}$ for each $x \in I$, we simply write $\prod^{D,\mathfrak{A}}\mathcal{B}$.

Let $a \in |\prod\{\mathcal{B}_x: x \in I\}|$, $x_0 \in I$. $a(x_0)^{red}$ denotes the element nt $b \in |\prod\{\mathcal{B}_x: x \in I\}|$ defined by

$$b(x) = \begin{cases} a(x_0) & \text{if } x = x_0; \\ 0 & \text{if } x \neq x_0. \end{cases}$$

To construct suitable Boolean algebras for a dense system we use some special subset systems.

Let $\varkappa$ be a cardinal, $A,B \subseteq pow(\varkappa)$. A is <u>almost disjoint</u> if for any $a,b \in A$ with $a \neq b$, $card(a \cap b) < \varkappa$. B is a <u>refinement</u> of A if there is a family $\{S_a: a \in A\}$ of subsets of B with $2 \leq card(S_a) < \omega_0$, $\bigcup S_a = a$ for each $a \in A$ and $\bigcup\{S_a: a \in A\} = B$.

A has <u>Property 1</u> if $card(A) = 2^{\varkappa}$ and a is almost disjoint.

A has <u>Property 2</u> if there is a sequence $\langle A_i: i<\omega\rangle$ with $A = \bigcup\{A_i: i<\omega\}$ so that $A_0$ is almost disjoint with $card(A_0) = 2^{\varkappa}$ and for $i<k<\omega$, $A_k$ is a refinement of $A_i$.

A has <u>Property 3</u> if there is a sequence $\langle A_i: i<\omega\rangle$ with $A = \bigcup\{A_i: i<\omega\}$ so that $card(A_0) \leq \omega_0$ and for $i<k<\omega$, $A_k$ is a refinement of $A_i$.

First assume $\varkappa > \omega_0$; the case $\varkappa = \omega_0$ is considered separately. Let $T_\varkappa$ be the set of all 0-1-sequences a with $lh(a) = \varkappa$ which satisfy

$$\forall\beta<\varkappa\,\exists\alpha<\varkappa(\beta<\alpha \wedge a(\alpha) = 1);$$

and let $C_\varkappa$ be the set of all 0-1-sequences a with $lh(a) < \varkappa$. Let $x \in T_\varkappa$, $y \in C_\varkappa$. Then we write $y < x$ for each $\beta < lh(y)$, $x(\beta) = y(\beta)$. Then $card(T_\varkappa) = 2^{\varkappa}$ and $card(C_\varkappa) = 2^{\underline{\varkappa}}$ $(= \sum\{2^\lambda: \lambda < \varkappa\})$.

Let f be any fixed one-one function from $C_\varkappa$ onto $2^{\underline{\varkappa}}$ satisfying the following condition:

for each $b \in C_{\varkappa}$ there are $\alpha, \beta < \varkappa$ and $j < \omega$ with $lh(b) = \alpha \cdot \omega + j$, $f(b) = \beta \cdot \omega + j$.

Let $k, l, m < \omega$ and $a \in T_{\varkappa}$. We set

$$C_{\varkappa}[k,l,m] = \{b \in C_{\varkappa}: \text{there are } \alpha < \varkappa \text{ and } j < \omega \text{ with } lh(b) = \alpha \cdot \omega^2 + l \cdot \omega + j \text{ and } j \equiv k (\mathrm{mod}\ m)\};$$

$$D_{\varkappa}[k,l,m](a) = \{f(b): b \in C_{\varkappa}[k,l,m] \text{ and } b < a\};$$

$$E_{\varkappa}[k,m] = \{D_{\varkappa}[k,i,m](b): b \in T_{\varkappa},\ i < \omega\};$$

$$F_{\varkappa}[k,m] = \{D_{\varkappa}[i,j,n \cdot m](b): b \in T_{\varkappa},\ i,j,n < \omega,\ i \equiv k (\mathrm{mod}\ m),\ n > 0\};$$

$$G_{\varkappa}[k,m] = \{f(b): \text{there are } \alpha < \varkappa,\ i,j < \omega \text{ with } lh(b) = \alpha \cdot \omega^2 + j \cdot \omega + i,\ j \equiv k (\mathrm{mod}\ 2^m)\};$$

$$G_{\varkappa} = \{G_{\varkappa}[i,j]: i,j < \omega\}.$$

Now assume $\varkappa = \omega_0$. Let $T_{\omega_0}$ be the set of all 0-1-sequences a with $lh(a) = \omega^2$ which satisfy

$$\forall \beta < \omega_0 \exists \alpha < \omega_0 (\beta < \alpha \wedge a(\alpha) = 1)$$

and let $C_{\omega_0}$ be the set of all 0-1-sequences a with $lh(a) < \omega^2$ and

$$\mathrm{card}(\{\alpha < lh(a): a(\alpha) = 1\}) < \omega_0.$$

Let f be any fixed one-one function from $C_{\omega_0}$ onto $\omega^{\omega}$ satisfying the following condition:

For each $b \in C_{\omega_0}$ there are $\alpha, \beta < \omega^{\omega}$ and $j < \omega$ with $lh(b) = \alpha \cdot \omega + j$, $f(b) = \beta \cdot \omega + j$.

Let be $k, l, m < \omega$ and $a \in T_{\omega_0}$. We set

$$C_{\omega_0}[k,l,m] = \{b \in C_{\omega_0}: \text{there are } \alpha < \omega^{\omega},\ j < \omega \text{ with } lh(b) = \alpha \cdot \omega^2 + l \cdot \omega + j \text{ and } j \equiv k (\mathrm{mod}\ m)\};$$

$D_{\omega_0}[k,l,m](a)$, $E_{\omega_0}[k,m]$ and $F_{\omega_0}[k,m]$ are defined as for $\varkappa > \omega_0$;

$$G_{\omega_0}[k,m] = \{f(b)\colon \text{there are } \alpha < \omega^\omega \text{ and } i,j < \omega \text{ with } lh(b) = \alpha\cdot\omega^2 + j\cdot\omega + i \text{ and } j \equiv k (\text{mod } 2^m)\};$$

$$G_{\omega_0} = \{G_{\omega_0}[i,j]\colon i,j < \omega\}.$$

Let $\varkappa$ be any infinite cardinal, $k,n < \omega$. Then it is easily seen that $E_\varkappa[k,m]$ has Property 1. Let be

$$A_n = \{D_\varkappa[i,j,n\cdot m](b)\colon b \in T_\varkappa,\ i,j < \omega,\ i \equiv k(\text{mod } m)\}.$$

Then it immediately follows that $F_\varkappa[k,m]$ has Property 2. Let be

$$A_n^+ = \{G_\varkappa[k,n]\colon k < n\}.$$

Then it follows immediately that G has Property 3.

For $\varkappa > \omega_0$, $k,n,m < \omega$ we set

$$B_\varkappa[k,n,m] = \mathrm{Pow}(2^\varkappa; \bigcup\{E_\varkappa[l,\ k+m]\colon k \le l < k+n\} \cup \bigcup\{F_\varkappa[l,\ k+m]\colon k+n \le l < k+m\});$$

$$B_\varkappa^+[k,n,m] = \mathrm{Pow}(2^\varkappa; \bigcup\{E_\varkappa[l,\ k+m]\colon k \le l < k+n\} \cup \bigcup\{F_\varkappa[l,\ k+m]\colon k+n \le l < k+m\} \cup G_\varkappa);$$

$$B_{\omega_0}[k,n,m] = \mathrm{Pow}(\omega^\omega; \bigcup\{E_{\omega_0}[l,\ k+m]\colon k \le l < k+n\} \cup \bigcup\{F_{\omega_0}[l,\ k+m]\colon k+n = l < k+m\});$$

$$B_{\omega_0}^+[k,n,m] = \mathrm{Pow}(\omega^\omega; \bigcup\{E_{\omega_0}[l,\ k+m]\colon k \le l < k+n\} \cup \bigcup\{F_{\omega_0}[l,\ k+m]\colon k+n \le l < k+m\} \cup G_{\omega_0}).$$

Let J be the ideal generated by the sets $\{E_\varkappa[l,\ k+m]\colon k \le l < k+n\}$, $\{F_\varkappa[l,\ k+m\colon k+n \le l < k+m\}$ and all ordinals less than $2^\varkappa$ if $\varkappa > \omega_0$ and all ordinals less than $\omega^\omega$ if $\varkappa = \omega_0$. Then

$$B_\varkappa[k,n,m]/J \cong B(2) \quad \text{and} \quad B_\varkappa^+[k,n,m]/J \cong B(\eta_0).$$

## § 3. Games for Boolean algebras

In Chapter I we gave a description of games that can be used to test structures for equivalence in languages with generalized quantifiers. Using these games the proofs in § 4 and § 5 would be very lengthy and complicated. Fortunately, it is possible to simplify the G-games in the case of Boolean algebras.

We define the $H^1$-games and the $H^2$-games. The $H^1$-games simplify the G-games and the $H^2$-games simplify the $H^1$-games. The $H^2$-games are used in § 4 to test density and in § 5 to show that the products used in § 4 preserve $Q_\alpha$-equivalence.

We start by showing that the G-games are equivalent to the $H^1$-games. To do this we reformulate the G-games in terms of partial isomorphisms. After having shown the equivalence of G-games and $H^1$-games we show that the $H^1$-games are equivalent to the $H^2$-games.

Definition. Let L be a language, $\mathfrak{A}$ and $\mathfrak{B}$ be structures for L, $\alpha,\beta \in On$, $n \in \omega$. Then

$$\mathfrak{A}\ {}^{\alpha}\underset{I}{\sim}{}^{\beta}_{n}\ \mathfrak{B}$$

iff there is a sequence $\Gamma$ with $lh(\Gamma) = n + 1$ so that

(I 1) for each $i \leq n$, $\Gamma(i)$ is a set of partial isomorphisms between $\mathfrak{A}$ and $\mathfrak{B}$;

(I 2) for each $i < n$, each $F \in \Gamma(i)$, and each $a \in |\mathfrak{A}|$ there is a $b \in |\mathfrak{B}|$ with $F \cup \{\langle a,b\rangle\} \in \Gamma(i + 1)$;

(I 3) for each $i < n$, each $F \in \Gamma(i)$, and each $b \in |\mathfrak{B}|$ there is an $a \in |\mathfrak{A}|$ with $F \cup \{\langle a,b\rangle\} \in \Gamma(i + 1)$;

(I 4) for each $i < n$, each $F \in \Gamma(i)$, and each $A \subseteq |\mathfrak{A}|$ with $card(A) = \omega_\alpha$ there exists a $B \subseteq |\mathfrak{B}|$ with $card(B) = \omega_\beta$ so that for each $b \in B$ there is an $a \in A$ with $F \cup \{\langle a,b\rangle\} \in \Gamma(i + 1)$;

(I 5) for each $i < n$, each $F \in \Gamma(i)$, and each $B \subseteq |\mathfrak{B}|$ with $card(B) = \omega_\beta$ there exists an $A \subseteq |\mathfrak{A}|$ with $card(A) = \omega_\alpha$ so that for each $a \in A$ there is a $b \in B$ with $F \cup \{\langle a,b\rangle\} \in \Gamma(i + 1)$;

(I 6) $\Gamma(0) \neq \emptyset$.

The sequence $\Gamma$ is called $(\alpha,\beta,n)$-I-sequence for $\mathfrak{A}$ and $\mathfrak{B}$.

By definition it easily follows that

Lemma 1. Let $\mathfrak{A}, \mathfrak{B} \in \mathrm{Ba}$, $\alpha, \beta \in \mathrm{On}$, $n \in \omega$. Then

$$\mathfrak{A} \;{}^{\alpha}\!\underset{I}{\sim}{}^{\beta}_{n}\; \mathfrak{B} \quad \text{iff} \quad \mathfrak{A} \;{}^{\alpha}\!\sim^{\beta}_{n}\; \mathfrak{B}.$$

Definition. Let $\mathfrak{A}, \mathfrak{B} \in \mathrm{Ba}$, $\alpha, \beta \in \mathrm{On}$, $n \in \omega$, $\overline{a^0} = \langle a^0_0, \ldots, a^0_{m-1} \rangle$ a partition of $\mathfrak{A}$, $\overline{b^0} = \langle b^0_0, \ldots, b^0_{m-1} \rangle$ a partition of $\mathfrak{B}$ so that for each $j < m$ $a^0_j = 0$ iff $b^0_j = 0$. Then

$$\langle \mathfrak{A}, \overline{a^0} \rangle \;{}^{\alpha}\!\underset{PS}{\sim}{}^{\beta}_{n}\; \langle \mathfrak{B}, \overline{b^0} \rangle$$

iff there is a sequence $\Gamma$ with $\mathrm{lh}(\Gamma) = n + 1$ so that

(PS 1) $\langle \overline{a^0}, \overline{b^0} \rangle \in \Gamma(0)$;

(PS 2) for each $i \leq n$ $\Gamma(i)$ is a set of pairs $\langle \bar{a}, \bar{b} \rangle$ so that $\bar{a}$ is a partition of $\mathfrak{A}$ and $\bar{b}$ is a partition of $\mathfrak{B}$ with $\mathrm{lh}(\bar{a}) = \mathrm{lh}(\bar{b})$ and for each $j < \mathrm{lh}(\bar{a})$ $a_j = 0$ iff $b_j = 0$;

(PS 3) if $i < n$, $\langle \bar{a}, \bar{b} \rangle \in \Gamma(i)$, and $\mathrm{lh}(\bar{a}) = m$, then for each $\bar{c} \in |\mathfrak{A}|^{<\omega}$ with $\mathrm{lh}(\bar{c}) = m$ and $c_j \leq a_j$ for each $j < m$, there exists a $\bar{d} \in |\mathfrak{B}|^{<\omega}$ with $\mathrm{lh}(\bar{d}) = m$ and $d_j \leq b_j$ for each $j < m$ so that $\langle \overline{a^*}, \overline{b^*} \rangle \in \Gamma(i + 1)$, where $\mathrm{lh}(\overline{a^*}) = 2m$, $a^*_{2j} = c_j$, $a^*_{2j+1} = a_j \cap \overline{\overline{c}}_j$, $b^*_{2j} = d_j$, and $b^*_{2j+1} = b_j \cap \overline{\overline{d}}_j$ for each $j < m$;

(PS 4) if $i < n$, $\langle \bar{a}, \bar{b} \rangle \in \Gamma(i)$ and $\mathrm{lh}(\bar{a}) = m$, then for each $\bar{d} \in |\mathfrak{B}|^{<\omega}$ with $\mathrm{lh}(\bar{d}) = m$ and $d_j \leq b_j$ for each $j < m$, there exists a $\bar{c} \in |\mathfrak{A}|^{<\omega}$ with $\mathrm{lh}(\bar{c}) = m$ and $c_j \leq d_j$ for each $j < m$ so that $\langle \overline{a^*}, \overline{b^*} \rangle \in \Gamma(i + 1)$, where $\mathrm{lh}(\overline{a^*}) = 2m$, $a^*_{2j} = c_j$, $a^*_{2j+1} = a_j \cap \overline{\overline{c}}_j$, $b^*_{2j} = d_j$, and $b^*_{2j+1} = b_j \cap \overline{\overline{d}}_j$ for each $j < m$;

(PS 5) if $i < n$, $\langle \bar{a}, \bar{b} \rangle \in \Gamma(i)$ with $\mathrm{lh}(\bar{a}) = m$, $j_0 < m$, $c_0, \ldots, c_{j_0-1}, c_{j_0+1}, \ldots, c_{m-1} \in |\mathfrak{A}|$ with $c_j \leq a_j$ for each $j < m$ with $j \neq j_0$, $A \subseteq |\mathfrak{A}|$ with $\mathrm{card}(A) = \omega_\alpha$ and $a \leq a_{j_0}$ for each $a \in A$, then there are $d_0, \ldots, d_{j_0-1}, d_{j_0+1}, \ldots, d_{m-1} \in |\mathfrak{B}|$ with $d_j \leq b_j$ for each $j < m$ with $j \neq j_0$, $B \subseteq |\mathfrak{B}|$ with $\mathrm{card}(B) = \omega_\beta$ and $b \leq b_{j_0}$ for each $b \in B$, so that for each $b \in B$ there is an $a \in A$ with $\langle \overline{a^*}, \overline{b^*} \rangle \in \Gamma(i + 1)$, where $\mathrm{lh}(\overline{a^*}) = 2m$, $a^*_{2j} = c_j$, $a^*_{2j+1} = a_j \cap \overline{\overline{c}}_j$, $b^*_{2j} = d_j$, $b^*_{2j+1} = b_j \cap \overline{\overline{d}}_j$ for $j < m$ with $j \neq j_0$, and $a^*_{2j_0} = a$, $a_{2j_0+1} = a_{j_0} \cap \overline{\overline{a}}$, $b^*_{2j_0} = b$, $b^*_{2j_0+1} = b_{j_0} \cap \overline{\overline{b}}$;

(PS 6) if $i < n$, $\langle \bar{a}, \bar{b} \rangle \in \Gamma(i)$ with $\mathrm{lh}(\bar{a}) = m$, $j_0 < m$, $d_0, \ldots, d_{j_0-1}, d_{j_0+1}, \ldots, d_{m-1} \in |\mathfrak{B}|$ with $d_j \leq b_j$ for each $j < m$ with $j \neq j_0$, $B \subseteq |\mathfrak{B}|$ with $\mathrm{card}(B) = \omega_\beta$ and $b \leq b_{j_0}$ for each $b \in B$, then there are $c_0, \ldots, c_{j_0-1}, c_{j_0+1}, \ldots, c_{m-1} \in |\mathfrak{A}|$

with $c_j \leq a_j$ for each $j < m$ with $j \neq j_0$, $A \subseteq |\mathfrak{A}|$ with card(A) = $\omega_\alpha$ and $a \leq a_{j_0}$ for each $a \in A$ so that for each $a \in A$ there is a $b \in B$ with $\langle \overline{a^*}, \overline{b^*} \rangle \in \Gamma(i + 1)$, where $lh(\overline{a^*}) = 2m$, $a^*_{2j} = c_j$, $a^*_{2j+1} = a_j \cap \overline{\overline{c_j}}$, $b^*_{2j} = d_j$, $b^*_{2j+1} = b_j \cap \overline{\overline{d_j}}$ for $j < m$ with $j \neq j_0$, and $a^*_{2j_0} = a$, $a^*_{2j_0+1} = a_{j_0} \cap \overline{\overline{a}}$, $b^*_{2j_0} = b$, $b^*_{2j_0+1} = b_{j_0} \cap \overline{\overline{b}}$.

The sequence $\Gamma$ is called <u>$(\alpha,\beta,n)$-PS-sequence</u> for $\langle \mathfrak{A}, \overline{a^0} \rangle$ and $\langle \mathfrak{B}, \overline{b^0} \rangle$.

Immediately by definition we get

<u>Lemma 2.</u> Let $\mathfrak{A}, \mathfrak{B} \in$ Ba, $\alpha, \beta \in$ On, $n \in \omega$, $\overline{a^0}$ a partition of $\mathfrak{A}$, $\overline{b^0}$ a partition of $\mathfrak{B}$ and $lh(\overline{a^0}) = lh(\overline{b^0})$, $\Gamma$ an $(\alpha,\beta,n)$-PS-sequence for $\langle \mathfrak{A}, \overline{a^0} \rangle$ and $\langle \mathfrak{B}, \overline{b^0} \rangle$. Then for each $k \leq n$ and each $\langle \overline{a^*}, \overline{b^*} \rangle \in \Gamma(k)$

$$\langle \mathfrak{A}, \overline{a^*} \rangle \;{}^{\alpha}\!\underset{PS}{\sim}{}^{\beta}_{n-k}\; \langle \mathfrak{B}, \overline{b^*} \rangle.$$

<u>Theorem 3.</u> Let $\mathfrak{A}, \mathfrak{B} \in$ Ba, $\alpha, \beta \in$ On, $n \in \omega$, $\overline{a^0}$ a partition of $\mathfrak{A}$, $\overline{b^0}$ a partition of $\mathfrak{B}$ with $lh(\overline{a^0}) = lh(\overline{b^0})$. Then

$$\langle \mathfrak{A}, \overline{a^0} \rangle \;{}^{\alpha}\!\underset{PS}{\sim}{}^{\beta}_{n}\; \langle \mathfrak{B}, \overline{b^0} \rangle \quad \text{iff} \quad \langle \mathfrak{A}, \overline{a^0} \rangle \;{}^{\alpha}\!\equiv^{\beta}_{n}\; \langle \mathfrak{B}, \overline{b^0} \rangle.$$

<u>Proof.</u> <u>($\longleftarrow$)</u>: Let $lh(\overline{a^0}) = m$ and let $\Gamma^*$ be an $(\alpha,\beta,n)$-I-sequence for $\langle \mathfrak{A}, \overline{a^0} \rangle$ and $\langle \mathfrak{B}, \overline{b^0} \rangle$. Let $\Gamma$ be the sequence which is determined by the following conditions:

(i) $lh(\Gamma) = n + 1$;

(ii) $\Gamma(0) = \{\langle \overline{a^0}, \overline{b^0} \rangle\}$;

(iii) for each $i \leq n$ $\Gamma(i)$ is the set of all pairs $\langle \bar{a}, \bar{b} \rangle = \langle\langle a_0,\ldots,a_{m-1}\rangle, \langle b_0,\ldots,b_{m-1}\rangle\rangle$ for which there are elements $a^0,\ldots,a^{i-1} \in |\mathfrak{A}|$, $b^0,\ldots,b^{i-1} \in |\mathfrak{B}|$ with $\langle\langle a^k,b^k\rangle : k < i\rangle \in \Gamma^*(i)$, so that $\bar{a}$ is a partition sequence for $\{a_0,\ldots,a_{m-1},a^0,\ldots,a^{i-1}\}$ and $\bar{b}$ is a partition sequence for $\{b_0,\ldots,b_{m-1},b^0,\ldots,b^{i-1}\}$.

Then $\Gamma$ is an $(\alpha,\beta,n)$-PS-sequence for $\langle \mathfrak{A}, \overline{a^0} \rangle$ and $\langle \mathfrak{B}, \overline{b^0} \rangle$.

($\longrightarrow$): Assume, on the contrary, that $\langle \mathfrak{A}, \overline{a^0} \rangle \;{}^{\alpha}\!\underset{PS}{\sim}{}^{\beta}_{n}\; \langle \mathfrak{B}, \overline{b^0} \rangle$, but $\langle \mathfrak{A}, \overline{a^0} \rangle \;{}^{\alpha}\!\not\equiv^{\beta}_{n}\; \langle \mathfrak{B}, \overline{b^0} \rangle$. Then, of course, $n \geq 1$. Let $lh(\overline{a^0}) = m$. Without loss of generality we assume that $\overline{a^0}$ and $\overline{b^0}$ are of such a form that n is minimal. Let $\varphi$ be a sentence in the language $L_{Ba}(Q)$ enriched by m new constants with $\langle \mathfrak{A}, \overline{a^0} \rangle \models_\alpha \varphi$ and $\langle \mathfrak{B}, \overline{b^0} \rangle \models_\beta \neg\varphi$. Without loss of generality we assume that $\varphi = \exists x \psi$ or $\varphi = Qx\psi$.

Let $\varphi = \exists x\psi$. Then there is an $a \in |\mathfrak{A}|$ with $\mathfrak{A} \models \varphi[a]$. Let $\bar{c}$ be a sequence with $lh(\bar{c}) = lh(a^0) = m$ so that for each $i < m$, $c_i = a \cap a_i$. Then it follows from (PS 3) that there is a $\bar{d}$ with $lh(\bar{d}) = m$ so that the pair $\langle a^*, b^* \rangle$ with $lh(a^*) = lh(b^*) = 2m$, $a^*_{2j} = c_j$, $a^*_{2j+1} = a_j \cap \overline{\overline{c_j}}$, $b^*_{2j} = d_j$, $b^*_{2j+1} = b_j \cap \overline{\overline{d_j}}$ for each $j < m$ is from $\Gamma(1)$. Now let $a = \bigcup[a^*_{2j}: j < m]$. From the minimality of n with respect to all partition sequences it follows, that $\langle \mathfrak{A}, a^* \rangle \,{}^{\alpha}{\equiv}^{\beta}_{n-1} \langle \mathfrak{B}, b^* \rangle$. Thus $\langle \mathfrak{A}, a^* \rangle \models_\alpha \psi[a]$ implies $\langle \mathfrak{B}, b^* \rangle \models_\beta \psi[\bigcup[b^*_{2j}: j < m]]$ and thus $\langle \mathfrak{B}, b^0 \rangle \models_\beta \varphi$, and so we got a contradiction.

Now let $\varphi = Qx\psi$. Then there is a set $A \subseteq |\mathfrak{A}|$ with $card(A) = \omega_\alpha$ so that for each $a \in A$, $\langle \mathfrak{A}, a^0 \rangle \models_\alpha \psi[a]$. Now there are only finitely many formulas $\chi$ with $fv(\chi) \subseteq \{x, x_0, \ldots, x_{m-1}\}$ and $qr(\chi) \leq n$. That means, there is a set $A' \subseteq A$ with $card(A') = \omega_\alpha$ so that for $a, a' \in A'$, $\langle \mathfrak{A}, a^0{}^\frown a \rangle \,{}^{\alpha}{\equiv}^{\alpha}_{n-1} \langle \mathfrak{A}, a^0{}^\frown a' \rangle$. Let $k < m$ be so that $card(\{a \cap a^0_k: a \in A'\}) = \omega_\alpha$. We choose some $a^* \in A'$. For $a \in A'$ let $\tilde{a} = (a \cap a^0_k) \cup (a^* \cap \overline{a^0_k})$. Let $A^* = \{\tilde{a}: a \in A'\}$. Then $card(A^*) = \omega_\alpha$, and the product property for $Q_\alpha$ implies that $\langle \mathfrak{A}, a^0{}^\frown a \rangle \,{}^{\alpha}{\equiv}^{\beta}_{n-1} \langle \mathfrak{A}, a^0{}^\frown a^* \rangle$ for each $a \in A^*$.

Furthermore, for each $a \in A^*$, $j < m$ with $j \neq k$ we have $a \cap a^0_j = a^* \cap a^0_j$. Let $c_j = a^* \cap a^0_j$ for $j < m$ with $j \neq k$. (PS 5) implies that there are $d_0, \ldots, d_{k-1}, d_{k+1}, \ldots, d_{m-1}$ with $d_j \leq b_j$ for each $j < m$ with $j \neq k$ and that there is a set $B \subseteq |\mathfrak{B}|$ with $card(B) = \omega_\beta$ with $b \leq b_k$ for each $b \in B$ and so that for each $a \in A^*$ and $b \in B$, $\langle \mathfrak{A}, a^0{}^\frown a \rangle \,{}^{\alpha}{\equiv}^{\alpha}_{n-1} \langle \mathfrak{B}, b^0{}^\frown b \rangle$. For $b \in B$ let $\tilde{b} = b \cup \bigcup[d_j: j < m, j \neq k]$. Then for each $b \in B$, $\langle \mathfrak{A}, a^0{}^\frown a^* \rangle \,{}^{\alpha}{\equiv}^{\beta}_{n-1} \langle \mathfrak{B}, b^0{}^\frown \tilde{b} \rangle$ and thus for each $b \in B$, $\langle \mathfrak{B}, b^0 \rangle \models_\beta \psi[\tilde{b}]$. This implies $\langle \mathfrak{B}, b^0 \rangle \models_\beta Qx\psi$ and so we got a contradiction.

Q.E.D.

The next lemma follows immediately by definition.

<u>Lemma 4.</u> Let $\mathfrak{A}, \mathfrak{B} \in Ba$, $\alpha, \beta \in On$, $n \in \omega$, and $\mathfrak{A} \,{}^{\alpha}{\underset{PS}{\sim}}{}^{\beta}_{n} \mathfrak{B}$. Then

(i) for any $a \in |\mathfrak{A}|$ ($b \in |\mathfrak{B}|$) there is a $b \in |\mathfrak{B}|$ ($a \in |\mathfrak{A}|$) with $\langle \mathfrak{A}, a \rangle \,{}^{\alpha}{\equiv}^{\beta}_{n-1} \langle \mathfrak{B}, b \rangle$;

(ii) for any $A \subseteq |\mathfrak{A}|$ with $card(A) = \omega_\alpha$ ($B \subseteq |\mathfrak{B}|$ with $card(B) = \omega_\beta$) there is a $B \subseteq |\mathfrak{B}|$ with $card(B) = \omega_\beta$ ($A \subseteq |\mathfrak{A}|$ with $card(A) = \omega_\alpha$) so that for each $b \in B$ ($a \in A$) there is an $a \in A$ ($b \in B$) with $\langle \mathfrak{A}, a \rangle \,{}^{\alpha}{\equiv}^{\beta}_{n-1} \langle \mathfrak{B}, b \rangle$.

It is possible to formulate the relation $\mathfrak{A} \,{}^{\alpha}{\underset{PS}{\sim}}{}^{\beta}_{n} \mathfrak{B}$ in game-theoretic terms:

The game $H^1(n,\mathfrak{A},\mathfrak{B},\alpha,\beta)$ is a two-person game over n rounds. For each $i \leq n$ the <u>result before the i-th round</u> is a pair $\langle \bar{a},\bar{b}\rangle$ with $lh(\bar{a}) = lh(\bar{b}) = 2^i$ so that $\bar{a}$ is a partition of $\mathfrak{A}$ and $\bar{b}$ is a partition of $\mathfrak{B}$ (therefore, the result before the 0-th round is the pair $\langle\langle 1_{\mathfrak{A}}\rangle,\langle 1_{\mathfrak{B}}\rangle\rangle$). The result before the n-th round is called the <u>result of the game</u>. For each $i < n$ the result before the (i+1)-th round is also called <u>the result after the i-th round</u>.

Let $i < n$, and $\langle \bar{a},\bar{b}\rangle$ be the result of the i-th round. The i-th round is as follows:

> Player I begins and decides if an $\exists$-round or a Q-round will be played.
>
> If player I decides to play an $\exists$-round, he chooses one of the two structures, for instance $\mathfrak{A}$, and for each $j < 2^i$ an element $c_j$ with $c_j \leq a_j$. Then player II chooses in the other structure an element $d_j$ with $d_j \leq b_j$ for each $j < 2^i$. Then the i-th round is finished and the pair $\langle \overline{a^*},\overline{b^*}\rangle$ with $lh(\overline{a^*}) = lh(\overline{b^*}) = 2^{i+1}$ and $a^*_{2j} = c_j$, $a^*_{2j+1} = a_j \cap \overline{c_j}$, $b^*_{2j} = d_j$, $b^*_{2j+1} = b_j \cap \overline{d_j}$ for each $j < 2^i$ is the result of the game before the (i+1)-th round.
>
> If player I decides to play a Q-round, he chooses one of the two structures, for instance $\mathfrak{B}$, $j_0 < 2^i$, for each $j < 2^i$ with $j \neq j_0$ an element $d_j \leq b_j$ and a set $B \subseteq |\mathfrak{B}|$ with $card(B) = \omega_\beta$ so that it is $b \leq b_{j_0}$ for each $b \in B$. Then player II chooses for each $j < 2^i$ with $j \neq j_0$ an element $c_j \leq a_j$ and a set $A \subseteq |\mathfrak{A}|$ with $card(A) = \omega_\alpha$ in the other structure so that for each $a \in A$ we have $a \leq a_{j_0}$. Now player I chooses an element $c_{j_0} \in A$ and at last player II chooses an element $d_{j_0} \in B$. Then the i-th round is finished and the pair $\langle \overline{a^*},\overline{b^*}\rangle$ with $lh(\overline{a^*}) = lh(\overline{b^*}) = 2^{i+1}$ and $a^*_{2j} = c_j$, $a^*_{2j+1} = a_j \cap \overline{c_j}$, $b^*_{2j} = d_j$, $b^*_{2j+1} = b_j \cap \overline{d_j}$ for each $j < 2^i$ is the result of the game before the (i+1)-th round.

After the (n-1)-th round the game is finished (remember that we start with the 0-th round). Player II wins the game if the result $\langle\langle a_0,\dots,a_{2n-1}\rangle,\langle b_0,\dots,b_{2n-1}\rangle\rangle$ satisfies the following condition:

> for each $i < 2^n$: $a_i = 0$ iff $b_i = 0$.

Now we have:

Lemma 5. Let $\mathfrak{A}, \mathfrak{B} \in$ Ba, $\alpha, \beta \in$ On, $n \in \omega$. Then

$$\mathfrak{A}\ {}^{\alpha}\!\sim_n^{\beta}\ \mathfrak{B}$$

iff player II has a winning strategy in the game $H^1(n, \mathfrak{A}, \mathfrak{B}, \alpha, \beta)$ (that means that he can ensure that he always wins).

Now we describe a new class of games, the $H^2$-games. Let $\mathfrak{A}, \mathfrak{B} \in$ Ba, $\alpha, \beta \in$ On, $n \in \omega$. The game $H^2(n, \mathfrak{A}, \mathfrak{B}, \alpha, \beta)$ is a two-person game over n rounds. For each $i \leq n$ the result before the i-th round is a pair $\langle a_i, b_i \rangle$ with $a_i \in |\mathfrak{A}|$ and $b_i \in |\mathfrak{B}|$. The result before the 0-th round is the pair $\langle 1^{\mathfrak{A}}, 1^{\mathfrak{B}} \rangle$. The result before the n-th round is called the result of the game. For each $i < n$ the result before the (i+1)-thround is also called the result after the i-th round. Let $i < n$, $\langle a_i, b_i \rangle$ be the result before the i-th round. The i-th round is as follows:

> Player I begins and decides if an $\exists$-round or a Q-round will be played.
>
> If player I decides to play an $\exists$-round, he chooses one of the two structures, for instance $\mathfrak{B}$, and an element d with $d \leq b_i$. Then player II chooses an element c with $c \leq a_i$ in the other structure. Now player I chooses one of the two pairs $\langle c, d \rangle$, $\langle a_i \cap \bar{c}, b_i \cap \bar{d} \rangle$, the result of the game before the (i+1)-th round. After that the i-th round is finished.
>
> If player I decides to play a Q-round, he chooses one of the two structures, for instance $\mathfrak{A}$, and a set $A \subseteq |\mathfrak{A}|$ with card(A) = $\omega_\alpha$ so that $a \leq a_i$ for each $a \in A$. Then player II chooses a set $B \subseteq |\mathfrak{B}|$ with card(B) = $\omega_\beta$ in the other structure so that $b \leq b_i$ for each $b \in B$. Now player I chooses an element $d \in B$ and then player II chooses an element $c \in A$. Finally, player I chooses one of the two pairs $\langle c, d \rangle$, $\langle a_i \cap \bar{c}, b_i \cap \bar{d} \rangle$, the result before the (i+1)-th round.

After the (n-1)-th round the game is finished. Player II wins if the result $\langle a_n, b_n \rangle$ of the game satisfies the following condition:

$$a_n = 0 \quad \text{iff} \quad b_n = 0.$$

Lemma 6. Let $\mathfrak{A}, \mathfrak{B} \in \mathrm{Ba}$, $\alpha, \beta \in \mathrm{On}$, $n \in \omega$. Then player II has a winning strategy in the game $H^1(n,\mathfrak{A},\mathfrak{B},\alpha,\beta)$ iff player II has a winning strategy in the game $H^2(n,\mathfrak{A},\mathfrak{B},\alpha,\beta)$.

Proof. The winning strategy for player II in the game $H^1(n,\mathfrak{A},\mathfrak{B},\alpha,\beta)$ immediately gives a winning strategy for player II in the game $H^2(n,\mathfrak{A},\mathfrak{B},\alpha,\beta)$. On the other hand it is possible to build up a winning strategy for player II in the game $H^1(n,\mathfrak{A},\mathfrak{B},\alpha,\beta)$ using a winning strategy for player II in the game $H^2(n,\mathfrak{A},\mathfrak{B},\alpha,\beta)$. Let us sketch the main idea: The pair $\langle a,b\rangle$ is a possible situation before the i-th round if $\langle a,b\rangle$ can occur as the result before the i-th round in the game $H^2(n,\mathfrak{A},\mathfrak{B},\alpha,\beta)$, where player II plays according to his winning strategy. In the game $H^1(n,\mathfrak{A},\mathfrak{B},\alpha,\beta)$ player II can ensure that for each $i \leq n$ the result $\langle \bar{a},\bar{b}\rangle = \langle\langle a_0^i,\dots,a_{2^i-1}^i\rangle,\langle b_0^i,\dots,b_{2^i-1}^i\rangle\rangle$ before the i-th round is so that $\langle a_j^i,b_j^i\rangle$ is a possible situation for each $j < 2^i$. Playing in such a way, player II will always win.

Q.E.D.

To prove $\mathfrak{A} \;{}^{\alpha}\!\underset{n}{\equiv}{}^{\beta}\; \mathfrak{B}$ we shall show that player II has a winning strategy in the game $H^2(n,\mathfrak{A},\mathfrak{B},\alpha,\beta)$.

We give some examples for the application of $H^2$-games.

Lemma 7. Let $\mathfrak{A}, \mathfrak{B} \in \mathrm{Ba}$, $\alpha, \beta \in \mathrm{On}$. Then

(i) if $\mathfrak{A}$ is $\omega_\alpha$-weakly-atomless with $\mathrm{card}(|\mathfrak{A}|) < \omega_\alpha$ and $\mathfrak{B}$ is $\omega_\beta$-weakly-atomless with $\mathrm{card}(|\mathfrak{B}|) < \omega_\beta$, then $\mathfrak{A} \;{}^{\alpha}\!\equiv^{\beta}\; \mathfrak{B}$;

(ii) if $\mathfrak{A}$ is $\omega_\alpha$-atomless and $\mathfrak{B}$ is $\omega_\beta$-atomless, then $\mathfrak{A} \;{}^{\alpha}\!\equiv^{\beta}\; \mathfrak{B}$.

Proof. Let be $n \in \omega$. We show that player II has a winning strategy in the game $H^2(n,\mathfrak{A},\mathfrak{B},\alpha,\beta)$. We show that player II can ensure that for each $i \leq n$ the result $\langle a_i,b_i\rangle$ before the i-th round satisfies the following condition

(*) $a_i = 0$ iff $b_i = 0$.

Of course, the result $\langle 1^{\mathfrak{A}},1^{\mathfrak{B}}\rangle$ before the 0-th round satisfies (*). Let $k < n$, and let the result $\langle a_k,b_k\rangle$ satisfy (*).

Ad (i): We have $\mathrm{card}(|\mathfrak{A}|) \geq \omega_\alpha$ and $\mathrm{card}(|\mathfrak{B}|) \geq \omega_\beta$; thus player I can only decide to play an $\exists$-round. Assume that player I chooses the structure $\mathfrak{A}$ and $a \in |\mathfrak{A}|$ with $a \leq a_k$. If $a = 0^{\mathfrak{A}}$, then player II chooses $0^{\mathfrak{B}}$. If $a = a_k$, then player II chooses $b_k$. If $0 < a < a_k$, then player II chooses any element b with $0 < b < b_k$. That is possible since $\mathfrak{B}$ is atomless.

Ad (ii): Assume player I decides to play an $\exists$-round. Then player II plays as described for (i). Assume player I decides to play a Q-round. Then he chooses a structure, for instance $\mathfrak{B}$, and a set $B \subseteq |\mathfrak{B}|$ with $\mathrm{card}(B) = \omega_\beta$. Then there is an element $b^* \in B$ with $0 < b^* < b_k$. Player II chooses a set $A \subseteq |\mathfrak{A}|$ with $\mathrm{card}(A) = \omega_\alpha$ so that for each $a \in A$, $0 < a < a_k$. This is possible because of $\mathrm{cd}(a_k) \geq \omega_\alpha$. For each $a \in A$ chosen by player I, player II chooses b. Thus the result $\langle a_{k+1}, b_{k+1} \rangle$ before the (k+1)-th round satisfies (*) for $i = k + 1$.

Q.E.D.

Let $\mathfrak{A}, \mathfrak{B} \in \mathrm{Ba}$, $\alpha, \beta \in \mathrm{On}$, $n \in \omega$. The game $H^2(n, \mathfrak{A}, \mathfrak{B})$ is the game which is played by the same rules as $H^2(n, \mathfrak{A}, \mathfrak{B}, \alpha, \beta)$, but with the restriction that player I always decides to play an $\exists$-round. Then it is easy to see that:

<u>Lemma 8.</u> Let $\mathfrak{A}, \mathfrak{B} \in \mathrm{Ba}$, $n \in \omega$. Then

$$\mathfrak{A} \equiv_n \mathfrak{B}$$

iff player II has a winning strategy in the game $H^2(n, \mathfrak{A}, \mathfrak{B})$.

<u>Lemma 9.</u> Let $n \in \omega$, $\mathfrak{A} \in \mathrm{Ba}$. If $\mathfrak{A}$ is atomic with at least $2^n$ atoms, then $\mathfrak{A} \equiv_n \mathrm{Pow}(2^n)$.

<u>Proof.</u> It is sufficient to show that player II has a winning strategy in the game $H^2(n, \mathfrak{A}, \mathrm{Pow}(2^n))$. Player II only has to ensure that the following condition is satisfied for each $i \leq n$:

> if $\langle a_i, b_i \rangle$ is the result before the i-th round, $k_i$ is the number of atoms $\leq a_i$ and $m_i$ is the number of atoms $\leq b_i$, then $k_i = m_i$ or $k_i, m_i \geq 2^{n-i}$.

Q.E.D.

For each $n \in \omega$ there are only finitely many $\varphi \in \mathrm{Sent}(L_{\mathrm{Ba}}(Q))$ with $\mathrm{qr}(\varphi) \leq n$ which are different with respect to Ba. Thus Lemma 9 together with the Feferman-Vaught-Theorem (see [Chang, Keisler 1973] or [Feferman, Vaught 1959]) implies:

Lemma 10. For each $n \in \omega$ there is a $k(n) \in \omega$ with: Let I be a set and for each $i \in I$ let $\mathfrak{A}_i \in Ba$ so that $\mathfrak{A}_i \ {}^{\alpha}{\equiv}_n^{\alpha} \ \mathfrak{A}_j$ for $i,j \in I$, where $\alpha$ is any ordinal. Let $J, K \subseteq I$ with $k(n) \leq \mathrm{card}(J) < \omega_0$, $k(n) \leq \mathrm{card}(K) < \omega_0$. Then

$$\prod\{\mathfrak{A}_i : i \in J\} \ {}^{\alpha}{\equiv}_n^{\alpha} \prod\{\mathfrak{A}_i : i \in K\}.$$

We use Lemma 10 to prove

Lemma 11. Let $\mathfrak{A}, \mathfrak{B}, \mathfrak{C} \in Ba$, $\alpha \in On$, $n \in \omega$. If $\mathfrak{A} \ {}^{\alpha}{\equiv}_n^{\alpha} \ \mathfrak{A} \times \mathfrak{B} \times \mathfrak{C}$, then $\mathfrak{A} \ {}^{\alpha}{\equiv}_n^{\alpha} \ \mathfrak{A} \times \mathfrak{B}$.

Proof. $Q_\alpha$ has the product property. Thus, replacing $\mathfrak{A}$ repeatedly by $\mathfrak{A} \times \mathfrak{B} \times \mathfrak{C}$ we get $\mathfrak{A} \ {}^{\alpha}{\equiv}_n \ \mathfrak{A} \times (\mathfrak{B} \times \mathfrak{C})^k$ for any $k \in \omega$. Let $k \geq k(n)$. Then $\mathfrak{B}^{k+1} \ {}^{\alpha}{\equiv}_n^{\alpha} \ \mathfrak{B}^k$ and thus

$$\mathfrak{A} \times \mathfrak{B} \ {}^{\alpha}{\equiv}_n^{\alpha} \ \mathfrak{A} \times \mathfrak{B}^{k+1} \times \mathfrak{C}^k \ {}^{\alpha}{\equiv}_n^{\alpha} \ \mathfrak{A} \times \mathfrak{B}^k \times \mathfrak{C}^k \ {}^{\alpha}{\equiv}_n^{\alpha} \ \mathfrak{A}.$$

Q.E.D.

Lemma 11 implies

Lemma 12. Let $\mathfrak{A} \in Ba$, $\alpha \in On$, $n \in \omega$, $a,b,c \in |\mathfrak{A}|$ with $a \leq c \leq b$. If $\mathfrak{A}[a] \ {}^{\alpha}{\equiv}_n^{\alpha} \ \mathfrak{A}[b]$, then $\mathfrak{A}[a] \ {}^{\alpha}{\equiv}_n^{\alpha} \ \mathfrak{A}[c]$.

Proof. $\mathfrak{A}[b] \cong \mathfrak{A}[a] \times \mathfrak{A}[\bar{\bar{a}} \cap c] \times \mathfrak{A}[\bar{\bar{c}} \cap b]$ and thus $\mathfrak{A}[a] \ {}^{\alpha}{\equiv}_n^{\alpha} \ \mathfrak{A}[b]$ implies

$$\mathfrak{A}[a] \ {}^{\alpha}{\equiv}_n^{\alpha} \ \mathfrak{A}[a] \times \mathfrak{A}[\bar{\bar{a}} \cap c] \cong \mathfrak{A}[c].$$

Q.E.D.

Theorem 13. Let $\mathfrak{A} \in Ba$, $\alpha \in On$, $n \in \omega$. Then $\mathrm{card}(|\mathfrak{A}|) \geq \omega_\alpha$ iff

$$\mathrm{card}(\{a \in |\mathfrak{A}| : \mathfrak{A}[a] \ {}^{\alpha}{\equiv}_n^{\alpha} \ \mathfrak{A}\}) \geq \omega_\alpha.$$

Proof. ($\longleftarrow$) is immediately clear.

($\longrightarrow$) Let $\mathrm{card}(|\mathfrak{A}|) = \omega_\alpha$. We set $X = \{a \in |\mathfrak{A}| : \mathrm{cd}(a) \geq \omega_\alpha\}$. Lemma 12 together with $\mathrm{card}(\mathrm{typ}(n,\alpha,Ba)) < \omega_0$ implies that there is an $a \in X$ with:

if $b \in X$ and $b \leq a$, then $\mathfrak{A}[a] \ {}^{\alpha}{\equiv}_n^{\alpha} \ \mathfrak{A}[b]$.

Now for each $c \in C$ we have $\mathrm{card}(\{a \in X: a \leq c\}) \geq \omega_\alpha$ and for each $b \in X$ with $b \leq a$ it is $\mathfrak{A} \; {}^{\alpha}\!\equiv_n^{\alpha} \; \mathfrak{A}[\bar{\bar{a}} \cap b]$. Thus

$$\mathrm{card}(\{a \in |\mathfrak{A}| : \mathfrak{A}[a] \; {}^{\alpha}\!\equiv_n^{\alpha} \; \mathfrak{A}\}) \geq \omega_\alpha .$$

Q.E.D.

## § 4. Dense systems

In this paragraph we shall describe dense systems for $Ba(Q_0)$, $Ba(Q_1)$, and $Ba(Q_\omega)$.

A term set T is a set of formal expressions with:

(i) $2, \eta_0, \eta \in T$;
(ii) if $t_1, t_2 \in T$, then $t_1 \times t_2 \in T$.

Let $\alpha$ be an ordinal number. The pair $\langle T, \mathcal{B} \rangle$ is a pre-density pair for $Q_\alpha$ if T is a term set and $\mathcal{B}$ is a function from T into the class of Boolean algebras with:

(i) $\mathcal{B}(2) = B(2)$;
(ii) $\mathcal{B}(\eta_0) = B(\eta_0)$;
(iii) $\mathcal{B}(\eta) = \mathcal{J}(\mathcal{H}_\alpha)$;
(iv) if $t_1, t_2 \in T$ (not necessarily $t_1 \neq t_2$), then $\mathcal{B}(t_1 \times t_2) = \mathcal{B}(t_1) \times \mathcal{B}(t_2)$.

$\mathcal{M}_T = \{\mathcal{B}(t) : t \in T\}$ is called pre-density system for $Q_\alpha$. Let $\langle T, \mathcal{B} \rangle$ be a pre-density pair for $Q_\alpha$, $\mathfrak{A} \in Ba$ and n be a natural number. Then we set

$$D_n(\mathfrak{A}) = \{a \in |\mathfrak{A}| : \text{for each } b \in |\mathfrak{A}| \text{ with } 0 < b \leq a \text{ there is a } t \in T \text{ with } \mathfrak{A}[b] \; {}^{\alpha}\!\equiv_n^{\alpha} \; \mathcal{B}(t)\};$$

for $a \in |\mathfrak{A}|$ let be

$$\mathrm{Typ}(n,\alpha,a) = \{T \in \mathrm{typ}(n,\alpha,Ba) : \text{there is a } b \in D_n(\mathfrak{A}) \text{ with } 0 < b \leq a \text{ and } \mathrm{typ}(n,\alpha,\mathfrak{A}[b]) = T\};$$

$$M_n(\mathfrak{A}) = \{a \in |\mathfrak{A}| \setminus D_n(\mathfrak{A}) : \text{if } b \leq a \text{ and } b \in |\mathfrak{A}| \setminus D_n(\mathfrak{A}), \text{ then } \mathrm{Typ}(n,\alpha,a) = \mathrm{Typ}(n,\alpha,b)\};$$

$$N_n^0(\mathfrak{A}) = \{a \in M_n(\mathfrak{A}) : a/D_n(\mathfrak{A}) \text{ is an atom}\};$$

$$N_n^1(\mathfrak{A}) = \{a \in M_n(\mathfrak{A}) : a/D_n(\mathfrak{A}) \text{ is } \omega_\alpha\text{-weakly-atomless and of cardinality } < \omega_\alpha\};$$

$$N_n^2(\mathfrak{A}) = \{a \in M_n(\mathfrak{A}) : a/D_n(\mathfrak{A}) \text{ is } \omega_\alpha\text{-atomless}\}.$$

If $a \in |\mathfrak{A}|$, then

$$D_n(\mathfrak{A}[a]) = D_n(\mathfrak{A}) \cap |\mathfrak{A}[a]|.$$

If $b \leq a$, then

$$\mathrm{Typ}(n,\alpha,b) = \mathrm{Typ}(n,\alpha,a).$$

It follows from $\mathrm{typ}(n,\alpha,Ba) < \omega_0$ that

$$D_n(\mathfrak{A}) = |\mathfrak{A}| \text{ iff } M_n(\mathfrak{A}) \neq \emptyset \text{ iff } N_n^0(\mathfrak{A}) \cup N_n^1(\mathfrak{A}) \cup N_n^2(\mathfrak{A}) = \emptyset.$$

Let $a \in M_n(\mathfrak{A})$. Then we set

$$R_n(a) = \{T \in \mathrm{Typ}(n,\alpha,a) : \mathrm{card}(\{b \in D_n(\mathfrak{A}) \setminus \{0\} : \mathrm{typ}(n,\alpha,\mathfrak{A}[b]) = T\} < \omega_\alpha\};$$

$$P_n(\mathfrak{A}) = \{a \in M_n(\mathfrak{A}) : \text{if } b \leq a,\ b \in M_n(\mathfrak{A}), \text{ then } R_n(a) = R_n(b)\}.$$

We set for each $i < 3$,

$$P_n^i(\mathfrak{A}) = P_n(\mathfrak{A}) \cap N_n^i(\mathfrak{A}).$$

If $b \leq a$, $b \in M_n(\mathfrak{A})$, then $R_n(b) \supseteq R_n(a)$. Thus

$$P_n^i(\mathfrak{A}) \neq \emptyset \quad \text{iff} \quad N_n^i(\mathfrak{A}) \neq \emptyset.$$

We have for each $i < 3$,

$$N_n^i(\mathfrak{A}) = \emptyset \quad \text{iff} \quad P_n^i(\mathfrak{A}) = \emptyset.$$

Thus

$$M_n(\mathfrak{A}) = \emptyset \quad \text{iff} \quad P_n^0(\mathfrak{A}) \cup P_n^1(\mathfrak{A}) \cup P_n^2(\mathfrak{A}) = \emptyset.$$

We set

$$S_n^0(\mathfrak{A}) = \{a \in P_n(\mathfrak{A}) : R_n(a) = \mathrm{Typ}(n,\alpha,a)\};$$

$$S_n^1(\mathfrak{A}) = \{a \in P_n(\mathfrak{A}) : R_n(a) = \emptyset\};$$

$$S_n^2(\mathfrak{A}) = P_n(\mathfrak{A}) \setminus (S_n^0(\mathfrak{A}) \cup S_n^1(\mathfrak{A})).$$

Thus if $a \in S_n^1(\mathfrak{A})$, then $cd(a) \geq \omega_\alpha$.

Then in order to show that $P_n^1(\mathfrak{A}) = \emptyset$ for any $i < 3$ it is sufficient to show that

$$P_n^i(\mathfrak{A}) \cap S_n^0(\mathfrak{A}) = \emptyset, \quad P_n^i(\mathfrak{A}) \cap S_n^1(\mathfrak{A}) = \emptyset \quad \text{and} \quad P_n^i(\mathfrak{A}) \cap S_n^2(\mathfrak{A}) = \emptyset.$$

<u>Lemma 14.</u> Let $\alpha \in \mathrm{On}$, $\langle T,\mathcal{B}\rangle$ be a pre-density pair for $Q_\alpha$, $\mathfrak{A} \in \mathrm{Ba}$, and $n \in \omega$. Then $D_n(\mathfrak{A})$ is an ideal and is dense in $\mathfrak{A}$.

<u>Proof.</u> It follows from Lemma 7 that all $\omega_\alpha$-atomless Boolean algebras and all $\omega_\alpha$-weakly-atomless Boolean algebras of cardinality $< \omega_\alpha$ are $Q_\alpha$-equivalent. But then $D_n(\mathfrak{A})$ contains all atoms, all $\omega_\alpha$-weakly-atomless elements of cardinality $< \omega_\alpha$, and all $\omega_\alpha$-atomless elements; thus $D_n(\mathfrak{A})$ is dense. Now we show that $D_n(\mathfrak{A})$ is an ideal on $\mathfrak{A}$. If $b \in D_n(\mathfrak{A})$ and $a \leq b$, then it follows by definition that $a \in D_n(\mathfrak{A})$. Let $a, b \in D_n(\mathfrak{A})$, $c \leq a \cup b$. If $c \leq a$ or $c \leq b$, then also $c \in D_n(\mathfrak{A})$. Thus let us assume that $c \not\leq a$ and $c \not\leq b$. We set $d = a \cap c$, $e = b \cap \bar{\bar{a}} \cap c$. Then we have $0 < d \leq a$ and $0 < e \leq b$. Thus there are $t_1, t_2 \in T$ with $\mathfrak{A}[d] \;{}^{\alpha}\!\!\equiv_n^{\alpha} \mathcal{B}(t_1)$ and $\mathfrak{A}[e] \;{}^{\alpha}\!\!\equiv_n^{\alpha} \mathcal{B}(t_2)$. Using the fact that $d \cup e = c$ and $d \cap e = 0$ we get

$$\mathfrak{A}[c] \cong \mathfrak{A}[d] \times \mathfrak{A}[e] \;{}^{\alpha}\!\!\equiv_n^{\alpha} \mathcal{B}(t_1) \times \mathcal{B}(t_2) = \mathcal{B}(t_1 \times t_2)$$

and thus $a \cup b \in D_n(\mathfrak{A})$.

Q.E.D.

Let $\alpha$ be an ordinal, $\langle T,\mathcal{B}\rangle$ be a pre-density pair for $Q_\alpha$, $\mathfrak{A} \in \mathrm{Ba}$, $a \in |\mathfrak{A}|$, and n be a natural number. A <u>representation</u> of $\mathrm{Typ}(n,\alpha,a)$

is a set $\{t_T: T \in \mathrm{Typ}(n,\alpha,a)\}$ so that for each $T \in \mathrm{Typ}(n,\alpha,a)$ we have $\mathrm{typ}(n,\alpha,\mathcal{B}(t_T)) = T$.

<u>Lemma 15.</u> Let $\alpha \in \mathrm{On}$, $\langle T,\mathcal{B}\rangle$ be a pre-density pair for $Q_\alpha$, $\mathfrak{A} \in \mathrm{Ba}$, $e \in M_n(\mathfrak{A})$ and $\{t_T: T \in \mathrm{Typ}(n,\alpha,e)\}$ be a representation of $\mathrm{Typ}(n,\alpha,e)$. Let $\mathcal{L} = \Pi\{\mathcal{B}(t_T): T \in \mathrm{Typ}(n,\alpha,e)\}$. Then for any $c \in |\mathcal{L}|$ there is a $d \in D_n(\mathfrak{A}[e])$ with $\mathfrak{A}[d] \;{}^{\alpha}{\equiv}^{\alpha}_{n-1}\; \mathcal{L}[c]$.

<u>Proof.</u> Let $\{b_T: T \in \mathrm{Typ}(n,\alpha,e)\}$ be a partition of $\mathcal{L}$ so that we have $\mathcal{L}[b_T] \cong \mathcal{B}(t_T)$ for each $T \in \mathrm{Typ}(n,\alpha,e)$. It follows from Lemma 14 that for each $T \in \mathrm{Typ}(n,\alpha,e)$ there is $T^* \in \mathrm{Typ}(n,\alpha,e)$ with

$$\mathrm{typ}(n-1,\alpha,\mathcal{L}[c \cap b_T]) = \mathrm{typ}(n-1,\alpha,\mathcal{B}(t_{T^*})).$$

Let $\mathrm{Typ}(n,\alpha,e) = \{T_i: i < l\}$. Then there is a $c_0 \in D_n(\mathfrak{A}[e])$ with $\mathfrak{A}[c_0] \;{}^{\alpha}{\equiv}^{\alpha}_{n-1}\; \mathcal{L}[e \cap b_{T_0}]$. But then $e \cap \overline{\overline{c_0}} \in M_n(\mathfrak{A}[e])$ and thus $\mathrm{Typ}(n,\alpha,e) = \mathrm{Typ}(n,\alpha,e \cap \overline{\overline{c_0}})$. That means, there is a $c_1 \in D_n(\mathfrak{A}[e \cap \overline{\overline{c_0}}])$ with $\mathfrak{A}[c_1] \;{}^{\alpha}{\equiv}^{\alpha}_{n-1}\; \mathcal{L}[c \cap b_{T_1}]$ and so on, at least there is $c_{l-1} \in D_n(\mathfrak{A}[e \cap \overline{\overline{c_0}} \cap \ldots \cap \overline{\overline{c_{l-2}}}])$ with $\mathfrak{A}[c_{l-1}] \;{}^{\alpha}{\equiv}^{\alpha}_{n-1}\; \mathcal{L}[c \cap b_{T_{l-1}}]$. Let $d = \bigcup[c_i: i < l]$. Now $\{c_i: i < l\}$ is a set of pairwise disjoint elements of $D_n(\mathfrak{A}[e])$. That means, $d \in D_n(\mathfrak{A}[e])$. Using the product property of $Q_\alpha$ we get $\mathfrak{A}[d] \;{}^{\alpha}{\equiv}^{\alpha}_{n-1}\; \mathcal{L}[c]$.

Q.E.D.

In the following we define pre-density pairs $\langle \mathrm{Term}_0, \mathcal{B}_0\rangle$, $\langle \mathrm{Term}_\omega, \mathcal{B}_\omega\rangle$ and $\langle \mathrm{Term}_1, \mathcal{B}_1\rangle$ for $Q_0$, $Q_\omega$ and $Q_1$, respectively. In order to show that the corresponding pre-density systems $\mathcal{M}_0$, $\mathcal{M}_\omega$ and $\mathcal{M}_1$ are dense for $\mathrm{Th}_0(\mathrm{Ba})$, $\mathrm{Th}_\omega(\mathrm{Ba})$ and $\mathrm{Th}_1(\mathrm{Ba})$, respectively, we show that for each $\alpha \in \{0,\omega,1\}$, each $i < 3$, each $\mathfrak{A} \in \mathrm{Ba}$ and each natural number $n$ we have $N^i_n(\mathfrak{A}) = \emptyset$. We do this by contradiction: Assume $e_0 \in N^i_n(\mathfrak{A})$. Then we choose a suitable Boolean algebra $\mathcal{L} \in \mathcal{M}_\alpha$ and define an ideal $D$ on $\mathcal{L}$ so that for each $e \leq e_0$ with $e \in N^i_n(\mathfrak{A})$ player II has such a strategy in the game $H^2(n,\mathfrak{A}[e],\mathcal{L},\alpha,\alpha)$ that for each $k \leq n$ the result $\langle a_k,b_k\rangle$ before the k-th round is such that

(1) $a_k \in D_n(\mathfrak{A})$ iff $b_k \in D$ and if $a_k \in D_n(\mathfrak{A})$, then $\mathfrak{A}[a_k] \;{}^{\alpha}{\equiv}^{\alpha}_{n-k}\; \mathcal{L}[b_k]$.

Thus $e_0 \in D_n(\mathfrak{A})$, contradicting $e_0 \in N^i_n$. Thus the density of $\mathcal{M}_\alpha$ is shown.

## § 4.1. A dense system for $Th_0(Ba)$

The case $\alpha = 0$ is the simplest case. Yet $Th_0(Ba)$, too, is more complicated than Th(Ba). For instance, Fin($\omega$) and Fin($\omega$) $\times$ Fin($\omega$) are elementary equivalent as can easily be seen by an Ehrenfeucht-game-argument. But Fin($\omega$) and Fin($\omega$) $\times$ Fin($\omega$) are not equivalent with respect to $Q_0$ as can be seen by the sentence

$$\exists xy(x \cap y = 0 \wedge Q_0 z(z \leq x) \wedge Q_0 z(z \leq y)).$$

Let $Term_0$ be the least class K with

(i) $2, \eta_0, \eta \in K$;
(ii) if $t_1, t_2 \in K$, then $p_1(t_1), p_2(t_1), t_1 \times t_2 \in K$ (not necessarily $t_1 \neq t_2$).

Let the function $\mathcal{B}_0$ from $Term_0$ into Ba be defined as follows:

$$\mathcal{B}_0(2) = B(2),$$

$$\mathcal{B}_0(\eta_0) = B(\eta_0),$$

$$\mathcal{B}_0(\eta) = \mathcal{J}(\mathcal{H}_0),$$

$$\mathcal{B}_0(p_1(t)) = \prod_{\omega}^{w} \mathcal{B}_0(t),$$

$$\mathcal{B}_0(p_2(t)) = \prod^{\mathcal{J}_1(\mathcal{H}_0)} \mathcal{B}_0(t),$$

$$\mathcal{B}_0(t_1 \times t_2) = \mathcal{B}_0(t_1) \times \mathcal{B}(t_2).$$

Then $\langle Term_0, \mathcal{B}_0 \rangle$ is a pre-density pair for $Q_0$. In the following we shall show that

$$\mathcal{M}_0 = \{\mathcal{B}_0(t) : t \in Term_0\}$$

is dense for $Th_0(Ba)$.

There does not exist any $\omega_0$-weakly-atomless Boolean algebra. Thus in order to show that $\mathcal{M}_0$ is dense for $Th_0(Ba)$ it is sufficient to prove that for any Boolean algebra $\mathfrak{A}$ and any natural number n, we have

$$N_n^0(\mathfrak{A}) = N_n^1(\mathfrak{A}) = \emptyset.$$

Lemma 16. Let $\mathfrak{A} \in \mathrm{Ba}$, $n \in \omega$. Then $N_n^0(\mathfrak{A}) = \emptyset$.

Proof. Assume, on the contrary, $e_0 \in N_n^0(\mathfrak{A})$. Let $\{t_T : T \in \mathrm{Typ}(n,0,e_0)\}$ be a representation of $\mathrm{Typ}(n,0,e_0)$. Let

$$\mathcal{L} = \prod\{\mathcal{B}_0(t_T) : T \in \mathrm{Typ}(n,0,e_0)\};$$

$$\mathfrak{B} = \prod_{\omega}^{w} \mathcal{L};$$

$$D = \{b \in |\mathfrak{B}| : \mathrm{card}(\{i \in \omega : b(i) \neq 0\}) < \omega_0\};$$

and let $e \in N_n^0(\mathfrak{A})$ with $e \leq e_0$.

In order to prove $\mathfrak{A}[e] \ {}^0{\equiv}_n^0 \ \mathfrak{B}$ we show that player II has such a strategy in the game $H^2(n, \mathfrak{A}[e], \mathfrak{B}, 0, 0)$ that for each $k \leq n$ the result $\langle a_k, b_k \rangle$ before the k-th round satisfies (1).

Assume $i < n$ and the result $\langle a_i, b_i \rangle$ before the i-th round satisfies (1) for $k = i$. (If $i = 0$, then, of course, $\langle a_0, b_0 \rangle = \langle e, 1^{\mathfrak{B}} \rangle$ satisfies (1) for $k = 0$.) We have to distinguish two cases:

Case 1. Player I decides to play an $\exists$-round.

Case 2. Player I decides to play a Q-round.

Ad case 1: First assume that $a_i \in D_n(\mathfrak{A})$. Then we have $\mathfrak{A}[a_i] \ {}^0{\equiv}_{n-i}^0 \ \mathfrak{B}[b_i]$ and thus for each $a \leq a_i$ there is a $b \leq b_i$ and for each $b \leq b_i$ there is an $a \leq a_i$ with

(2) $\langle \mathfrak{A}[a_i], a \rangle \ {}^0{\equiv}_{n-i-1}^0 \ \langle \mathfrak{B}[b_i], b \rangle$.

Playing in such a way that (2) is satisfied, player II ensures that the result $\langle a_{i+1}, b_{i+1} \rangle$ after the i-th round satisfies (1) for $k = i + 1$.

Assume now that $a_i \notin D_n(\mathfrak{A})$. Let $a \leq a_i$ with $a \in D_n(\mathfrak{A})$. Now $b_i \notin D$, thus there is a $j \in \omega$ with $b_i(j) = 1$. But then $\mathfrak{B}[b_i(j)^{\mathrm{red}}] \cong \mathcal{L}$ and Lemma 15 ensures that there is a $b \leq b_i(j)^{\mathrm{red}}$ with

$$\langle \mathfrak{A}[a_i], a \rangle \ {}^0{\equiv}_{n-i-1}^0 \ \langle \mathfrak{B}[b_i], b \rangle.$$

Let $a \leq a_i$ with $a \notin D_n(\mathfrak{A})$. Then $a_i \cap \bar{a} \in D_n(\mathfrak{A})$ and similar to the above consideration there is a $b \leq b_i$ with $b \notin D$ and

$$\langle \mathfrak{A}[a_i], a_i \cap \bar{a}\rangle \ {}^{0}{\equiv}^{0}_{n-i-1} \ \langle \mathfrak{B}[b_i], b_i \cap \bar{b}\rangle.$$

That means, if player I chooses an element from $\mathfrak{A}[a_i]$, then player II can ensure that the result $\langle a_{i+1}, b_{i+1}\rangle$ after the i-th round satisfies (1) for $k = i + 1$.

Let $b \leq b_i$ with $b \in D$. Now $a_i \notin D_n(\mathfrak{A})$ and thus $\mathrm{Typ}(n,0,a_i) = \mathrm{Typ}(n,0,e_0)$. Thus there is an $a \leq a_i$, $a \in D_n(\mathfrak{A})$ with $\mathrm{typ}(n,0,\mathfrak{A}[a]) = \mathrm{typ}(n,0,\mathfrak{B}[b])$. But then

$$\langle \mathfrak{A}[a_i], a\rangle \ {}^{0}{\equiv}^{0}_{n-i-1} \ \langle \mathfrak{B}[b_i], b\rangle.$$

Assume $b \leq b_i$ with $b \notin D$. Then $b_i \cap \bar{b} \in D$ and similar to the above consideration there is an $a \leq a_i$ with $a \notin D_n(\mathfrak{A})$ and

$$\langle \mathfrak{A}[a_i], a_i \cap \bar{a}\rangle \ {}^{0}{\equiv}^{0}_{n-i-1} \ \langle \mathfrak{B}[b_i], b_i \cap \bar{b}\rangle.$$

That means, if player I chooses an element from $\mathfrak{B}[b_i]$, then player II can ensure that the result $\langle a_{i+1}, b_{i+1}\rangle$ after the i-th round satisfies (1) for $k = i + 1$.

Ad case 2: First assume $a_i \in D_n(\mathfrak{A})$. Then $\mathfrak{A}[a_i] \ {}^{0}{\equiv}^{0}_{n-i} \ \mathfrak{B}[b_i]$ and thus for each $A \subseteq |\mathfrak{A}[a_i]|$ with $\mathrm{card}(A) = \omega_0$ there is a set $B \subseteq |\mathfrak{B}[b_i]|$ with $\mathrm{card}(B) = \omega_0$ so that for each $b \in B$ there is an $a \in A$ satisfying (2). On the other hand, also for each set $B \subseteq |\mathfrak{B}[b_i]|$ with $\mathrm{card}(B) = \omega_0$ there is a set $A \subseteq |\mathfrak{A}[a_i]|$ with $\mathrm{card}(A) = \omega_0$ so that for each $a \in A$ there is a $b \in B$ satisfying (2). Thus player II can ensure that the result $\langle a_{i+1}, b_{i+1}\rangle$ after the i-th round satisfies (1) for $k = i + 1$.

Assume now $a_i \notin D_n(\mathfrak{A})$. Let $A \subseteq |\mathfrak{A}[a_i]|$ with $\mathrm{card}(A) = \omega_0$. Let $A_0 = A \cap D_n(\mathfrak{A})$, $A_1 = A \setminus A_0$. Then $\mathrm{card}(A_0) = \omega_0$ or $\mathrm{card}(A_1) = \omega_0$. Let us assume that $\mathrm{card}(A_0) = \omega_0$ (the other case is similar). Then there are a set $A' \subseteq A_0$ with $\mathrm{card}(A') = \omega_0$ and $T \in \mathrm{Typ}(n,0,e_0)$ so that for each $a \in A'$, $\mathrm{typ}(n,0,\mathfrak{A}[a]) = T$. Let $I^* = \{j \in I: b_i(j) = 1\}$. From $b_i \in D$ it follows that $\mathrm{card}(I^*) = \omega_0$. For each $j \in I^*$ there is a $d_j \leq b_i(j)^{red}$ with $\mathrm{typ}(n,0,\mathfrak{A}[d_j]) = T$. If $j_0, j_1 \in I^*$ and $j_0 \neq j_1$, then $d_{j_0} \neq d_{j_1}$. Let $B = \{d_j: j \in I^*\}$. Then $\mathrm{card}(B) = \omega_0$, $B \subseteq D$ and for each $b \in B$, $\mathrm{typ}(n,0,\mathfrak{A}[b]) = T$. Assuming that player I chooses a set in the structure $\mathfrak{A}[a_i]$, player II can ensure that the result $\langle a_{i+1}, b_{i+1}\rangle$ after the i-th round satisfies (1) for $k = i + 1$.

Let be $B \subseteq |\mathfrak{B}[b_i]|$ with $\mathrm{card}(B) = \omega_0$. Let be $B_0 = B \cap D$, $B_1 = B \setminus D$.

Then $\mathrm{card}(B_0) = \omega_0$ or $\mathrm{card}(B_1) = \omega_0$. Assume $\mathrm{card}(B_1) = \omega_0$ (the other case is similar). Then there are a set $B' \subseteq B_1$ with $\mathrm{card}(B') = \omega_0$ and $T \in \mathrm{Typ}(n,0,e_0)$ so that for each $b \in B'$, $\mathrm{typ}(n,0,\mathfrak{B}[b_1 \cap \bar{b}]) = T$. Now $a_i \in M_n(\mathfrak{A})$. Thus there is a set $A \subseteq |\mathfrak{A}[a_i]| \cap D_n(\mathfrak{A})$ with $\mathrm{card}(A) = \omega_0$ so that $\mathrm{typ}(n,0,\mathfrak{A}[a]) = T$ for each $a \in A$. Let $A' = \{a_i \cap \bar{a}: a \in A\}$. Assuming that player I chooses a set in the structure $\mathfrak{B}$, player II can ensure that the result $\langle a_{i+1}, b_{i+1}\rangle$ after the i-th round satisfies (1) for $k = i + 1$.

Thus $\mathfrak{A}[e] \;{}^0{\equiv}^0_n\; \mathfrak{B}$. Remember that $\mathfrak{L}$ is the direct product of finitely many algebras from $\mathcal{M}_0$. Thus it follows by definition that $\mathfrak{L} \in \mathcal{M}_0$. Let $\mathfrak{L} = \mathcal{B}_0(t)$. Then $\mathfrak{B} = \mathcal{B}_0(p_1(t)) \in \mathcal{M}_0$; thus $e_0 \in D_n(\mathfrak{A})$, contradicting $e_0 \in N^0_n(\mathfrak{A})$.

Q.E.D.

<u>Lemma 17.</u> Let $\mathfrak{A} \in \mathrm{Ba}$, n be a natural number. Then $N^1_n(\mathfrak{A}) = \emptyset$.

<u>Proof.</u> Assume $e_0 \in N^1_n(\mathfrak{A})$. Let $\{t_T: T \in \mathrm{Typ}(n,0,e_0)\}$ be a representation of $\mathrm{Typ}(n,0,e_0)$. Let

$$\mathfrak{L} = \prod \{\mathcal{B}_0(t_T): T \in \mathrm{Typ}(n,0,e_0)\},$$

$$\mathfrak{B} = \prod^{\mathcal{J}_1(\mathcal{H}_0)} \mathfrak{L},$$

$$D = \{b \in |\mathfrak{B}|: \mathrm{card}(\{s \in |\mathcal{H}_0| : b(s) \neq 0\}) < \omega_0\},$$

and let $e \in N^1_n(\mathfrak{A})$ with $e \leq e_0$. In order to prove $\mathfrak{A}[e] \;{}^0{\equiv}^0_n\; \mathfrak{L}$ we show that player II has such a strategy in the game $H^2(n,\mathfrak{A}[e],\mathfrak{B},0,0)$ that for each $k \leq n$ the result $\langle a_k, b_k\rangle$ before the k-th round satisfies (1).

Assume $i < n$ and the result $\langle a_i, b_i\rangle$ before the i-th round satisfies (1) for $k = i$. If $a_i \in D_n(\mathfrak{A})$, everything is as in the proof of Lemma 16. So let us assume that $a_i \notin D_n(\mathfrak{A})$. We distinguish the following two cases:

<u>Case 1.</u> Player I decides to play an $\exists$-round.

<u>Case 2.</u> Player I decides to play a Q-round.

<u>Ad case 1:</u> Let $a \leq a_i$. If $a \in D_n(\mathfrak{A})$ or $a_i \cap \bar{a} \in D_n(\mathfrak{A})$, everything goes as in the proof of Lemma 16. So let us assume $a \notin D_n(\mathfrak{A})$ and $a_i \cap \bar{a} \notin D_n(\mathfrak{A})$. Now $b_i \in D$. Thus there is an open nonempty interval $\mathcal{T} \in |\mathcal{H}_0|$ so that for each $s \in \mathcal{T}$, $b_i(s) = 1$. Let $\mathcal{T}_1$ be a nonempty open subinterval of $\mathcal{T}$, $\mathcal{T}_1 \neq \mathcal{T}$. Let $b \in |\mathfrak{B}|$ be defined by

$$b(s) = \begin{cases} 0, & \text{if } s \notin \mathcal{T}_1, \\ 1, & \text{if } s \in \mathcal{T}_1. \end{cases}$$

Then $b \leq b_i$, $b \notin D$ and $b_i \cap \bar{b} \notin D$. That means, if player I chooses an element from $\mathfrak{A}[a_i]$, then player II can ensure that the result $\langle a_{i+1}, b_{i+1}\rangle$ **after the i-th round satisfies (1) for k = i + 1.**

Let $b \leq b_i$. If $b \in D$ or $b_i \cap \bar{b} \in D$, everything goes as in the proof **of Lemma 16. So let us assume that** $b \notin D$, $b_i \cap \bar{b} \notin D$. Now $a_i \in N_n^1(\mathfrak{A})$ and thus $a_i/D_n(\mathfrak{A})$ is an atomless element. Thus there is an $a \leq a_i$ with $a \notin D_n(\mathfrak{A})$, $a_i \cap \bar{a} \notin D_n(\mathfrak{A})$.

That means, if player I chooses an element from $\mathfrak{B}[b_i]$, then player II **can ensure that the result** $\langle a_{i+1}, b_{i+1}\rangle$ **after the i-th round satisfies (1) for k = i + 1.**

Ad case 2. Let $A \subseteq |\mathfrak{A}[a_i]|$ with $\mathrm{card}(A) = \omega_0$. Let

$$A_0 = A \setminus D_n(\mathfrak{A}), \quad A_1 = \{a \in A: a_i \cap \bar{a} \in D_n(\mathfrak{A})\}, \quad A_2 = A \setminus (A_0 \cup A_1).$$

If $\mathrm{card}(A_0) = \omega_0$ or $\mathrm{card}(A_1) = \omega_0$, everything goes as in Lemma 16.

Thus let us assume that $\mathrm{card}(A_0) < \omega_0$ and $\mathrm{card}(A_1) < \omega_0$. But then $\mathrm{card}(A_2) = \omega_0$. Now $b_i \notin D$. Thus there is an open nonempty interval $\mathcal{T} \subseteq |\mathcal{H}_0|$ **so that we have** $b_i(s) = 1$ **for each** $s \in \mathcal{T}$. **Let** $\{\mathcal{T}_k: k \in \omega\}$ be a family of nonempty open subintervals of $\mathcal{T}$, each different from $\mathcal{T}$. For each $k \in \omega$ let $d_k$ be such that

$$d_k(s) = \begin{cases} 1 & \text{if } s \in \mathcal{T}_k, \\ 0 & \text{if } s \notin \mathcal{T}_k. \end{cases}$$

Let $B = \{d_k: k \in \omega\}$. Then for each $b \in B$ it is $b \notin D$ and $b_i \cap \bar{b} \notin D$. Assuming that player I chooses a set in the structure $\mathfrak{A}[e]$, player II **can ensure that the result** $\langle a_{i+1}, b_{i+1}\rangle$ **after the i-th round satisfies (1) for k = i + 1.**

Let $B \subseteq |\mathfrak{B}[b_i]|$ with $\mathrm{card}(B) = \omega_0$. Let

$$B_0 = B \cap D, \quad B_1 = \{b \in B: b_i \cap \bar{b} \in D\}, \quad B_2 = B \setminus (B_1 \cup B_2).$$

If $\mathrm{card}(B_0) = \omega_0$ or $\mathrm{card}(B_1) = \omega_0$, everything goes as in Lemma 16.

Thus let us assume that $\mathrm{card}(B_0) < \omega_0$ and $\mathrm{card}(B_1) < \omega_0$. But then $\mathrm{card}(B_2) = \omega_0$. Now $a_i \in N_n^1(\mathfrak{A})$ and thus $a_i/D_n(\mathfrak{A})$ is atomless. Let $A^* \subseteq |\mathfrak{A}[a_i]/D_n(\mathfrak{A})|$ with $\mathrm{card}(A^*) = \omega_0$, $0/D_n(\mathfrak{A}) \notin A^*$, $a_i/D_n(\mathfrak{A}) \notin A^*$. For each $a^* \in A^*$ let $a \in |\mathfrak{A}[a_i]|$ be an element with $a/D_n(\mathfrak{A}) = a^*$. Let $A = \{a: a^* \in A^*\}$. Then for each $a \in A$, $a \notin D_n(\mathfrak{A})$ and $a_i \cap \bar{a} \in D_n(\mathfrak{A})$. Assuming that player I chooses a set in the structure $\mathfrak{B}$, player II can ensure that the result $\langle a_{i+1}, b_{i+1}\rangle$ after the i-th round satisfies (1) for $k = i + 1$.

Thus $\mathfrak{A}[e] \equiv_n^0 \mathfrak{B}$. Now $\mathfrak{L} \in \mathcal{M}_0$, let $\mathfrak{L} = \mathcal{B}_0(t)$. But then $\mathfrak{B} \cong B_0(p_2(t)) \in \mathcal{M}_0$. Thus $e_0 \in D_n(\mathfrak{A})$, contradicting the assumption.

Q.E.D.

Now it follows from the Lemmas 16 and 17

<u>Theorem 18.</u> $\mathcal{M}_0$ is dense for $\mathrm{Th}_0(\mathrm{Ba})$.

## § 4.2. A dense system for $\mathrm{Th}_\omega(\mathrm{Ba})$

We assume in the following that for each $n \in \omega$ we have $2^{\omega_n} < \omega_\omega$. Without this assumption $\mathrm{Th}_\omega(\mathrm{Ba})$ is not uniquely determined and can become equal to $\mathrm{Th}_1(\mathrm{Ba})$ (see § 4.4).

$\mathrm{Th}_\omega(\mathrm{Ba})$ is more complicated than $\mathrm{Th}_0(\mathrm{Ba})$. The reason is, that we can distinguish atomless elements, whereas in $\mathrm{Th}_0(\mathrm{Ba})$ this cannot be done. But, on the other hand, it follows from Lemma 1 that for each Boolean algebra $\mathfrak{A}$ with $\mathrm{card}(\mathfrak{A}) \geq \omega_\omega$, each dense subset D must also have $\mathrm{card}(D) \geq \omega_\omega$. This fact is the reason that the construction of a dense system for $\mathrm{Th}_\omega(\mathrm{Ba})$ is not to complicated.

Let $\mathrm{Term}_\omega$ be the least class K with

(i) $2, \eta_0, \eta \in K$;
(ii) if $t_1, t_2 \in K$, then $p_1(t_1), p_2(t), t_1 \times t_2, p_3(t_1), p_4(t_1, t_2), p_5(t_1, t_2) \in K$ (not necessarily $t_1 \neq t_2$).

Remember that $\mathcal{H}_\omega$ is a dense linear ordering with first element and $\mathcal{J}(\mathcal{H}_\omega)$ is $\omega_\omega$-atomless (see § 2).

Let $f_\omega : 2^{<\omega} \longrightarrow pow(\omega_\omega)$ be a function with

(i) $f_\omega(\varphi) = \omega_\omega$,
(ii) if $s \in 2^{<\omega}$, then $f_\omega(s^\frown 0) \cup f_\omega(s^\frown 1) = f_\omega(s)$ and
$f_\omega(s^\frown 0) \cap f_\omega(s^\frown 1) = \emptyset$,
(iii) if $s \in 2^{<\omega}$, then $card(f_\omega(s)) = \omega_\omega$ and $card(f_\omega(s) \cap \omega_0) = \omega_0$.

Let $B(f_\omega) = Pow(\omega_\omega; \{f_\omega(s): s \in 2^{<\omega}\})$, that means $B(f_\omega)$ is the subalgebra of $Pow(\omega_\omega)$ generated by $\{f_\omega(s): s \in 2^{<\omega}\}$ and the finite subsets of $\omega_\omega$.

Let the function $\mathcal{B}_\omega$ from $Term_\omega$ into Ba be defined as follows:

$$\mathcal{B}_\omega(2) = B(2),$$

$$\mathcal{B}_\omega(\eta_0) = B(\eta_0),$$

$$\mathcal{B}_\omega(\eta) = \mathcal{J}(\mathcal{H}_\omega),$$

$$\mathcal{B}_\omega(t_1 \times t_2) = \mathcal{B}_\omega(t_1) \times \mathcal{B}_\omega(t_2),$$

$$\mathcal{B}_\omega(p_1(t_1)) = \prod_\omega^w \mathcal{B}_\omega(t_1),$$

$$\mathcal{B}_\omega(p_2(t_1)) = \prod^{\mathcal{J}_1(\mathcal{H}_\omega)} \mathcal{B}_\omega(t_1),$$

$$\mathcal{B}_\omega(p_3(t_1)) = \prod^{\mathcal{J}_1(\mathcal{H}_0)} \mathcal{B}_\omega(t_1),$$

$$\mathcal{B}_\omega(p_4(t_1, t_2)) = \prod^w \{\mathcal{L}_\alpha: \alpha < \omega_\omega\}$$

with

$$\mathcal{L}_\alpha = \begin{cases} \mathcal{B}_\omega(t_1) & \text{for } \alpha < \omega, \\ \mathcal{B}_\omega(t_2) & \text{for } \omega \leq \alpha < \omega_\omega, \end{cases}$$

$$\mathcal{B}_\omega(p_5(t_1, t_2)) = \prod^{B(f_\omega)} \{\mathcal{L}_\alpha: \alpha < \omega_\omega\}$$

with

$$\mathcal{L}_\alpha = \begin{cases} \mathcal{B}_\omega(t_1) & \text{for } \alpha < \omega, \\ \mathcal{B}_\omega(t_2) & \text{for } \omega \leq \alpha < \omega. \end{cases}$$

Then $\langle \mathrm{Term}_\omega, \mathcal{B}_\omega \rangle$ is a pre-density pair for $Q_\omega$. In the following we show that

$$\mathcal{M}_\omega = \{\mathcal{B}_\omega(t) : t \in \mathrm{Term}_\omega\}$$

is dense for $\mathrm{Th}_\omega(\mathrm{Ba})$.

We have to show that for each $\mathfrak{A} \in \mathrm{Ba}$ and each $n \in \omega$

$$N_n^0(\mathfrak{A}) = N_n^1(\mathfrak{A}) = N_n^2(\mathfrak{A}) = \emptyset.$$

<u>Lemma 19.</u> Let $\mathfrak{A} \in \mathrm{Ba}$, $n \in \omega$. Then $P_n^0(\mathfrak{A}) \cap S_n^0(\mathfrak{A}) = \emptyset$.

<u>Proof.</u> Assume, on the contrary, $e_0 \in P_n^0(\mathfrak{A}) \cap S_n^0(\mathfrak{A})$ and let $\{t_T : T \in \mathrm{Typ}(n,\omega,e_0)\}$ be a representation of $\mathrm{Typ}(n,\omega,e_0)$. Let

$$\mathcal{L} = \prod\{\mathcal{B}_\omega(t_T) : T \in \mathrm{Typ}(n,\omega,e_0)\},$$

$$\mathcal{L}' = \prod_\omega^w \mathcal{L},$$

$$D = \{b \in |\mathcal{L}'| : \mathrm{card}(\{i < \omega_0 : b(i) \neq 0\}) < \omega_0\}.$$

Let $e \in P_n^0(\mathfrak{A}) \cap S_n^0(\mathfrak{A})$ with $e \leq e_0$. In order to prove $\mathfrak{A}[e] \;{}^\omega\!\equiv_n^\omega \mathcal{L}'$ we show that player II has such a strategy in the game $H^2(n, \mathfrak{A}[e], \mathcal{L}', \omega, \omega)$ that for each $k \leq n$ the result $\langle a_k, b_k \rangle$ before the k-th round satisfies (1).

Now we have $\mathrm{card}(|\mathcal{B}_\omega(t_T)|) < \omega_\omega$ for each $T \in \mathrm{Typ}(n,\omega,e_0)$. Thus $\mathrm{card}(|\mathcal{L}|) < \omega_\omega$ and also $\mathrm{card}(|\mathcal{L}'|) < \omega_\omega$. Thus player I has always to decide to play an $\exists$-round. Then the proof goes as the proof of Lemma 16.

Now $\mathcal{L} \in \mathcal{M}_\omega$, let $\mathcal{L} \cong \mathcal{B}_\omega(t_1)$. Then $\mathcal{L}' \cong \mathcal{B}_\omega(p_1(t_1)) \in \mathcal{M}_\omega$ and thus $e_0 \in D_n(\mathfrak{A})$, contradicting the assumption.

Q.E.D.

<u>Lemma 20.</u> Let $\mathfrak{A} \in \mathrm{Ba}$, $n \in \omega$. Then $P_n^0(\mathfrak{A}) \cap S_n^1(\mathfrak{A}) = \emptyset$.

<u>Proof.</u> Assume, on the contrary, $e_0 \in P_n^0(\mathfrak{A}) \cap S_n^1(\mathfrak{A})$ and let $\{t_T : T \in \mathrm{Typ}(n,\omega,e_0)\}$ be a representation of $\mathrm{Typ}(n,\omega,e_0)$. Let

$$\mathcal{L} = \prod\{\mathcal{B}_\omega(t_T) : T \in \mathrm{Typ}(n,\omega,e_0)\},$$

$\mathcal{L} = \prod^{w}_{\omega_\omega} \mathcal{L}$,

$D = \{b \in |\mathcal{L}|: \mathrm{card}(\{i < \omega_\omega : b(i) \neq 0\}) < \omega_0\}$.

Let $e \in P^0_n(\mathfrak{A}) \cap S^1_n(\mathfrak{A})$ with $e \leq e_0$. In order to prove $\mathfrak{A}[e] \;{}^{\omega}\!\equiv^{\omega}_{n}\; \mathcal{L}$ we show that player II has such a strategy in the game $H^2(n,\mathfrak{A}[e],\mathcal{L},\omega,\omega)$ that for each $k \leq n$ the result $\langle a_k,b_k\rangle$ before the k-th round satisfies (1).

Assume $i < n$ and the result $\langle a_i,b_i\rangle$ before the i-th round satisfies (1) for $k = i$. If $a_i \in D_n(\mathfrak{A})$ or if player I decided to play an $\exists$-round, then everything goes as in the proof of Lemma 16.

Thus let us assume that $a_i \notin D_n(\mathfrak{A})$ and player I decided to play a Q-round. Let $A \subseteq |\mathfrak{A}[a_i]|$ with $\mathrm{card}(A) = \omega_\omega$. We set $A_0 = A \cap D_n(\mathfrak{A})$, $A_1 = A \setminus D_n(\mathfrak{A})$. Then $\mathrm{card}(A_0) = \omega_\omega$ or $\mathrm{card}(A_1) = \omega_\omega$. Without loss of generality let us assume that $\mathrm{card}(A_0) = \omega_\omega$. Then there are $A' \subseteq A_0$ with $\mathrm{card}(A') = \omega_\omega$ and $T \in \mathrm{Typ}(n,\omega,e_0)$ so that it is $\mathrm{typ}(n,\omega,\mathfrak{A}[a]) = T$ for each $a \in A'$.

Let $I = \{\alpha < \omega_\omega : b_i(\alpha) = 1\}$. From $b_i \notin D$ it follows that $\mathrm{card}(I) = \omega_\omega$. For each $\alpha \in I$ there is $d_\alpha = b_i(\alpha)^{\mathrm{red}}$ with $\mathrm{typ}(n,\omega,\mathcal{L}[d_\alpha]) = T$. Let $B = \{d_\alpha : \alpha \in I\}$. Then $\mathrm{card}(B) = \omega_\omega$. Assuming that player I chooses a set in the structure $\mathfrak{A}[a_i]$, player II can ensure that the result $\langle a_{i+1},b_{i+1}\rangle$ after the i-th round satisfies (1) for $k = i + 1$.

Let $B \subseteq |\mathcal{L}[b_i]|$ with $\mathrm{card}(B) = \omega_\omega$. We set $B_0 = B \cap D$, $B_1 = B \setminus D$. Then $\mathrm{card}(B_0) = \omega_\omega$ or $\mathrm{card}(B_1) = \omega_\omega$. Without loss of generality let us assume that $\mathrm{card}(B_0) = \omega_\omega$. Then there are $B' \subseteq B_0$ with $\mathrm{card}(B') = \omega_\omega$ and $T \in \mathrm{Typ}(n,\omega,e_0)$ so that for each $b \in B'$ it is $\mathrm{typ}(n,\omega,\mathcal{L}[b]) = T$.

It follows from $a_i \in P^0_n(\mathfrak{A})$ that there is a set $A \subseteq |\mathfrak{A}[a_i]|$ with $\mathrm{card}(A) = \omega_\omega$ so that for each $a \in A$ we have $\mathrm{typ}(n,\omega,\mathfrak{A}[a]) = T$. Assuming that player I chooses a set in the structure $\mathcal{L}[b_i]$, player II can ensure that the result $\langle a_{i+1},b_{i+1}\rangle$ after the i-th round satisfies (1) for $k = i + 1$.

Now $\mathcal{L} \in \mathcal{M}_\omega$ let be $\mathcal{L} \cong \mathcal{B}_\omega(t_1)$. But then $\mathcal{L} \cong \mathcal{B}_\omega(p_4(t_1,t_1)) \in \mathcal{M}_\omega$, and thus $e_0 \in D_n(\mathfrak{A})$, contradicting the assumption.

Q.E.D.

Lemma 21. Let $\mathfrak{A} \in Ba$, $n \in \omega$. Then $P_n^0(\mathfrak{A}) \cap S_n^2(\mathfrak{A}) = \emptyset$.

Proof. Assume, on the contrary, $e_0 \in P_n^0(\mathfrak{A}) \cap S_n^2(\mathfrak{A})$ and let $\{t_T : T \in Typ(n,\omega,e_0)\}$ be a representation of $Typ(n,\omega,e_0)$. We start by showing:

Claim. Let $X \subseteq |\mathfrak{A}[e_0]|$ with $card(X) = \omega_\omega$. Then

$$card(\{a \in |\mathfrak{A}[e_0]| : a \text{ is an atom and there is a } b \in X \text{ with } a \leq b\}) = \omega_\omega$$

or

$$card(\{a \in |\mathfrak{A}[e_0]| : a \text{ is } \omega_\omega\text{-weakly-atomless and there is a } b \in X \text{ with } a \leq b\}) = \omega_\omega$$

or

there are an $a \in |\mathfrak{A}[e_0]|$, a $b \in X$ with $a \leq b$ and a is $\omega_\omega$-atomless.

Proof of the claim. Assume that

$$card(\{a \in |\mathfrak{A}[e_0]| : a \text{ is an atom and there is a } b \in X \text{ with } a \leq b\}) < \omega_\omega$$

and

$$card(\{a \in |\mathfrak{A}[e_0]| : a \text{ is } \omega_\omega\text{-weakly atomless and there is a } b \in X \text{ with } a \leq b\}) < \omega_\omega.$$

If there are no $a \in |\mathfrak{A}[e_0]|$, $b \in X$ with $a \leq b$ and a being $\omega_\omega$-atomless, then two elements $c_0, c_1 \in X$ are equal if they contain the same set of atoms and the same set of $\omega_\omega$-weakly-atomless elements. But then it follows by cardinality reasons that $card(X) < \omega_\omega$, contradicting the assumption.

We distinguish the following four cases:

Case 1. $card(\{d \leq e_0 : d \text{ is an atom}\}) < \omega_\omega$ and $card(\{d \leq e_0 : d \text{ is } \omega_\omega\text{-weakly-atomless}\}) < \omega_\omega$;

Case 2. $card(\{d \leq e_0 : d \text{ is an atom}\}) \geq \omega_\omega$ and $card(\{d \leq e_0 : d \text{ is } \omega_\omega\text{-weakly-atomless}\}) < \omega_\omega$;

Case 3. card($\{d \leq e_0$: d is an atom$\}) < \omega_\omega$ and card($\{d \leq e_0$: d is $\omega_\omega$-weakly-atomless$\}) \geq \omega_\omega$;

Case 4. card($\{d \leq e_0$: d is an atom$\}) \geq \omega_\omega$ and card($\{d \leq e_0$: d is $\omega_\omega$-weakly-atomless$\}) \geq \omega_\omega$.

Ad case 1: Let

$$\mathcal{L} = \mathcal{T}(\mathcal{H}_\omega) \times \prod\{\mathcal{B}_\omega(t_T) : T \in Typ(n,\omega,e_0)\},$$

$$\mathcal{B} = \prod_\omega^w \mathcal{L},$$

$$D = \{b \in |\mathcal{B}| : card(\{i < \omega : b(i) \neq 0\}) < \omega_0,$$

and let $e \in P_n^0(\mathfrak{A}) \cap S_n^2(\mathfrak{A})$ with $e = e_0$. In order to prove $\mathfrak{A}[e] \equiv_n^{\omega\omega} \mathcal{B}$ we show that player II has such a strategy in the game $H^2(n,\mathfrak{A}[e],\mathcal{B},\omega,\omega)$ that for each $k \leq n$ the result $\langle a_k,b_k\rangle$ before the k-th round satisfies (1).

Assume $i < n$ and the result $\langle a_i,b_i\rangle$ before the i-th round satisfies (1) for $k = i$.

If $a_i \in D_n(\mathfrak{A})$, everything goes as in the proof of Lemma 16. Thus let us assume that $a_i \notin D_n(\mathfrak{A})$. If $a \leq a_i$, then it follows as in the proof of Lemma 16 that there is a $b \leq b_i$ with $a \in D_n(\mathfrak{A})$ iff $b \in D_n(\mathfrak{A})$ and with $\langle \mathfrak{A}[a_i],a\rangle \equiv_{n-i-1}^{\omega\omega} \langle \mathcal{B}[b_i],b\rangle$.

Let $b \leq b_i$. Without loss of generality we assume that $b \in D$. Let $I = \{j < \omega : b(j) \neq 0\}$. For $j \in I$ let $c_j^0, c_j^1 \leq 1(j)^{red}$ be such that $c_j^0 \cap c_j^1 = 0$, $c_j^0 \cup c_j^1 = 1(j)^{red}$, $\mathcal{B}[c_j^0] \cong \mathcal{T}(\mathcal{H}_\omega)$, $\mathcal{B}[c_j^1] \cong \prod\{\mathcal{B}_\omega(t_T) : T \in Typ(n,\omega,e_0)\}$. Let $b^0 = b \cap (\bigcup[c_j^0 : j \in I])$, $b^1 = b \cap (\bigcup[c_j^1 : j \in I])$, then $b^0$ is $\omega_\omega$-weakly-atomless and $typ(n,\omega,\mathcal{B}[b^1]) \in Typ(n,\omega,e_0)$ and thus there is an $a^1 \leq a_i$ with $a^1 \in D_n(\mathfrak{A})$ and $typ(n,\omega,\mathcal{B}[b^1]) = typ(n,\omega,\mathfrak{A}[a^1])$. It follows from the claim (with $X = \{c \in |\mathfrak{A}[a_i]| : c \leq a_i \cap \overline{a^1}\}$) that there is an $a^0 \leq a_i \cap \overline{a^1}$ so that $a^0$ is $\omega_\omega$-atomless. But then $a^0 \in D_n(\mathfrak{A})$ and thus $a = a^0 \cup a^1 \in D_n(\mathfrak{A})$.

Thus assuming that player I decides to play an $\exists$-round, player II can ensure that the result $\langle a_{i+1},b_{i+1}\rangle$ after the i-th round satisfies (1) for $k = i + 1$.

Let $A \subseteq |\mathfrak{A}[a_i]|$ with $card(A) = \omega_\omega$. Let $A_0 = A \cap D_n(\mathfrak{A})$, $A_1 = A \setminus D_n(\mathfrak{A})$. Then $card(A_0) = \omega_\omega$ or $card(A_1) = \omega_\omega$. Let $card(A_0) = \omega_\omega$. Then it follows by the claim that there are $a^0 \in A_0$ and $a^1 \leq a^0$ so that $a^1$ is $\omega_\omega$-atomless. Let $T = typ(n,\omega,\mathfrak{A}[a^0])$. Let $j_0 < \omega$ be such that $b_i(j_0) = 1$. Let $b^0, b^1 \leq b_i(j_0)^{red}$ be such that $\mathfrak{B}[b^0] \cong \mathcal{J}(\mathcal{H}_\omega)$, $\mathfrak{B}[b^1] \cong \prod\{\mathcal{B}_\omega(t_T): T \in Typ(n,\omega,e_0)\}$. Then there is a set $B^*$ with $card(B^*) = \omega_\omega$ so that we have $b \leq b^0$ for each $b \in B^*$, $b$ is $\omega_\omega$-weakly-atomless. Let $B = \{b \cup b^1: b \in B^*\}$. Then $card(B) = \omega_\omega$ and for each $b \in B$ we have $b \in D$ and $typ(n,\omega,\mathfrak{B}[b]) = typ(n,\omega,\mathfrak{A}[a^0])$.

Assuming that player I chooses a set in the structure $\mathfrak{A}[e]$, player II can ensure that the result $\langle a_{i+1}, b_{i+1}\rangle$ after the i-th round satisfies (1) for $k = i + 1$.

Let $B \subseteq |\mathfrak{B}[b_i]|$ with $card(B) = \omega_\beta$. Let $B_0 = B \cap D$, $B_1 = B \setminus D$. Let $card(B_0) = \omega_\omega$. For each $j < \omega$ let $c_j^0, c_j^1 \leq 1(j)^{red}$ be such that $c_j^0 \cap c_j^1 = 0$, $c_j^0 \cup c_j^1 = 1(j)^{red}$, $\mathfrak{B}[c_j^0] \cong \mathcal{J}(\mathcal{H}_\omega)$, $\mathfrak{B}[b_j^1] \cong \prod\{\mathcal{B}_\omega(t_T): T \in Typ(n,\omega,e_0)\}$.

Assume there are $b^* \in B_0$ and $j_0 < \omega$ with $b^* \cap c_{j_0}^0 \neq 0$. Let $I = \{j < \omega: b^*(j) \neq 0\}$, $b^0 = b^* \cap (\bigcup[c_j^0: j \in I])$, $b^1 = b^* \cap (\bigcup[c_j^1: j \in I])$. Then $b^0$ is an $\omega_\omega$-atomless element and $typ(n,\omega,\mathfrak{B}[b^1]) \in Typ(n,\omega,e_0)$. Let $a^0 \leq a_i$ be an $\omega_\omega$-atomless element. Then $a^0 \in D_n(\mathfrak{A})$, thus $a_i \cap \overline{\overline{a^0}} \notin D_n(\mathfrak{A})$. Consequently, there is an $a^1 \leq a_i \cap \overline{\overline{a^0}}$ so that $a^1 \in D_n(\mathfrak{A})$ and $\mathfrak{A}[a^1] \;{}^{\omega}{\equiv}^{\omega}_{n}\; \mathfrak{B}[b^1]$. Let $A^*$ be a set with $card(A^*) = \omega_\omega$ so that we have $a \leq a^0$ for each $a \in A^*$, $a$ is $\omega_\omega$-atomless. Let $A = \{a \cup a^1: a \in A^*\}$. Then $card(A) = \omega_\omega$ and for each $a \in A$ we have $a \in D_n(\mathfrak{A})$ and $typ(n,\omega,\mathfrak{A}[a]) = typ(n,\omega,\mathfrak{B}[b^*])$.

Assume that for each $b \in B_0$ and each $j < \omega$ we have $b \cap c_j^0 = 0$. Then there is $j_0 < \omega$ so that $card(\{b \cap c_{j_0}^1 : b \in B_0\}) = \omega_\omega$. Let $\{d_T: T \in Typ(n,\omega,e_0)\}$ be a partition of $c_{j_0}^1$ so that for each $T \in Typ(n,\omega,e_0)$ it is $\mathfrak{B}[d_T] \cong \mathcal{B}_\omega(t_T)$. Then there is $T_0 \in Typ(n,\omega,e_0)$ so that $card(\{b \cap d_{T_0} : b \in B_0\}) = \omega_\omega$. Let $C = \{b \cap d_{T_0} : b \in B_0\}$. Then there are $T_1 \in Typ(n,\omega,e_0)$ and $C_0 \subseteq C$ with $card(C_0) = \omega_\omega$ so that for each $c \in C_0$, $typ(n,\omega,\mathfrak{B}[c]) = T_1$. Let $b^* \in B_0$ be such that $b^* \cap d_{T_0} \in C_0$. We then can choose disjoint elements $a^0, a^1 \in D_n(\mathfrak{A}[a_i])$ so that $\mathfrak{A}[a^0] \;{}^{\omega}{\equiv}^{\omega}_{n}\; \mathfrak{B}[b \cap \overline{\overline{d_{T_0}}}]$ and $\mathfrak{A}[a^1] \;{}^{\omega}{\equiv}^{\omega}_{n}\; \mathfrak{B}[d_{T_0}]$. But then it follows from Theorem 13 that there is a set $A^* \subseteq |\mathfrak{A}[a^1]|$ with

$\mathrm{card}(A^*) = \omega_\omega$ so that we have $\mathrm{typ}(n,\omega,\mathfrak{A}[a]) = T_1$ for each $a \in A^*$. Let $A = \{a^0 \cup a: a \in A^*\}$. Then $\mathrm{card}(A) = \omega_\omega$ and for each $a \in A$ we have $\mathrm{typ}(n,\omega,\mathfrak{A}[a]) = \mathrm{typ}(n,\omega,\mathfrak{L}[b^*])$.

Thus we have shown: Assuming that player I chooses a set in the structure $\mathfrak{L}$, player II can ensure that the result $\langle a_{i+1}, b_{i+1}\rangle$ after the i-th round satisfies (1) for $k = i + 1$.

Now $\mathcal{L} \in \mathcal{M}_\omega$, let $\mathcal{L} \cong \mathcal{B}_\omega(t_1)$. Then $\mathfrak{L} \cong \mathcal{B}_\omega(p_1(t)) \in \mathcal{M}_\omega$. Thus $e_0 \in D_n(\mathfrak{A})$, contradicting the assumption.

Ad case 2: Let

$$\mathcal{L} = \prod\{\mathcal{B}_\omega(t_T): T \in \mathrm{Typ}(n,\omega,e_0)\},$$

$$\mathfrak{L} = \prod^{w}\{\mathcal{L}_\alpha : \alpha < \omega_\omega\}$$

with

$$\mathcal{L}_\alpha = \begin{cases} \mathcal{L} \text{ for } \alpha < \omega, \\ B(2) \text{ for } \omega \le \alpha < \omega_\omega, \end{cases}$$

$$D = \{b \in |\mathfrak{L}|: \mathrm{card}(\{\alpha < \omega_\omega: b(\alpha) \neq 0\}) < \omega_0\},$$

and let $e \in P_n^0(\mathfrak{A}) \cap S_n^2(\mathfrak{A})$ with $e \le e_0$. In order to prove $\mathfrak{A}[e] \; {}^{\omega}{\equiv}_n^{\omega} \; \mathfrak{L}$ we show that player II has such a strategy in the game $H^2(n,\mathfrak{A}[e],\mathfrak{L},\omega,\omega)$ that for each $k = n$ the result $\langle a_k, b_k\rangle$ before the k-th round satisfies (1).

Assume $i < n$ and the result $\langle a_i, b_i\rangle$ before the i-th round satisfies (1) for $k = i$.

If $a_i \in D_n(\mathfrak{A})$, everything goes as in the proof of Lemma 16. Thus let us assume that $a_i \notin D_n(\mathfrak{A})$. If $a \le a_i$, then it follows as in the proof of Lemma 16 that there is a $b \le b_i$ with $a \in D_n(\mathfrak{A})$ iff $b \in D$ and $\langle \mathfrak{A}[a_i], a\rangle \; {}^{\omega}{\equiv}_{n-i-1}^{\omega} \; \langle \mathfrak{L}[b_i], b\rangle$.

Let $b \le b_i$. Without loss of generality we assume that $b \in D$. Let $I = \{\alpha < \omega_\omega: b(\alpha) \neq 0\}$. We set $b^0 = \bigcup[b(\alpha)^{red}: \alpha \in I \cap \omega]$, $b^1 = \bigcup[b(\alpha)^{red}: \alpha \in I \setminus \omega]$. Then there is an $a^0 \le a_i$ with $a^0 \in D_n(\mathfrak{A})$ and $\mathrm{typ}(n,\omega,\mathfrak{A}[a^0]) = \mathrm{typ}(n,\omega,\mathfrak{L}[b^0])$. $b^1$ is a finite union of atoms.

Using the fact that $a_i \cap \overline{\overline{a^0}} \in D_n(\mathfrak{A})$ we see that there is an $a^1 \leq a_i \cap \overline{\overline{a^0}}$ with $cd(a^1) = cd(b^1)$, that means, $a^1$ is a finite union of atoms and contains as many atoms as $b^1$. Thus $\mathfrak{A}[a^1] \cong \mathfrak{B}[b^1]$ and $typ(n,\omega,\mathfrak{A}[a^0 \cup a^1]) = typ(n,\omega,\mathfrak{B}[b])$.

Thus assuming that player I decided to play an $\exists$-round, player II can ensure that the result $\langle a_{i+1}, b_{i+1} \rangle$ after the i-th round satisfies (1) for $k = i + 1$.

Let $A \subseteq |\mathfrak{A}[a_i]|$ with $card(A) = \omega_\omega$. Let $A_0 = A \cap D_n(\mathfrak{A})$, $A_1 = A \setminus D_n(\mathfrak{A})$. Let us assume that $card(A_0) = \omega_\omega$.

Assume there is an $a^* \in A_0$ which contains an atom. Let $a^0 \leq a^*$ be an atom and let $b^* \leq b_i$, $b^* \in D$ be such that $typ(n,\omega,\mathfrak{A}[a \cap \overline{\overline{a^0}}]) = typ(n,\omega,\mathfrak{B}[b^*])$ and $b(\alpha) = 0$ for $\omega \leq \alpha < \omega_\omega$.

Let $B = \{b^* \cup 1(\alpha)^{red}: \omega \leq \alpha < \omega_\omega\}$. Then $card(B) = \omega_\omega$ and for each $b \in B$ we have $typ(n,\omega,\mathfrak{A}[a^*]) = typ(n,\omega,\mathfrak{B}[b])$.

Assume that no element of $A_0$ contains an atom. By the claim it then follows that there are $a^0 \in A_0$ and $a^1 \leq a^0$ so that $a^1$ is $\omega_\omega$-atomless. Thus $cd(a^0) \geq \omega_\omega$. Let $b^* \leq b_i$ be such that $b^* \in D$ and $typ(n,\omega,\mathfrak{A}[a^0]) = typ(n,\omega,\mathfrak{B}[b^*])$. Then $cd(b^*) \geq \omega_\omega$ and it follows from Theorem 13 that there is a set $B \subseteq \mathfrak{B}[b^*]$ with $card(B) = \omega_\omega$ so that for each $b \in B$ we have $typ(n,\omega,\mathfrak{B}[b^*]) = typ(n,\omega,\mathfrak{B}[b])$. Thus $typ(n,\omega,\mathfrak{A}[a^0]) = typ(n,\omega,\mathfrak{B}[b])$ for each $b \in B$.

Let $B \subseteq |\mathfrak{B}[b_i]|$ with $card(B) = \omega_\omega$. Let $B_0 = B \cap D$, $B_1 = B \setminus D$. Assume $card(B_0) = \omega_\omega$. Assume there are $b^0 \in B_0$ and $b^1 \leq b^0$ so that $b^1$ is an atom. Let $a^* \in D_n(\mathfrak{A}[a_i])$ be such that $typ(n,\omega,\mathfrak{A}[a^*]) = typ(n,\omega,\mathfrak{B}[b^0 \cap \overline{\overline{b^1}}])$. Then there is a set $A^* \subseteq |\mathfrak{A}[a_i \cap \overline{\overline{a}}]|$ with $card(A^*) = \omega_\omega$ so that each $a \in A^*$ is an atom. Let $A = \{a^* \cup a: a \in A^*\}$. Then $card(A) = \omega_\omega$ and for each $a \in A$, $typ(n,\omega,\mathfrak{A}[a]) = typ(n,\omega,\mathfrak{B}[b^0])$.

Assume that no element of $B_0$ contains an atom. By the claim it then follows that there is an element in $B_0$ which contains an $\omega_\omega$-atomless element. Thus there is $b^* \in B_0$ with $cd(b^*) = \omega_\omega$. Let $a^* \in D_n(\mathfrak{A}[a_i])$ be such that $typ(n,\omega,\mathfrak{A}[a^*]) = typ(n,\omega,\mathfrak{B}[b^*])$. Then $cd(a^*) \geq \omega_\omega$ and it follows from Theorem 13 that there is a set A with $card(A) = \omega_\omega$ so that for each $a \in A$, $a \leq a^*$ and $typ(n,\omega,\mathfrak{A}[a]) = typ(n,\omega,\mathfrak{B}[b^*])$.

Thus assuming that player I decided to play a Q-round, player II can ensure that the result $\langle a_{i+1}, b_{i+1}\rangle$ after the i-th round satisfies (1) for k = i + 1.

Now $\mathcal{L} \in \mathcal{M}_\omega$, let $\mathcal{L} \cong \mathcal{B}_\omega(t_1)$. Then $\mathcal{L}' \cong \mathcal{B}_\omega(p_4(t_1,2)) \in \mathcal{M}_\omega$. Thus $e_0 \in D_n(\mathfrak{A})$, contradicting the assumption.

Ad case 3: Let

$$\mathcal{L} = \prod\{\mathcal{B}_\omega(t_T): T \in Typ(n,\omega,e_0)\},$$

$$\mathcal{L}' = \prod^W\{\mathcal{L}_\alpha: \alpha < \omega_\omega\}$$

with

$$\mathcal{L}_\alpha = \begin{cases} \mathcal{L} \text{ for } \alpha < \omega, \\ B(\eta_0) \text{ for } \omega \le \alpha < \omega_\omega, \end{cases}$$

$$D = \{b \in |\mathcal{L}'|: card(\{\alpha < \omega_\omega: b(\alpha) \neq 0\}) < \omega_0\},$$

and let $e \in P_n^0(\mathfrak{A}) \cap S_n^2(\mathfrak{A})$ with $e \le e_0$. Similar as in case 2 it can be shown that player II has such a strategy in the game $H^2(n,\mathfrak{A}[e],\mathcal{L}',\omega,\omega)$ that for each $k \le n$ the result $\langle a_k, b_k\rangle$ before the k-th round satisfies (1). This implies $\mathfrak{A}[e] \overset{\omega}{\equiv}{}_n^{\omega} \mathcal{L}'$.

Now $\mathcal{L} \in \mathcal{M}_\omega$, let $\mathcal{L} \cong \mathcal{B}_\omega(t_1)$. Then $\mathcal{L}' \cong \mathcal{B}_\omega(p_4(t_1,\eta_0)) \in \mathcal{M}_\omega$. Thus $e_0 \in D_n(\mathfrak{A})$, contradicting the assumption.

Ad case 4: Let

$$\mathcal{L} = \prod\{\mathcal{B}_\omega(t_T): T \in Typ(n,\omega,e_0)\},$$

$$\mathcal{L}' = \prod^W\{\mathcal{L}_\alpha: \alpha < \omega_\omega\}$$

with

$$\mathcal{L}_\alpha = \begin{cases} \mathcal{L} \text{ for } \alpha < \omega, \\ B(2) \times B(\eta_0) \text{ for } \omega \le \alpha < \omega_\omega, \end{cases}$$

$$D = \{b \in |\mathcal{L}'|: card(\{\alpha < \omega_\omega: b(\alpha) \neq 0\}) < \omega_0\},$$

and let $e \in P_n^0(\mathfrak{A}) \cap S_n^2(\mathfrak{A})$ with $e \leq e_0$. Similar as in case 2 it can be shown that player II has such a strategy in the game $H^2(n,\mathfrak{A}[e],\mathfrak{L},\omega,\omega)$ that for each $k \leq n$ the result $\langle a_k,b_k\rangle$ before the k-th round satisfies (1). This implies $\mathfrak{A}[e] \ {}^{\omega}{\equiv}_n^{\omega}\ \mathfrak{L}$.

Now $\mathcal{L} \in \mathcal{M}_\omega$, let $\mathcal{L} \cong \mathcal{B}_\omega(t_1)$. But then $\mathfrak{L} \cong \mathcal{B}_\omega(p_3(t_1, 2\times\eta_0)) \in \mathcal{M}_\omega$. Thus $e_0 \in D_n(\mathfrak{A})$, contradicting the assumption.

Q.E.D.

Now it follows from Lemmas 19, 20 and 21 that

<u>Corollary 22.</u> Let $\mathfrak{A} \in Ba$, $n \in \omega$. Then $N_n^0(\mathfrak{A}) = \emptyset$.

<u>Lemma 23.</u> Let $\mathfrak{A} \in Ba$, $n \in \omega$. Then $P_n^1(\mathfrak{A}) \cap S_n^0(\mathfrak{A}) = \emptyset$.

<u>Proof.</u> Assume, on the contrary, that $e_0 \in P_n^1(\mathfrak{A}) \cap S_n^0(\mathfrak{A})$ and let $\{t_T\colon T \in Typ(n,\omega,e_0)$ be a representation of $Typ(n,\omega,e_0)$. Let

$$\mathcal{L} = \prod\{\mathcal{B}_\omega(t_T)\colon T \in Typ(n,\omega,e_0)\},$$

$$\mathfrak{L} = \prod^{\mathcal{I}_1(\mathcal{H}_0)} \mathcal{L},$$

$$D = \{b \in |\mathfrak{L}|\colon card(\{s \in |\mathcal{H}_0|\colon b(s) \neq 0\}) < \omega_0\}),$$

and let $e \in P_n^1(\mathfrak{A}) \cap S_n^0(\mathfrak{A})$ with $e \leq e_0$. In order to prove $\mathfrak{A}[e] \ {}^{\omega}{\equiv}_n^{\omega}\ \mathfrak{L}$ we show that player II has such a strategy in the game $H^2(n,\mathfrak{A}[e],\mathfrak{L},\omega,\omega)$ that for each $k \leq n$ the result $\langle a_k,b_k\rangle$ before the k-th round satisfies (1).

Now we have $card(|\mathcal{B}_\omega(t_T)|) < \omega_\omega$ for each $T \in Typ(n,\omega,e_0)$. Thus $card(|\mathcal{L}|) < \omega_\omega$ and also $card(|\mathfrak{L}|) < \omega_\omega$. Thus player I has always to decide to play an $\exists$-round. Then the proof goes as the proof of Lemma 17.

Now $\mathcal{L} \in \mathcal{M}_\omega$, let $\mathcal{L} \cong \mathcal{B}_\omega(t_1)$. Then $\mathfrak{L} \cong \mathcal{B}_\omega(p_3(t_1)) \in \mathcal{M}_\omega$ and thus $e_0 \in D_n(\mathfrak{A})$, contradicting the assumption.

Q.E.D.

<u>Lemma 24.</u> Let $\mathfrak{A} \in Ba$, $n \in \omega$. Then $P_n^1(\mathfrak{A}) \cap S_n^1(\mathfrak{A}) = \emptyset$.

<u>Proof.</u> Assume, on the contrary, that $e_0 \in P_n^1(\mathfrak{A}) \cap S_n^1(\mathfrak{A})$ and let

$\{t_T : T \in Typ(n,\omega,e_0)\}$ be a representation of $Typ(n,\omega,e_0)$. Let

$$\mathcal{L} = \prod\{\mathcal{B}_\omega(t_T) : T \in Typ(n,\omega,e_0)\},$$

$$\mathcal{B} = \prod^{B(f_\omega)} \mathcal{L},$$

$$D = \{b \in |\mathcal{B}| : card(\{\alpha < \omega_\omega : b(\alpha) \neq 0\}) < \omega_0\},$$

and let $e \in P^1_n(\mathfrak{A}) \cap S^1_n(\mathfrak{A})$ with $e \leq e_0$. In order to prove $\mathfrak{A}[e] \ {}^{\omega}\!\equiv^{\omega}_n \mathcal{B}$ we show that player II has such a strategy in the game $H^2(n,\mathfrak{A}[e],\mathcal{B},\omega,\omega)$ that for each $k \leq n$ the result $\langle a_k,b_k\rangle$ before the k-th round satisfies (1).

If $a_i \in D_n(\mathfrak{A})$, everything goes as in the proof of Lemma 17.

Thus let us assume that $a_i \notin D_n(\mathfrak{A})$. Let $a \leq a_i$. If $a \in D_n(\mathfrak{A})$ or $a_i \cap \bar{a} \in D_n(\mathfrak{A})$, then it follows as in Lemma 17 that there is a $b \leq b_i$ so that $\langle \mathfrak{A}[a_i],a\rangle \ {}^{\omega}\!\equiv^{\omega}_n \langle \mathcal{B}[b_i],b\rangle$ and $a \in D_n(\mathfrak{A})$ iff $b \in D$, $a_i \cap \bar{a} \in D_n(\mathfrak{A})$ iff $b_i \cap \bar{b} \in D$.

Let $a \notin D_n(\mathfrak{A})$ and $a_i \cap \bar{a} \notin D_n(\mathfrak{A})$. Now $b_i \notin D$ and thus there is an $s \in 2^{<\omega^n}$ so that we have $b_i(\alpha) = 1$ for each $\alpha \in f_\omega(s)$. Let $b^*$ be such that

$$b^*(\alpha) = \begin{cases} 1 \text{ if } \alpha \in f_\omega(s^\frown 0), \\ 0 \text{ otherwise.} \end{cases}$$

Then $b^* \leq b_i$, $b^* \notin D$ and $b_i \cap \bar{\bar{b^*}} \notin D$.

Let $b \leq b_i$. If $b \in D$ or $b_i \cap \bar{b} \in D$, then it follows as in Lemma 17 that there is an $a \leq a_i$ so that $\langle \mathfrak{A}[a_i],a\rangle \ {}^{\omega}\!\equiv^{\omega}_n \langle \mathcal{B}[b_i],b\rangle$ and $a \in D_n(\mathfrak{A})$ iff $b \in D$, $a_i \cap \bar{a} \in D_n(\mathfrak{A})$ iff $b_i \cap \bar{b} \in D$.

Let $b \notin D$, $b_i \cap \bar{b} \notin D$. From $a_i \in N^1_n(\mathfrak{A})$ it follows that there is an $a \leq a_i$ with $a \in N^1_n(\mathfrak{A})$, $a_i \cap \bar{a} \in N^1_n(\mathfrak{A})$ and thus $a \notin D_n(\mathfrak{A})$, $a_i \cap \bar{a} \notin D_n(\mathfrak{A})$. That means, if player I decides to play an $\exists$-round, then player II can ensure that the result $\langle a_{i+1},b_{i+1}\rangle$ after the i-th round satisfies (1) for $k = i + 1$.

Let $A \subseteq |\mathfrak{A}[a_i]|$ with $card(A) = \omega_\omega$. Let $A_0 = A \cap D_n(\mathfrak{A})$, $A_1 = \{a \in A : a_i \cap \bar{a} \in D_n(\mathfrak{A})\}$, $A_2 = A \setminus (A_0 \cup A_1)$.

If $\mathrm{card}(A_0) = \omega_\omega$ or $\mathrm{card}(A_1) = \omega_\omega$, everything goes as in Lemma 20. Hence, let us assume that $\mathrm{card}(A_0) < \omega_\omega$ and $\mathrm{card}(A_1) < \omega_\omega$. Then $\mathrm{card}(A_2) = \omega_\omega$. Now $b_i \notin D$. Thus there is an $s \in 2^{<\omega}$ so that we have $b_i(\alpha) = 1$ for each $\alpha \in f_\omega(s)$. For each $\alpha \in f_\omega(s^\frown 0)$ let $c_\alpha$ be defined by

$$c_\alpha(l) = \begin{cases} 1 \text{ if } l = j \text{ or } l \in f_\omega(s^\frown 1), \\ 0 \text{ otherwise.} \end{cases}$$

Let $B = \{c_\alpha\colon \alpha \in f_\omega(s^\frown 0)\}$. Then $\mathrm{card}(B) = \omega_\omega$ and for each $b \in B$, $b \notin D$ and $b_i \cap \bar{b} \notin D$. That means if player I chooses a set in the structure $\mathfrak{A}[e]$, player II can ensure that the result $\langle a_{i+1}, b_{i+1}\rangle$ after the i-th round satisfies (1) for k = i + 1.

Let $B \subseteq |\mathfrak{B}[b_i]|$ with $\mathrm{card}(B) = \omega_\omega$. Let $B_0 = B \cap D$, $B_1 = \{b \in B\colon b_i \cap \bar{b} \in D\}$, $B_2 = B \setminus (B_0 \cup B_1)$.

If $\mathrm{card}(B_0) = \omega_\omega$ or $\mathrm{card}(B_1) = \omega_\omega$, everything goes as in Lemma 20. Thus let us assume $\mathrm{card}(B_0) < \omega_\omega$ and $\mathrm{card}(B_1) < \omega_\omega$. Then $\mathrm{card}(B_2) = \omega_\omega$. Now $a_i \in N_n^1(\mathfrak{A})$. Consequently, there are pairwise disjoint $a^0, a^1, a^2 \in N_n^1(\mathfrak{A})$ with $a^0 \cup a^1 \cup a^2 = a_i$. It follows from $a_i \in S_n^1(\mathfrak{A})$ that $\mathrm{cd}(a^0) \geq \omega_\omega$. Let $A^* \subseteq |\mathfrak{A}[a^0]|$ with $\mathrm{card}(A^*) = \omega_\omega$. We set $A = \{a \cup a^1\colon a \in A^*\}$. Then $\mathrm{card}(A) = \omega_\omega$, and for each $a \in A$ we have $a \notin D_n(\mathfrak{A})$, $a_i \cap \bar{a} \notin D_n(\mathfrak{A})$. That means if player I chooses a set in the structure $\mathfrak{B}$, player II can ensure that the result $\langle a_{i+1}, b_{i+1}\rangle$ after the i-th round satisfies (1) for k = i + 1.

Now $\mathfrak{L} \in \mathcal{M}_\omega$, let $\mathfrak{L} \cong \mathcal{B}_\omega(t_1)$. But then $\mathfrak{B} \cong \mathcal{B}_\omega(p_5(t_1, t_1)) \in \mathcal{M}_\omega$. Thus $e_0 \in D_n(\mathfrak{A})$, contradicting the assumption.

Q.E.D.

<u>Lemma 25.</u> Let $\mathfrak{A} \in \mathrm{Ba}$, $n \in \omega$. Then $P_n^1(\mathfrak{A}) \cap S_n^2(\mathfrak{A}) = \emptyset$.

<u>Proof.</u> Assume, on the contrary, that $e_0 \in P_n^1(\mathfrak{A}) \cap S_n^2(\mathfrak{A})$ and let $\{t_T\colon T \in \mathrm{Typ}(n,\omega,e_0)\}$ be a representation of $\mathrm{Typ}(n,\omega,e_0)$. It is easy to see that the claim stated in Lemma 21 is valid here, too.

We have to distinguish the following four cases:

<u>Case 1.</u> $\mathrm{card}(\{d \leq e_0\colon d \text{ is an atom}\}) < \omega_\omega$ and $\mathrm{card}(\{d \leq e_0\colon d \text{ is } \omega_\omega\text{-weakly-atomless}\}) < \omega_\omega$,

Case 2. card($\{d \le e_0$: d is an atom$\}) \ge \omega_\omega$ and card($\{d \le e_0$: d is $\omega_\omega$-weakly-atomless$\}) < \omega_\omega$,

Case 3. card($\{d \le e_0$: d is an atom$\}) < \omega_\omega$ and card($\{d \le e_0$: d is $\omega_\omega$-weakly-atomless$\}) \ge \omega_\omega$,

Case 4. card($\{d \le e_0$: d is an atom$\}) \ge \omega_\omega$ and card($\{d \le e_0$: d is $\omega_\omega$-weakly-atomless$\}) \ge \omega_\omega$.

Ad case 1: Let

$$\mathcal{L} = \mathcal{T}(\mathcal{H}_\omega) \dot{\times} \prod\{\mathcal{B}_\omega(t_T): T \in \mathrm{Typ}(n,\omega,e_0)\},$$

$$\mathcal{B} = \prod^{\mathcal{T}_1(\mathcal{H}_0)} \mathcal{L},$$

$$D = \{b \in |\mathcal{B}|: \mathrm{card}(\{s \in |\mathcal{H}_0|: b(s) \ne 0\}) < \omega_0\},$$

and let $e \in P_n^1(\mathfrak{A}) \cap S_n^2(\mathfrak{A})$ with $e \le e_0$. In order to prove $\mathfrak{A}[e] \;{}^{\omega}\!\equiv_n^{\omega}\; \mathcal{B}$ we show that player II has such a strategy in the game $H^2(n,\mathfrak{A}[e],\mathcal{B},\omega,\omega)$ that for each $k \le n$ the result $\langle a_k,b_k\rangle$ before the k-th round satisfies (1).

Assume $i < n$ and the result $\langle a_i,b_i\rangle$ before the i-th round satisfies (1) for $k = i$.

If $a_i \in D_n(\mathfrak{A})$ or if player I decides to play an $\exists$-round, everything goes as in the proofs of Lemmas 16 and 17.

Let $a_i \notin D_n(\mathfrak{A})$. Let $A \subseteq |\mathfrak{A}[a_i]|$ with card(A) $= \omega_\omega$. We set $A_0 = A \cap D_n(\mathfrak{A})$, $A_1 = \{a \in A: a_i \cap a \in D_n(\mathfrak{A})\}$, $A_2 = A \setminus (A_0 \cup A_1)$.

Assume card($A_0$) $= \omega_\omega$. By the claim (see Lemma 21) it then follows that there are $a^0, a^1$ with $a^0 \in A_0$, $a^1 \le a^0$ so that $a^1$ is $\omega_\omega$-weakly-atomless. Let $T_0 = \mathrm{typ}(n,\omega,\mathfrak{A}[a^0])$. Let $j_0 \in |\mathcal{R}^+|$ be such that $b_i(j_0) = 1$. Let $b^0, b^1 \le b_i(j_0)^{\mathrm{red}}$ be such that $b^0 \cap b^1 = 0$, $\mathcal{B}[b^0] \cong \mathcal{T}(\mathcal{H}_\omega)$, $\mathcal{B}[b^1] \cong \prod\{\mathcal{B}_\omega(t_T): T \in \mathrm{Typ}(n,\omega,e_0)\}$. Then there is a set $B^*$ with card($B^*$) $= \omega_\omega$ so that we have $b \le b^0$ for each $b \in B^*$, b is $\omega_\omega$-atomless. Let $B = \{b \cup b^1: b \in B^*\}$. Then card(B) $= \omega_\omega$ and for each $b \in B$ we have $b \in D$, $\mathrm{typ}(n,\omega,\mathcal{B}[b]) = \mathrm{typ}(n,\omega,\mathfrak{A}[a^0])$.

For card($A_1$) $= \omega_\omega$ the considerations are similar.

Assume now $\mathrm{card}(A_0) < \omega_\omega$ and $\mathrm{card}(A_1) < \omega_\omega$. Then $\mathrm{card}(A_2) = \omega_\omega$. Now $b_i \notin D$. Thus there is a nonempty open interval $\mathcal{T} \in |\mathcal{H}_0|$ so that $b_i(s) = 1$ for each $s \in \mathcal{T}$. Let $\mathcal{T}_0, \mathcal{T}_1, \mathcal{T}_2$ be pairwise disjoint nonempty open subintervals of $\mathcal{T}$. Let $b^0, b^1 \in |\mathcal{B}|$ be such that

$$b^0(s) = \begin{cases} 1 \text{ if } s \in \mathcal{T}_0, \\ 0 \text{ otherwise}; \end{cases} \qquad b^1(s) = \begin{cases} 1 \text{ if } s \in \mathcal{T}_1, \\ 0 \text{ otherwise}. \end{cases}$$

Then $b^1$ contains an $\omega_\omega$-atomless element. Thus there is a set $B^*$ with $\mathrm{card}(B^*) = \omega_\omega$ so that we have $b \le b^1$ for each $b \in B^*$, $b$ is $\omega_\omega$-atomless. Let $B = \{b^0 \cup b: b \in B^*\}$. Then $\mathrm{card}(B) = \omega_\omega$ and for each $b \in B$: $b \notin D$ and $b_i \cap \bar{b} \notin D$.

Thus, assuming that player I chooses a set in $\mathfrak{A}[e]$, player II can ensure that the result $\langle a_{i+1}, b_{i+1}\rangle$ after the i-th round satisfies (1) for $k = i + 1$.

Let $B \subseteq |\mathcal{B}[b_i]|$, $\mathrm{card}(B) = \omega_\omega$. Let $B_0 = B \cap D$, $B_1 = \{b \in B: b_i \cap \bar{b} \in D\}$, $B_2 = B \setminus (B_0 \cup B_1)$.

Assume $\mathrm{card}(B_0) = \omega_\omega$. For each $s \in |\mathcal{H}_0|$ let $c_s^0, c_s^1 \le 1(s)^{red}$ be such that $c_s^0 \cap c_s^1 = 0$, $c_s^0 \cup c_s^1 = 1(s)^{red}$, $\mathcal{B}[c_s^0] = \mathcal{I}(\mathcal{H}_\omega)$, $\mathcal{B}[c_s^1] \cong \prod\{\mathcal{B}_\omega(t_T): T \in \mathrm{Typ}(n,\omega,e_0)\}$.

Assume there are $b^* \in B_0$ and $s_0 \in |\mathcal{H}_0|$ with $b^* \cap c_{s_0}^0 = 0$. Let $I = \{s \in |\mathcal{H}_0|: b^*(s) \ne 0\}$, $b^0 = b^* \cap (\bigcup[c_s^0: s \in I])$, $b^1 = b^* \cap (\bigcup[c_s^1: s \in I])$. Then $b^0$ is an $\omega_\omega$-atomless element and $\mathrm{typ}(n,\omega,\mathcal{B}[b^1]) \in \mathrm{Typ}(n,\omega,e_{\underline{0}})$. Let $a^0 \le a_i$ be an $\omega_\omega$-atomless element. Then $a^0 \in D_n(\mathfrak{A})$, thus $a_i \cap \overline{a^0} \notin D_n(\mathfrak{A})$. Hence, there is $a^1 \le a_i \cap \overline{a^0}$ so that $a^1 \in D_n(\mathfrak{A})$ and $\mathfrak{A}[a^1] \; {}^{\omega}{\equiv}_n^{\omega} \; \mathcal{B}[b^1]$. Let $A^*$ be a set with $\mathrm{card}(A^*) = \omega_\omega$ so that for each $a \in A^*$: $a \le a^0$ and $a$ is $\omega_\omega$-atomless. Let $A = \{a \cup a^1: a \in A^*\}$. Then $\mathrm{card}(A) = \omega_\omega$ and for each $a \in A$ it is $a \in D_n(\mathfrak{A})$ and $\mathrm{typ}(n,\omega,\mathfrak{A}[a]) = \mathrm{typ}(n,\omega,\mathcal{B}[b^*])$.

Assume that $b \cap c_s^0 = 0$ for each $b \in B_0$ and each $s \in |\mathcal{H}_0|$. Then there is an $s_0 \in |\mathcal{R}^+|$ so that $\mathrm{card}(\{b \cap c_{s_0}^1 : b \in B_0\}) = \omega_\omega$. Let $\{d_T: T \in \mathrm{Typ}(n,\omega,e_0)\}$ be a partition of $c_{s_0}^1$ so that for each $T \in \mathrm{Typ}(n,\omega,e_0)$ it is $\mathfrak{A}[d_T] \cong \mathcal{B}_\omega(t_T)$. Then there is a $T_0 \in \mathrm{Typ}(n,\omega,e_0)$ so that $\mathrm{card}(\{b \cap d_{T_0}\}: b \in B_0\}) = \omega_\omega$. Let $C = \{b \cap d_{T_0} : b \in B_0\}$. Then there are $T_1 \in \mathrm{Typ}(n,\omega,e_0)$ and $C_0 \subseteq C$ with $\mathrm{card}(C_0) = \omega_\omega$ so that for each $c \in C_0$, $\mathrm{typ}(n,\omega,\mathcal{B}[c]) = T_1$. Let $b^* \in B_0$ be such that

$b^* \cap d_{T_0} \in C_0$. Then we can choose disjoint elements $a^0, a^1 \in D_n(\mathfrak{A}[a_i])$ so that $\mathfrak{A}[a^0] \ {}^{\omega}\!\equiv_n^{\omega} \mathfrak{B}[b^* \cap \overline{d_{T_0}}]$ and $\mathfrak{A}[a^1] \ {}^{\omega}\!\equiv_n^{\omega} \mathfrak{B}[d_{j_0}]$. But then it follows from Theorem 13 that there is a set $A^* \subseteq |\mathfrak{A}[a^1]|$ with $\mathrm{card}(A^*) = \omega_\omega$ so that we have $\mathrm{typ}(n,\omega,\mathfrak{A}[a]) = T_1$ for each $a \in A^*$. Let $A = \{a^0 \cup a : a \in A^*\}$. Then $\mathrm{card}(A) = \omega_\omega$ and for each $a \in A$, $\mathrm{typ}(n,\omega,\mathfrak{A}[a]) = \mathrm{typ}(n,\omega,\mathfrak{B}[b^*])$.

If $\mathrm{card}(A_1) = \omega_\omega$, the considerations are similar.

Assume $\mathrm{card}(A_0) < \omega_\omega$ and $\mathrm{card}(A_1) < \omega_\omega$. Then $\mathrm{card}(A_2) = \omega_\omega$. Now $b_i \notin D$. Thus there is an open interval $\mathcal{T} \subseteq |\mathcal{H}_0|$ so that for each $s \in \mathcal{T}$, $b_i(s) = 1$. Let $\mathcal{T}_0, \mathcal{T}_1$ be disjoint nonempty open intervals contained in $\mathcal{T}$. Let $s_0 \in \mathcal{T}_0$ and let $b^0 = 1(s_0)^{\mathrm{red}}$ be an $\omega_\omega$-atomless element. Let $b^1$ be an element with

$$b^1(s) = \begin{cases} 1 \text{ if } s \in \mathcal{T}_0 \text{ and } s \neq s_0, \\ 0 \text{ otherwise.} \end{cases}$$

Let $B^* \subseteq |\mathfrak{B}[b^1]|$ be a set with $\mathrm{card}(B^*) = \omega_\omega$ and each $b \in B^*$ is $\omega_\omega$-atomless. Let $B = \{b^1 \cup b : b \in B^*\}$. Then $\mathrm{card}(B) = \omega_\omega$, if $b \in B$, then $b \leq b_i$, $b \notin D$, and $b_i \cap \overline{b} \notin D$.

Hence, assuming that player I chooses a set in $\mathfrak{B}$, player II can ensure that the result $\langle a_{i+1}, b_{i+1} \rangle$ after the i-th round satisfies (1) for $k = i + 1$.

Now $\mathfrak{L} \in \mathcal{M}_\omega$, let $\mathfrak{L} \cong \mathcal{B}_\omega(t_1)$. Then $\mathfrak{B} \cong \mathcal{B}_\omega(p_2(t_1)) \in \mathcal{M}_\omega$. Thus $e_0 \in D_n(\mathfrak{A})$, contradicting the assumption.

Ad case 2: Let

$$\mathfrak{L} = \prod\{\mathcal{B}_\omega(t_T) : T \in \mathrm{Typ}(n,\omega,e_0)\},$$

$$\mathfrak{B} = \prod^{B(f_\omega)} \{\mathfrak{L}_\alpha : \alpha < \omega_\omega\}$$

with

$$\mathfrak{L}_\alpha = \begin{cases} \mathfrak{L} \text{ for } \alpha < \omega, \\ B(2) \text{ for } \omega \leq \alpha < \omega_\omega, \end{cases}$$

$$D = \{b \in |\mathfrak{B}| : \mathrm{card}(\{\alpha < \omega_\omega : b(\alpha) \neq 0\}) < \omega_0\},$$

and let $e \in P^1_n(\mathfrak{A}) \cap S^2_n(\mathfrak{A})$ with $e \leq e_0$. In order to prove $\mathfrak{A}[e] \,{}^{\omega}\!\!\equiv^{\omega}_{n}\, \mathfrak{B}$ we show that player II has such a strategy in the game $H^2(n, \mathfrak{A}[e], \mathfrak{B}, \omega, \omega)$ that for each $k \leq n$ the result $\langle a_k, b_k \rangle$ before the k-th round satisfies (1).

Assume $i < n$ and the result $\langle a_i, b_i \rangle$ before the i-th round satisfies (1) for $k = i$.

If $a_i \in D_n(\mathfrak{A})$ everything goes as in the proof of Lemma 16. If player I decides to play an $\exists$-round, then the considerations are similar to those of Lemma 17.

Let $A \subseteq |\mathfrak{A}[a_i]|$ with $\mathrm{card}(A) = \omega_\omega$. Let $A_0 = A \cap D_n(\mathfrak{A})$, $A_1 = \{a \in A: a_i \cap \bar{a} \in D_n(\mathfrak{A})\}$, $A_2 = A \setminus (A_0 \cup A_1)$. If $\mathrm{card}(A_0) = \omega_\omega$ or $\mathrm{card}(A_1) = \omega_\omega$, everything goes as in Lemma 17.

Let $\mathrm{card}(A_0) < \omega_\omega$ and $\mathrm{card}(A_1) < \omega_\omega$. Then $\mathrm{card}(A_2) = \omega_\omega$. Now $b_i \notin D$, thus there is an $s \in 2^{<\omega}$ so that $b_i(\alpha) = 1$ for each $\alpha \in f_\omega(s)$. Let $b^0, b^1$ be such that

$$b^0(\alpha) = \begin{cases} 1 \text{ if } \alpha \in f_\omega(s^\frown 0), \\ 0 \text{ otherwise}; \end{cases} \qquad b^1(\alpha) = \begin{cases} 1 \text{ if } \alpha \in f_\omega(s^\frown 1), \\ 0 \text{ otherwise}. \end{cases}$$

Then $b^0, b^1$ are disjoint, $b^0 \cup b^1 \leq b_i$. Let $B^*$ be a set with $\mathrm{card}(B^*) = \omega_\omega$, for each $b \in B^*$, $b$ is an atom and $b \leq b^1$. Let $B = \{b^0 \cap b: b \in B^*\}$. Then $\mathrm{card}(B) = \omega_\omega$, we have $b \notin D$ and $b_i \cap \bar{b} \notin D$ for each $b \in B$.

Thus, assuming that player I chooses a set in the structure $\mathfrak{A}[e]$, player II can ensure that the result $\langle a_{i+1}, b_{i+1} \rangle$ after the i-th round satisfies (1) for $k = i + 1$.

Let $B \subseteq |\mathfrak{B}[b_i]|$ with $\mathrm{card}(B) = \omega_\omega$. Let $B_0 = B \cap D$, $B_1 = \{b \in B: b_i \cap \bar{b} \in D\}$, $B_2 = B \setminus (B_0 \cup B_1)$.

If $\mathrm{card}(B_0) = \omega_\omega$ or $\mathrm{card}(B_1) = \omega_\omega$, everything goes as in Lemma 17.

Let $\mathrm{card}(B_0) < \omega_\omega$ and $\mathrm{card}(B_1) < \omega_\omega$. Thus $\mathrm{card}(B_2) = \omega_\omega$. Now $a_i \notin D_n(\mathfrak{A})$. It follows from $a_i \in N^1_n(\mathfrak{A})$ that there are pairwise disjoint elements $a^0, a^1, a^2$ with $a^0 \cup a^1 \cup a^2 = a_i$, $a^0, a^1, a^2 \notin D_n(\mathfrak{A})$. From $a_i \in S^2_n(\mathfrak{A})$ it follows that $\mathrm{cd}(a_i) \geq \omega_\omega$.

Let $A^* \subseteq |\mathfrak{A}[a^0]|$ with $\mathrm{card}(A^*) = \omega_\omega$. Let $A = \{a^1 \cup a: a \in A^*\}$.

Then card(A) $= \omega_\omega$ and for each $a \in A$ we have $a \notin D_n(\mathfrak{A})$ and $a_i \cap \bar{a} \notin D_n(\mathfrak{A})$.

Hence, assuming that player I chooses a set in the structure $\mathcal{L}$, player II can ensure that the result $\langle a_{i+1}, b_{i+1}\rangle$ after the i-th round satisfies (1) for $k = i + 1$.

Now $\mathcal{L} \in \mathcal{M}_\omega$, let $\mathcal{L} \cong \mathcal{B}_\omega(t_1)$. Then $\mathcal{L} \cong \mathcal{B}_\omega(p_5(t_1,2)) \in \mathcal{M}_\omega$ and thus $e_0 \in D_n(\mathfrak{A})$, contradicting the assumption.

<u>Ad case 3:</u> Let

$$\mathcal{L} = \prod\{\mathcal{B}_\omega(t_T)\colon T \in \mathrm{Typ}(n,\omega,e_0)\},$$

$$\mathcal{L} = \prod^{B(f_\omega)}\{\mathcal{L}_\alpha\colon \alpha < \omega_\omega\}$$

with

$$\mathcal{L}_\alpha = \begin{cases} \mathcal{L} \text{ for } \alpha < \omega, \\ B(\eta_0) \text{ for } \omega \leq \alpha < \omega_\omega, \end{cases}$$

$$D = \{b \in |\mathcal{L}|\colon \mathrm{card}(\{\alpha < \omega_\omega\colon b(\alpha) \neq 0\}) < \omega_0\},$$

and let $e \in P_n^1(\mathfrak{A}) \cap S_n^2(\mathfrak{A})$ with $e \leq e_0$.

Similar as in case 2 it can be shown that player II has such a strategy in the game $H^2(n,\mathfrak{A}[e],\mathcal{L},\omega,\omega)$ that for each $k \leq n$ the result $\langle a_k,b_k\rangle$ before the k-th round satisfies (1). This implies $\mathfrak{A}[e] \;{}^{\omega}{\equiv}_n^{\omega}\; \mathcal{L}$.

Now $\mathcal{L} \in \mathcal{M}_\omega$, let $\mathcal{L} \cong \mathcal{B}_\omega(t_1)$. Then $\mathcal{L} \cong \mathcal{B}_\omega(p_5(t_1,\eta_0)) \in \mathcal{M}_\omega$. Thus $e_0 \in D_n(\mathfrak{A})$, contradicting the assumption.

<u>Ad case 4:</u> Let

$$\mathcal{L} = \prod\{\mathcal{B}_\omega(t_T)\colon T \in \mathrm{Typ}(n,\omega,e_0)\},$$

$$\mathcal{L} = \prod^{B(f_\omega)}\{\mathcal{L}_\alpha\colon \alpha < \omega_\omega\}$$

with

$$\mathcal{L}_\alpha = \begin{cases} \mathcal{L} \text{ for } \alpha < \omega, \\ B(2) \times B(\eta_0) \text{ for } \omega \leq \alpha < \omega_\omega, \end{cases}$$

$$D = \{b \in |\mathcal{L}|: \text{card}(\{\alpha < \omega_\omega: b(\alpha) \neq 0\}) < \omega_0\},$$

and let $e \in P_n^1(\mathfrak{A}) \cap S_n^2(\mathfrak{A})$ with $e \leq e_0$.

Similar as in case 2 it can be shown that player II has such a strategy in the game $H^2(n,\mathfrak{A}[e],\mathcal{L},\omega,\omega)$ that for each $k \leq n$ the result $\langle a_k,b_k\rangle$ before the k-th round satisfies (1). This implies $\mathfrak{A}[e] \; {}^{\omega}\!\equiv_n^{\omega} \mathcal{L}$

Now $\mathcal{L} \in \mathcal{M}_\omega$, let $\mathcal{L} \cong \mathcal{B}_\omega(t_1)$. Then $\mathcal{L} \cong \mathcal{B}_\omega(p_5(t_1,2\times\eta_0)) \in \mathcal{M}_\omega$.
Thus $e_0 \in D_n(\mathfrak{A})$, contradicting the assumption.

Q.E.D.

Now it follows from Lemmas 23, 24 and 25 that

__Corollary 26.__ Let $\mathfrak{A} \in$ Ba, $n \in \omega$. Then $N_n^1(\mathfrak{A}) = \emptyset$.

__Lemma 27.__ Let $\mathfrak{A} \in$ Ba, $n \in \omega$. Then $P_n^2(\mathfrak{A}) \cap S_n^0(\mathfrak{A}) = \emptyset$.

__Proof.__ $a \in P_n^2(\mathfrak{A})$ implies $\text{cd}(a) \geq \omega_\omega$, whereas $a \in S_n^0(\mathfrak{A})$ implies $\text{cd}(a) < \omega_\omega$.

Q.E.D.

__Lemma 28.__ Let $\mathfrak{A} \in$ Ba, $n \in \omega$. Then $P_n^2(\mathfrak{A}) \cap S_n^1(\mathfrak{A}) = \emptyset$.

__Proof.__ Similar to the proof of Lemma 24.

Q.E.D.

__Lemma 29.__ Let $\mathfrak{A} \in$ Ba, $n \in \omega$. Then $P_n^2(\mathfrak{A}) \cap S_n^2(\mathfrak{A}) = \emptyset$.

__Proof.__ Similar to the proof of Lemma 25.

Q.E.D.

Now it follows from Lemmas 27, 28 and 29 that

__Corollary 30.__ Let $\mathfrak{A} \in$ Ba, $n \in \omega$. Then $N_n^2(\mathfrak{A}) = \emptyset$.

Using Corollaries 22, 26 and 30 we obtain

__Theorem 31.__ $\mathcal{M}_\omega$ is dense for $\text{Th}_\omega(\text{Ba})$.

## § 4.3. A dense system for $Th_1(Ba)$

This is the most complicated case. The reason for this is that there are Boolean algebras of power $\omega_1$ which possess countable dense subsets (see the examples in § 2). Thus we nave to consider the case that $\mathfrak{A}$ is a Boolean algebra with $card(|\mathfrak{A}|) = \omega_1$ and $card(D_n(\mathfrak{A})) = \omega_0$. This fact makes it very complicated to describe a dense system for $Th_1(Ba)$.

$Term_1$ denotes the least class K with

(i) $2, \eta_0, \eta \in K$,
(ii) if $t_1, t_2 \in K$, then $p_1(t_1)$, $p_2(t_1)$, $t_1 \times t_2$, $p_3(t_1)$, $p_4(t_1,t_2)$, $p_5(t_1,t_2) \in K$ (not necessarily $t_1 \neq t_2$),
(iii) if $m_0 > 0$, $t_0,\dots,t_{m_0-1} \in K$, $X \in (pow(\{t_i : i < m_0\}) \setminus \{\emptyset\})^{<\omega}$, $n \leq lh(X)$, then $p_6(t_0,\dots,t_{m_0-1}; n, X)$, $p_7(t_0,\dots,t_{m_0-1}; n, X) \in K$.

Let $f_1 : 2^{<\omega} \longrightarrow pow(\omega_1)$ be a function with

(i) $f_1(\varphi) = \omega_1$,
(ii) if $s \in 2^{<\omega}$, then $f_1(s^\frown 0) \cup f_1(s^\frown 1) = f_1(s)$ and $f_1(s^\frown 0) \cap f_1(s^\frown 1) = \emptyset$,
(iii) if $s \in 2^{<\omega}$, then $card(f_1(s)) = \omega_1$ and $card(f_1(s)) \cap \omega) = \omega_0$.

Let $B(f_1) = Pow(\omega_1; \{f_1(s) : s \in 2^{<\omega}\})$, that means, $B(f_1)$ is the subalgebra of $Pow(\omega_1)$ generated by $\{f_1(s) : s \in 2^{<\omega}\}$ and the finite subsets of $\omega_1$.

Let the function $\mathcal{B}_1$ from $Term_1$ into Ba be defined as follows:

$$\mathcal{B}_1(2) = B(2),$$

$$\mathcal{B}_1(\eta_0) = B(\eta_0),$$

$$\mathcal{B}_1(\eta) = \mathcal{J}(\mathcal{H}_1),$$

$$\mathcal{B}_1(t_1 \times t_2) = \mathcal{B}_1(t_1) \times \mathcal{B}_1(t_2),$$

$$\mathcal{B}_1(p_1(t_1)) = \prod{}^{w} \mathcal{B}_1(t_1),$$

$$\mathcal{B}_1(p_2(t_1)) = \prod{}^{\mathcal{J}_1(\mathcal{H}_1)} \mathcal{B}_1(t_1),$$

$$\mathcal{B}_1(p_3(t_1)) = \prod{}^{\mathcal{J}_1(\mathcal{H}_0)} \mathcal{B}_1(t_1),$$

$$\mathcal{B}_1(p_4(t_1,t_2)) = \prod^{W}\{\mathcal{L}_\alpha : \alpha < \omega_1\}$$

with

$$\mathcal{L}_\alpha = \begin{cases} \mathcal{B}_1(t_1) & \text{for } \alpha < \omega, \\ \mathcal{B}_1(t_2) & \text{for } \omega \leq \alpha < \omega_1, \end{cases}$$

$$\mathcal{B}_1(p_5(t_1,t_2)) = \prod^{B(f_1)}\{\mathcal{L}_\alpha : \alpha < \omega_1\}$$

with

$$\mathcal{L}_\alpha = \begin{cases} \mathcal{B}_1(t_1) & \text{for } \alpha < \omega, \\ \mathcal{B}_1(t_2) & \text{for } \omega \leq \alpha < \omega_1, \end{cases}$$

$$\mathcal{B}_1(p_6(t_0,\ldots,t_{m_0-1};n,X)) = \prod^{\mathcal{L}}\{\mathcal{L}_\alpha : \alpha < \omega^\omega\}$$

with

$$\mathcal{L} = B_{\omega_0}[m_0,n,lh(X)],$$

$$\mathcal{L}_{\alpha\cdot\omega+i} = \begin{cases} \mathcal{B}_1(t_k) \text{ if } i \equiv k(\mathrm{mod}(m_0+lh(X))) \text{ and } 0 \leq k < m_0, \\ \prod\{\mathcal{B}_1(t) : t \in X(k-m_0)\} \text{ if } i \equiv k(\mathrm{mod}(m_0+lh(X))) \\ \qquad\qquad \text{and } m_0 \leq k < m_0 + lh(X), \end{cases}$$

$$\mathcal{B}_1(p_7(t_0,\ldots,t_{m_0-1};n,X) = \prod^{\mathcal{L}}\{\mathcal{L}_\alpha : \alpha < \omega^\omega\}$$

with

$$\mathcal{L} = B^{+}_{\omega_0}[m_0,n,lh(X)],$$

$$\mathcal{L}_{\alpha\cdot\omega+i} = \begin{cases} \mathcal{B}_1(t_k) \text{ if } i \equiv k(\mathrm{mod}(m_0+lh(X))) \text{ and } 0 \leq k < m_0, \\ \prod\{\mathcal{B}_1(t) : t \in X(k-m_0)\} \text{ if } i \equiv k(\mathrm{mod}(m_0+lh(X))) \\ \qquad\qquad \text{and } m_0 \leq k < m_0 + lh(X). \end{cases}$$

Remark. It is easy to see that for each $t \in \mathrm{Term}_1$ we have

$$\mathcal{B}_1(p_1(t)) = \mathcal{B}_1(p_5(t;0,\varphi)) \quad \text{and} \quad \mathcal{B}_1(p_2(t)) = \mathcal{B}_1(p_6(t;0,\varphi)).$$

Then $\langle \mathrm{Term}_1, \mathcal{B}_1 \rangle$ is a pre-density pair for $Q_1$. In the following we show that

$$\mathcal{M}_1 = \{\mathcal{B}_1(t) : t \in \mathrm{Term}_1\}$$

is dense for $\mathrm{Th}_1(\mathrm{Ba})$.

We have to show that for each $\mathfrak{A} \in \mathrm{Ba}$ and each $n \in \omega$

$$N_n^0(\mathfrak{A}) = N_n^1(\mathfrak{A}) = N_n^2(\mathfrak{A}) = \emptyset.$$

Lemma 32. Let $R \subseteq \mathrm{Term}_1$ with $\mathrm{card}(R) < \omega_0$, $\mathfrak{A} \in \mathrm{Ba}$ with $\mathrm{card}(|\mathfrak{A}|) \leq \omega_0$ such that we have for each $a \in |\mathfrak{A}|$:

$$\{t \in R: \text{there is a } d \leq a \text{ with } \mathfrak{A}[d] \ {}^1{\equiv}_n^1\ \mathcal{B}_1(t)\} = R$$

or

$$\{t \in R: \text{there is a } d \leq \bar{\bar{a}} \text{ with } \mathfrak{A}[d] \ {}^1{\equiv}_n^1\ \mathcal{B}_1(t)\} = R.$$

Let

$$D = \{d \in |\mathfrak{A}|: \text{there is a } t \in R \text{ with } \mathfrak{A}[d] \ {}^1{\equiv}_n^1\ \mathcal{B}_1(t)\}.$$

If D is an ideal of $\mathfrak{A}$, then:

(i) if $\mathfrak{A}/D = B(2)$, then $\mathfrak{A} \ {}^1{\equiv}_n^1 \prod_\omega^w \mathcal{L}$ with $\mathcal{L} = \prod\{\mathcal{B}_1(t): t \in R\}$,

(ii) if $\mathfrak{A}/D = B(\eta_0)$, then $\mathfrak{A} \ {}^1{\equiv}_n^1 \prod^{\mathcal{T}_1(\mathcal{H}_0)} \mathcal{L}$ with $\mathcal{L} = \prod\{\mathcal{B}_1(t): t \in R\}$.

Proof. From $\mathrm{card}(|\mathfrak{A}|) \leq \omega_0$ it follows that we have $\mathrm{card}(|\mathcal{B}_1(t)|) \leq \omega_0$ for each $t \in R$. Thus $\mathrm{card}(|\prod_\omega^w \mathcal{L}|) \leq \omega_0$ and $\mathrm{card}(|\prod^{\mathcal{T}_1(\mathcal{H}_0)} \mathcal{L}|) \leq \omega_0$ with $\mathcal{L} = \prod\{\mathcal{B}_1(t): t \in R\}$. That means, each round in the game $H^2(n, \mathfrak{A}, \prod_\omega^w \mathcal{L}, 1, 1)$ and in the game $H^2(n, \mathfrak{A}, \prod^{\mathcal{T}_1(\mathcal{H}_0)} \mathcal{L}, 1, 1)$ must be an $\exists$-round, and the proof is similar to the proof of Lemmas 16 and 17.

Q.E.D.

The proofs of the following two lemmas are similar to the proofs of Lemmas 19 and 20, respectively.

Lemma 33. Let $\mathfrak{A} \in \mathrm{Ba}$, $n \in \omega$. Then $P_n^0(\mathfrak{A}) \cap S_n^0(\mathfrak{A}) = \emptyset$.

Lemma 34. Let $\mathfrak{A} \in \mathrm{Ba}$, $n \in \omega$. Then $P_n^0(\mathfrak{A}) \cap S_n^1(\mathfrak{A}) = \emptyset$.

Lemma 35. Let $\mathfrak{A} \in \mathrm{Ba}$, $n \in \omega$. Then $P_n^0(\mathfrak{A}) \cap S_n^2(\mathfrak{A}) = \emptyset$.

Proof. Assume, on the contrary, that $e_0 \in P_n^0(\mathfrak{A}) \cap S_n^2(\mathfrak{A})$ and let $\{t_T: T \in \mathrm{Typ}(n,1,e_0)\}$ be a representation of $\mathrm{Typ}(n,1,e_0)$.

We distinguish the following four cases:

Case 1. $\mathrm{card}(\{d \leq e_0: d \text{ is an atom}\}) < \omega_1$ and $\mathrm{card}(\{d \leq e_0: d \text{ is } \omega_1\text{-weakly-atomless}\}) < \omega_1$,

Case 2. $\mathrm{card}(\{d \leq e_0: d \text{ is an atom}\}) \geq \omega_1$ and $\mathrm{card}(\{d \leq e_0: d \text{ is } \omega_1\text{-weakly-atomless}\}) < \omega_1$,

Case 3. $\mathrm{card}(\{d \leq e_0: d \text{ is an atom}\}) < \omega_1$ and $\mathrm{card}(\{d \leq e_0: d \text{ is } \omega_1\text{-weakly-atomless}\}) \geq \omega_1$,

Case 4. $\mathrm{card}(\{d \leq e_0: d \text{ is an atom}\}) \geq \omega_1$ and $\mathrm{card}(\{d \leq e_0: d \text{ is } \omega_1\text{-weakly-atomless}\}) \geq \omega_1$.

Ad case 1: Let

$$R = \{T \in \mathrm{Typ}(n,1,e_0): \mathrm{card}(|\mathcal{B}_1(t_T)|) < \omega_1,\ \mathrm{card}(\{a \leq e_0: \mathrm{typ}(n,1,\mathfrak{A}[a]) = T\}) \geq \omega_1\},$$

$$D^0 = \{a \leq e_0: \mathrm{cd}(a) < \omega_1\},$$

$$E^0 = \{a \leq e_0: \text{there is a } T \in R \text{ with } \mathrm{typ}(n,1,\mathfrak{A}[a]) = T\},$$

$$E^1 = \{a \in E^0: \text{if } b \in E^0 \text{ and } b \leq a, \text{then } \mathrm{typ}(n,1,\mathfrak{A}[a]) = \mathrm{typ}(n,1,\mathfrak{A}[b])\}.$$

We have $E^1 \subseteq E^0$ and $D^0 \cap E^0 = \emptyset$. $D^0$ is an ideal on $\mathfrak{A}[e_0]$. Let

$$E^2 = \{a \in E^1: a/D^0 \text{ is an atom}\},$$

$$E^3 = \{a \in E^1: a/D^0 \text{ is atomless}\},$$

then $\{a/D^0: a \in E^2 \cup E^3\}$ is dense in $\mathfrak{A}[e_0]/D^0$.

Let $a \in E^2$,

$$\{T \in R\colon \text{there is a } b \leq a \text{ with } b \in D^0,\ typ(n,1,\mathfrak{A}[b]) = T\} = \{T_0,\ldots,T_{m-1}\},$$

$$t_a = (\ldots(t_{T_0} \times t_{T_1}) \times \ldots) \times t_{T_{m-1}}.$$

Then it follows from Lemma 32 (i) that $\mathfrak{A}[a] \ {}^1\!\equiv^1_n \mathcal{B}_1(p_1(t_a))$.

Let $a \in E^3$,

$$\{T \in R\colon \text{there is a } b \leq a \text{ with } b \in D^0,\ typ(n,1,\mathfrak{A}[b]) = T\} = \{T_0,\ldots,T_{m-1}\},$$

$$t_a = (\ldots(t_{T_0} \times t_{T_1}) \times \ldots) \times t_{T_{m-1}}.$$

Then $\mathfrak{A}[a]/D^0 = B(\eta_0)$, and thus it follows from Lemma 32 (ii) that $\mathfrak{A}[a] \ {}^1\!\equiv^1_n \mathcal{B}_1(p_3(t_a))$.

For $a \in E^2 \cup E^3$ let

$$R_n(a) = \{T \in Typ(n,1,e_0)\colon \text{there is a } b \leq a \text{ with } typ(n,1,\mathfrak{A}[b]) = T,$$

$$Y_0 = \{R_n(a)\colon a \in E^2\}, \quad Y_1 = \{R_n(a)\colon a \in E^3\}.$$

Let $Z_0$, $Z_1$ be sequences with $\{Z_0(i)\colon i < lh(Z_0)\} = Y_0$, $\{Z_1(i)\colon i < lh(Z_1)\} = Y_1$. Let $m = lh(Y_0)$, $m_1 = lh(Y_0) + lh(Y_1)$. For $i < m_1$ let

$$X(i) = \begin{cases} \{t_T\colon T \in Z_0(i)\} & \text{if } i < m, \\ \{t_T\colon T \in Z_1(i-n) & \text{if } m \leq i < m_1. \end{cases}$$

Let $X = \langle X(0),\ldots,X(m_1-1)\rangle$, and let $\{t_T\colon T \in Typ(n,1,e_0)\} = \{t_0,\ldots,t_{m_0-1}\}$. We set

$$\mathcal{L} = B_{\omega_0}[m_0,m,m_1].$$

We set

$$\mathcal{L}_{\alpha\cdot\omega+i} = \begin{cases} \mathcal{B}_1(t_k) \text{ if } i \equiv k(\mathrm{mod}(m_0+m_1)) \text{ and } 0 \leq k < m_0, \\ \prod\{\mathcal{B}_1(t)\colon t \in X(k-m_0) \text{ if } i \equiv k(\mathrm{mod}(m_0+m_1)) \\ \qquad\qquad\qquad\qquad \text{and } m_0 \leq k < m_0 + m_1. \end{cases}$$

Let

$$\mathcal{L} = \prod^{\mathcal{L}}\{\mathcal{L}_\alpha : \alpha < \omega^\omega\}.$$

Let $I^0$ be the ideal on Pow$(\omega^\omega)$ generated by the finite subsets of $\omega^\omega$ and the sets $\{E_{\omega_0}[i,m_0+m_1]: m_0 \leq i < m_0 + m\}$ and $\{F_{\omega_0}[i,m_0+m_1]: m_0 + m \leq i < m_0 + m_1\}$.

Let

$$D = \{b \in |\mathcal{L}|: \{i < \omega_0: b(i) \neq 0\} \in I^0\}$$

and let $e \in P^0_n(\mathfrak{A}) \cap S^2_n(\mathfrak{A})$ with $e \leq e_0$. In order to prove $\mathfrak{A}[e] \,{}^1\!\equiv^1_n \mathcal{L}$ we show that player II has such a strategy in the game $H^2(n,\mathfrak{A}[e],\mathcal{L},1,1)$ that for each $k \leq n$ the result $\langle a_k,b_k\rangle$ before the k-th round satisfies (1).

Assume $i < n$ and the result $\langle a_i,b_i\rangle$ before the i-th round satisfies (1) for $k = i$.

If $a_i \in D_n(\mathfrak{A})$, everything goes as in the proof of Lemma 16.

Thus let us assume $a_i \notin D_n(\mathfrak{A})$. Assume that player I decided to play an $\exists$-round. Let $a \leq a_i$. Without loss of generality we assume $a \in D_n(\mathfrak{A})$. Let $j_0 < m_0$ be such that $\mathrm{typ}(n,1,\mathfrak{A}[a]) = \mathrm{typ}(n,1,\mathcal{B}_1(t_{T_{j_0}}))$. Now it follows from $b_i \notin D$ that there is a $k_0 < \omega_0$ with $b_i(k_0) = 1$ and $j_0 \equiv k_0(\mathrm{mod}(m_0+m_1))$. But then $\langle \mathfrak{A}[a_i],a\rangle \,{}^1\!\equiv^1_n \langle \mathcal{L}[b_i],(b_i(k_0))^{red}\rangle$.

Let $b \leq b_i$. Without loss of generality we assume that $b \in D$. From the definition of D it follows that there are pairwise disjoint sets $J_0,\ldots,J_{m_2},K$ (where $m_2$ is a suitable number) with

(i) for each $k < m_2$ there is an $X \in \bigcup\{E_{\omega_0}[i,m_0+m_1]: m_0 \leq i < m_0 + m\} \cup \bigcup\{F_{\omega_0}[i,m_0+m_1]: m_0 + m \leq i < m_0 + m_1\}$ with $J_k \subseteq X$ and $\mathrm{card}(X \setminus J_k) < \omega_0$,
(ii) $\mathrm{card}(K) < \omega_0$,
(iii) if $\alpha \in \bigcup\{J_k: k < m_2\}$, then $b(\alpha) = 1$,
(iv) $\{\alpha < \omega^\omega: b(\alpha) \neq 0\} \subseteq \bigcup\{J_k: k < m_2\} \cup K$.

For $k < m_2$ let $b^k$ be defined by

$$b^k(\alpha) = \begin{cases} 0 \text{ if } \alpha \notin J_k, \\ 1 \text{ if } \alpha \in J_k. \end{cases}$$

Then $\{b^k: k < m_2\} \cup \{b(\alpha)^{red}: \alpha \in K\}$ is a partition of b. For each $k < m_2$ we have $typ(n,1,\mathcal{B}[b^k]) \in R$ and thus for each $b^* \in \{b^k: k < m_2\} \cup \{b(\alpha)^{red}: \alpha \in K\}$ we have $typ(n,1,\mathcal{B}[b^*]) \in Typ(n,1,e_0)$. Hence, for each $k < m_2$ there is an $a^k \leq a_i$ with $a^k \in D_n(\mathfrak{A})$, and for each $\alpha \in K$ there is a $c^\alpha \leq a_i$ with $c^\alpha \in D_n(\mathfrak{A})$ so that for each $k < m_2$, $\mathfrak{A}[a^k] \ {}^1{\equiv}_n^1 \ \mathcal{B}[b^*]$, for each $\alpha \in K$, $\mathfrak{A}[c^\alpha] \ {}^1{\equiv}_n^1 \ \mathcal{B}[b(\alpha)^{red}]$, and the $a^k$'s $(k < m_2)$ and the $c^\alpha$'s $(\alpha \in K)$ are pairwise disjoint elements.

Let $a = \bigcup[a^k: k < m_2] \cup \bigcup[c^\alpha: \alpha \in K]$. Then $\langle\mathfrak{A}[a_i],a\rangle \ {}^1{\equiv}_n^1 \ \langle\mathcal{B}[b_i],b\rangle$. Thus assuming that player I decides to play an $\exists$-round, player II can ensure that the result $\langle a_{i+1},b_{i+1}\rangle$ after the i-th round satifies (1) for $k = i + 1$.

Assume now that player I decides to play a Q-round. Let $A \subseteq |\mathfrak{A}[a_i]|$ with $card(A) = \omega_1$. Let $A_0 = A \cap D_n(\mathfrak{A})$, $A_1 = A \setminus A_0$. Let $card(A_0) = \omega_1$. Then there are a set $A' \subseteq A_0$ with $card(A') = \omega_1$ and $T' \in Typ(n,1,e_0)$ so that for each $a \in A'$, $typ(n,1,\mathfrak{A}[a]) = T'$.

Assume there is an $a^0 \in A'$ with $cd(a^0) \geq \omega_1$ (thus for each $a \in A'$, $cd(a) \geq \omega_1$). Then it follows from $b_i \notin D$ that there is a $b^0 \leq b_i$ with $b^0 \in D$ and $cd(b^0) = \omega_1$. From Theorem 13 it follows that there is a set $B \subseteq |\mathcal{B}[b^0]|$ with $card(B) = \omega_1$ so that for each $b \in B$, $typ(n,1,\mathcal{B}[b]) = T'$.

Assume that we have $cd(a) < \omega_1$ for each $a \in A'$. Then $T' \in R$. Let $a^0 \in A'$. Then there is an $a^1 \leq a^0$ with $a^1 \in E^2 \cup E^3$. Assume $a^1 \in E^2$. Let $\{T \in R:$ there is a $c \leq a^1$ with $typ(n,1,\mathfrak{A}[c]) = T\} = X(k)$, where $k < m$. It follows from $a^0 \in D_n(\mathfrak{A})$ that there are $b^0,b^1 \in D$ with $b^1 \leq b^0$ so that $\langle\mathcal{B}[b^0],b^1\rangle \ {}^1{\equiv}_{n-1}^1 \ \langle\mathfrak{A}[a^0],a^1\rangle$. Let $Z = \{\alpha < \omega^\omega: b^0(\alpha) \neq 0\}$. It follows from $b^0 \in D$ that there is an $E \subseteq pow(\omega^\omega)$ with $card(E) = \omega_1$ so that we have for each $X \in E$: $X \cap Z = \emptyset$ and there is a $Y \in E_{\omega_0}[m_0+k,m_0+m_1]$ with $X \subseteq Y$, $card(Y \setminus X) < \omega_0$. For each $X \in E$ let $b_X$ be defined by

$$b_X(\alpha) = \begin{cases} 1 \text{ if } \alpha \in X, \\ 0 \text{ otherwise.} \end{cases}$$

Then for each $X \in E$, $\mathfrak{A}[a^1] \;{}^1\!\equiv^1_{n-1} \mathcal{B}[b_X]$. Let $B = \{(b^0 \cap \overline{b^1}) \cup b_X : X \in E\}$. Then $card(B) = \omega_1$ and for each $b \in B$, $\mathfrak{A}[a^0] \;{}^1\!\equiv^1_{n-1} \mathcal{B}[b]$. Thus assuming that player I chooses a set in the structure $\mathfrak{A}[e]$, player II can ensure that the result $\langle a_{i+1}, b_{i+1}\rangle$ after the i-th round satisfies (1) for $k = i + 1$.

Let $B \subseteq |\mathcal{B}[b_i]|$ with $card(B) = \omega_1$. Let $B_0 = B \cap D$, $B_1 = B \setminus D$. Let $card(B_0) = \omega_1$. Assume that there is a $b^* \in B_0$ with $card(\{\alpha < \omega^\omega : b^*(\alpha) \neq 0\}) = \omega_0$. Then there are $k^* < m_1$,

$$X^* \in \begin{cases} E_{\omega_0}[m_0+k^*, m_0+m_1] & \text{if } k^* < m, \\ F_{\omega_0}[m_0+k^*, m_0+m_1] & \text{if } m \leq k^* < m_1, \end{cases}$$

and $X \subseteq \omega_0$ with $X \subseteq X^*$, $card(X^* \setminus X) < \omega_0$ so that for each $j \in X$, $b^*(j) = 1$. Let $b_X$ be defined by

$$b_X(j) = \begin{cases} 1 & \text{if } j \in X, \\ 0 & \text{otherwise}. \end{cases}$$

Then $card(X^* \setminus X) < \omega_0$ implies $b_X \in |\mathcal{B}|$. Let $T' \in R$ be such that $typ(n,1,\mathcal{B}[b_X]) = T'$. Let $a^0 \in D_n(\mathfrak{A}[a_i])$ be such that $\mathfrak{A}[a^0] \;{}^1\!\equiv^1_{n-1} \mathcal{B}[b^* \cap \overline{b_X}]$. $T' \in R$ implies that there is a set $A^* \subseteq D_n(\mathfrak{A}[a_i])$ so that for each $c \in A^*$: $c \cap a^0 = 0$ and $typ(n,1,\mathfrak{A}[c]) = T'$. Let $A = \{a^0 \cup c : c \in A^*\}$. Then $card(A) = \omega_1$ and for each $a \in A$, $typ(n,1,\mathfrak{A}[a]) = typ(n,1,\mathcal{B}[b^*])$.

Assume for each $b \in B_0$ we have $card(\{\alpha < \omega^\omega : b(\alpha) \neq 0\}) < \omega_0$. Then there is an $\alpha_0 < \omega^\omega$ with $card(\{b(\alpha_0) : b \in B_0\}) = \omega_1$. It follows that there are $B^* \subseteq B_0$ and $T_0, T_1 \in Typ(n,1,e_0)$ so that $typ(n,1,\mathcal{B}[b \cap \overline{b(\alpha_0)^{red}}]) = T_0$, $typ(n,1,\mathcal{B}[b(\alpha_0)^{red}]) = T_1$ for each $b \in B^*$. Then there are $a^0, a^1 \in D_n(\mathfrak{A}[a_i])$ with $a^0 \cap a^1 = 0$, $typ(n,1,\mathfrak{A}[a^0]) = T_0$, $typ(n,1,\mathfrak{A}[a^1]) = T_1$. Assume $card(\{b(\alpha_0)^{red} : b \in B^*\}) = \omega_1$. Then there is a set $A^* \subseteq |\mathfrak{A}[a_i]|$ with $card(A^*) = \omega_1$ so that for each $a \in A^*$ it is $a \cap a^1 = 0$, $typ(n,1,\mathfrak{A}[a]) = T_0$. Let $A = \{a^0 \cup a : a \in A^*\}$. Then $card(A) = \omega_1$ and for each $a \in A$, $typ(n,1,\mathfrak{A}[a]) = typ(n,1,\mathcal{B}_1(t_{T_0} \times t_{T_1}))$.

Assume $card(\{b(\alpha_0)^{red} : b \in B^*\}) < \omega_1$. Then $card(\{b \cap \overline{b(\alpha_0)^{red}} : b \in B^*\}) = \omega_1$ and thus there is a set $A^{**} \subseteq |\mathfrak{A}[a_i]|$ with $card(A^{**}) = \omega_1$ so that for each $a \in A^{**}$ it is $a \cap a^0 = 0$, $typ(n,1,\mathfrak{A}[a]) = T_1$. Let $A' = \{a^1 \cup a : a \in A^{**}\}$. Then $card(A') = \omega_1$ and for each $a \in A'$,

$typ(n,1,\mathfrak{A}[a]) = typ(n,1,\mathcal{B}_1(t_{T_0} \times t_{T_1}))$. Thus assuming that player I chooses a set in the structure $\mathcal{L}$, player II can ensure that the result $\langle a_{i+1}, b_{i+1} \rangle$ after the i-th round satisfies (1) for $k = i + 1$.

Now we have $\mathfrak{A}[e] \ {}^1{\equiv}^1_n \ \mathcal{B}_1(p_6(t_0,\ldots,t_{m_0-1};m,X)) \in \mathcal{M}_1$. Thus $e_0 \in D_n(\mathfrak{A})$, contradicting the assumption.

Ad cases 2 - 4: For these cases the desired result can be proved in a similar way.

Q.E.D.

Now it follows from Lemmas 33, 34 and 35

Corollary 36. Let $\mathfrak{A} \in Ba$, $n \in \omega$. Then $N^0_n(\mathfrak{A}) = \emptyset$.

Similar to Lemmas 23 and 24 it can be shown:

Lemma 37. Let $\mathfrak{A} \in Ba$, $n \in \omega$. Then $P^1_n(\mathfrak{A}) \cap S^0_n(\mathfrak{A}) = \emptyset$.

Lemma 38. Let $\mathfrak{A} \in Ba$, $n \in \omega$. Then $P^1_n(\mathfrak{A}) \cap S^1_n(\mathfrak{A}) = 0$.

Using $p_6$ instead of $p_5$ in the proof of Lemma 35 it can be shown:

Lemma 39. Let $\mathfrak{A} \in Ba$, $n \in \omega$. Then $P^1_n(\mathfrak{A}) \cap S^2_n(\mathfrak{A}) = \emptyset$.

Now it follows from Lemmas 37, 38 and 39

Corollary 40. Let $\mathfrak{A} \in Ba$, $n \in \omega$. Then $N^1_n(\mathfrak{A}) = \emptyset$.

Similar to Lemmas 27 and 28 it can be shown:

Lemma 41. Let $\mathfrak{A} \in Ba$, $n \in \omega$. Then $P^2_n(\mathfrak{A}) \cap S^0_n(\mathfrak{A}) = \emptyset$.

Lemma 42. Let $\mathfrak{A} \in Ba$, $n \in \omega$. Then $P^2_n(\mathfrak{A}) \cap S^1_n(\mathfrak{A}) = \emptyset$.

Using $p_6$ instead of $p_5$ in the proof of Lemma 35 it can be shown:

Lemma 43. Let $\mathfrak{A} \in Ba$, $n \in \omega$. Then $P^2_n(\mathfrak{A}) \cap S^2_n(\mathfrak{A}) = \emptyset$.

Now it follows from Lemmas 41, 42 and 43

Corollary 44. Let $\mathfrak{A} \in$ Ba, $n \in \omega$. Then $N_n^2(\mathfrak{A}) = \emptyset$.

Using Corollaries 36, 40 and 44 we obtain

Theorem 45. $\mathcal{M}_1$ is dense for $Th_1(Ba)$.

## § 4.4. The general case

In §§ 4.1, 4.2, and 4.3 we described dense systems for $Th_0(Ba)$, $Th_\omega(Ba)$, and $Th_1(Ba)$, respectively, assuming the generalized continuum hypothesis (it was needed only for $Th_\omega(Ba)$ in the weaker form $\forall n < \omega(2^{\omega_n} < \omega_\omega)$). Here we show how we can get dense systems for each $\alpha \in On$ even without assuming the generalized continuum hypothesis.

We set

$$Th^0 = .Th_0(\mathcal{M}_0)., \quad Th^\omega = .Th_\omega(\mathcal{M}_\omega)., \quad Th^1 = .Th_1(\mathcal{M}_1).,$$

where $\mathcal{M}_0, \mathcal{M}_\omega, \mathcal{M}_1$ are the systems described in §§ 4.1, 4.2, and 4.3, respectively.

From now on we are working in any fixed model of ZFC.

In order to show that $\mathcal{M}_0$ is dense for $Th_0(Ba)$ in § 4.1 we did not need anything arising from the generalized continuum hypothesis; thus $\mathcal{M}_0$ is dense here, too, that means $Th^0 = .Th_0(Ba)..$

Let $\alpha \in On$, $\alpha \neq 0$. $\alpha$ is of the first kind if $\forall \beta < \alpha(2^{\omega_\beta} < \omega_\alpha)$; otherwise $\alpha$ is of the second kind.

Let $\alpha \in On$ be of the first kind. Let $Term^\alpha$ be the least class K with:

(i) $2, \eta_0, \eta \in K$,
(ii) if $t_1, t_2 \in K$, then $p_1(t_1)$, $p_2(t_1)$, $t_1 \times t_2$, $p_3(t_1)$, $p_4(t_1, t_2)$, $p_5(t_1, t_2) \in K$ (not necessarily $t_1 \neq t_2$).

Let $f_\alpha : 2^{<\omega} \longrightarrow pow(\omega_\alpha)$ be a function with

(i) $f_\alpha(\varphi) = \omega_\alpha$,

(ii) if $s \in 2^{<\omega}$, then $f_\alpha(s^\frown 0) \cup f_\alpha(s^\frown 1) = f_\alpha(s)$ and $f_\alpha(s^\frown 0) \cap f_\alpha(s^\frown 1) = \emptyset$,
(iii) if $s \in 2^{<\omega}$, then $\mathrm{card}(f_\alpha(s)) = \omega_\alpha$ and $\mathrm{card}(f_\alpha(s) \cap \omega_0) = \omega_0$.

Let $B(f_\alpha) = \mathrm{Pow}(\omega_\alpha; \{f_\alpha(s): s \in 2^{<\omega}\})$, that means, $B(f_\alpha)$ is the subalgebra of $\mathrm{Pow}(\omega_\alpha)$ generated by $\{f_\alpha(s): s \in 2^{<\omega}\}$ and the finite subsets of $\omega_\alpha$.

Let the function $\mathcal{B}^\alpha$ from $\mathrm{Term}^\alpha$ into Ba be defined as follows:

$$\mathcal{B}^\alpha(2) = B(2),$$

$$\mathcal{B}^\alpha(\eta_0) = B(\eta_0),$$

$$\mathcal{B}^\alpha(\eta) = \mathcal{J}(\mathcal{H}_\alpha),$$

$$\mathcal{B}^\alpha(t_1 \times t_2) = \mathcal{B}^\alpha(t_1) \times \mathcal{B}^\alpha(t_2),$$

$$\mathcal{B}^\alpha(p_1(t_1)) = \prod_\omega{}^{w} \mathcal{B}^\alpha(t_1),$$

$$\mathcal{B}^\alpha(p_2(t_1)) = \prod^{\mathcal{J}_1(\mathcal{H}_\alpha)} \mathcal{B}^\alpha(t_1),$$

$$\mathcal{B}^\alpha(p_3(t_1)) = \prod^{\mathcal{J}_1(\mathcal{H}_0)} \mathcal{B}^\alpha(t_1),$$

$$\mathcal{B}^\alpha(p_4(t_1,t_2)) = \prod^{w}\{\mathcal{L}_\beta: \beta < \omega_\alpha\}$$

with

$$\mathcal{L}_\beta = \begin{cases} \mathcal{B}^\alpha(t_1) & \text{for } \beta < \omega, \\ \mathcal{B}^\alpha(t_2) & \text{for } \omega \le \beta < \omega_\alpha, \end{cases}$$

$$\mathcal{B}^\alpha(p_5(t_1,t_2)) = \prod^{B(f_\alpha)}\{\mathcal{L}_\beta: \beta < \omega_\alpha\}$$

with

$$\mathcal{L}_\beta = \begin{cases} \mathcal{B}^\alpha(t_1) & \text{for } \beta < \omega, \\ \mathcal{B}^\alpha(t_2) & \text{for } \omega \le \beta < \omega_\alpha. \end{cases}$$

Then $\langle \mathrm{Term}^\alpha, \mathcal{B}^\alpha \rangle$ is a pre-density pair for $\Omega_\alpha$. We set

$$\mathcal{M}_\alpha = \{\mathcal{B}^\alpha(t): t \in \mathrm{Term}^\alpha\}.$$

Then it can be shown completely analogously as in § 4.2 that $\mathcal{M}_\alpha$ is dense for $Th_\alpha(Ba)$.

Let $\alpha \in On$, $\alpha \neq 0$, be of the second kind. Let $Term^\alpha$ be the least class K with:

(i) $2, \eta_0, \eta \in K$,
(ii) if $t_1, t_2 \in K$, then $p_1(t_1)$, $p_2(t_1)$, $t_1 \times t_2$, $p_3(t_1)$, $p_4(t_1,t_2)$, $p_5(t_1,t_2) \in K$ (not necessarily $t_1 \neq t_2$),
(iii) if $m_0 > 0$, $t_0,\dots,t_{m_0-1} \in K$, $X \in (pow(\{t_i : i < m_0\}) \setminus \{\emptyset\})^{<\omega}$, $n \leq lh(X)$, then $p_6(t_0,\dots,t_{m_0-1};n,X)$, $p_7(t_0,\dots,t_{m_0-1};n,X) \in K$.

Let $f_\alpha : 2^{<\omega} \longrightarrow pow(\omega_\alpha)$ be a function with

(i) $f_\alpha(\varnothing) = \omega_\alpha$,
(ii) if $s \in 2^{<\omega}$, then $f_\alpha(s^\frown 0) \cup f_\alpha(s^\frown 1) = f_\alpha(s)$ and $f_\alpha(s^\frown 0) \cap f_\alpha(s^\frown 1) = \emptyset$,
(iii) if $s \in 2^{<\omega}$, then $card(f_\alpha(s)) = \omega_\alpha$ and $card(f_\alpha(s) \cap \omega) = \omega_0$.

Let $B(f_\alpha) = Pow(\omega_\alpha; \{f_\alpha(s) : s \in 2^{<\omega}\})$. Let $\beta$ be the least ordinal with $2^{\omega_\beta} \geq \omega_\alpha$. Then $2^{\underset{\smile}{\omega_\beta}} < \omega_\alpha$.

Let the function $\mathcal{B}^\alpha$ from $Term^\alpha$ into Ba be defined as follows:

$$\mathcal{B}^\alpha(2) = B(2),$$

$$\mathcal{B}^\alpha(\eta_0) = B(\eta_0),$$

$$\mathcal{B}^\alpha(\eta) = \mathcal{I}(\mathcal{H}_\alpha),$$

$$\mathcal{B}^\alpha(t_1 \times t_2) = \mathcal{B}^\alpha(t_1) \times \mathcal{B}^\alpha(t_2),$$

$$\mathcal{B}^\alpha(p_1(t_1)) = \prod_\omega{}^{w} \mathcal{B}^\alpha(t_1),$$

$$\mathcal{B}^\alpha(p_2(t_1)) = \prod{}^{\mathcal{I}_1(\mathcal{H}_\alpha)} \mathcal{B}^\alpha(t_1),$$

$$\mathcal{B}^\alpha(p_3(t_1)) = \prod{}^{\mathcal{I}_1(\mathcal{H}_0)} \mathcal{B}^\alpha(t_1),$$

$$\mathcal{B}^\alpha(p_4(t_1,t_2)) = \prod{}^{w} \{\mathcal{L}_\gamma : \gamma < \omega_\alpha\}$$

with

$$\mathcal{L}_\gamma = \begin{cases} \mathcal{B}^\alpha(t_1) \text{ for } \gamma < \omega, \\ \mathcal{B}^\alpha(t_2) \text{ for } \omega \leq \gamma < \omega_\alpha, \end{cases}$$

$$\mathcal{B}^\alpha(p_5(t_1,t_2)) = \prod{}^{B(f_\alpha)}\{\mathcal{L}_\gamma : \gamma < \omega_\alpha\}$$

with

$$\mathcal{L}_\gamma = \begin{cases} \mathcal{B}^\alpha(t_1) \text{ for } \gamma < \omega, \\ \mathcal{B}^\alpha(t_2) \text{ for } \omega \leq \gamma < \omega_\alpha, \end{cases}$$

$$\mathcal{B}^\alpha(p_6(t_0,\dots,t_{m_0-1};n,X)) = \prod{}^{\mathcal{L}}\{\mathcal{L}_\gamma : \gamma < 2^{\omega_\alpha}\}$$

with

$$\mathcal{L} = B_{\omega_\alpha}[m_0,n,lh(X)],$$

$$\mathcal{L}_{\delta\cdot\omega+i} = \begin{cases} \mathcal{B}^\alpha(t_k) \text{ if } i \equiv k(\mathrm{mod}(m_0+lh(X)) \text{ and } k < m_0, \\ \prod\{\mathcal{B}^\alpha(t) : t \in X(k-m_0)\} \text{ if } i \equiv k(\mathrm{mod}(m_0+lh(X)) \text{ and} \\ \qquad m_0 \leq k < m_0 + lh(X), \end{cases}$$

$$\mathcal{B}^\alpha(p_7(t_0,\dots,t_{m_0-1};n,X)) = \prod{}^{\mathcal{L}}\{\mathcal{L}_\gamma : \gamma < 2^{\omega_\alpha}\}$$

with

$$\mathcal{L} = B^+_{\omega_\alpha}[m_0,n,lh(X)],$$

$$\mathcal{L}_{\delta\cdot\omega+i} = \begin{cases} \mathcal{B}^\alpha(t_k) \text{ if } i \equiv k(\mathrm{mod}(m_0+lh(X)) \text{ and } k < m_0, \\ \prod\{\mathcal{B}^\alpha(t) : t \in X(k-m_0)\} \text{ if } i \equiv k(\mathrm{mod}(m_0+lh(X)) \text{ and} \\ \qquad m_0 \leq k < m_0 + lh(X). \end{cases}$$

Then $\langle \mathrm{Term}^\alpha, \mathcal{B}^\alpha \rangle$ is a pre-density pair for $\mathbb{Q}_\alpha$. We set

$$\mathcal{M}_\alpha = \{\mathcal{B}^\alpha(t) : t \in \mathrm{Term}^\alpha\}.$$

Then it can be shown completely analogously as in § 4.3 that $\mathcal{M}_\alpha$ is dense for $\mathrm{Th}_\alpha(\mathrm{Ba})$.

Thus we get

Theorem 46. Let $\alpha \in On$. Then
(i) if $\alpha = 0$, then $.Th_{\alpha}(Ba). = Th^{0}$,
(ii) if $\alpha$ is of the first kind, then $.Th_{\alpha}(Ba). = Th^{\omega}$,
(iii) if $\alpha$ is of the second kind, then $.Th_{\alpha}(Ba). = Th^{1}$.

## § 5. The decidability of $Th_{\alpha}(Ba)$

We start by showing that for each $\mathfrak{A} \in \mathcal{M}_1$, $Th_1(\mathfrak{A})$ is decidable. This is used to show the decidability of $Th_1(Ba)$. Then the decidability of $Th_1(Ba)$ implies the decidability of $Th_{\alpha}(Ba)$ for any $\alpha \in On$. It follows from Lemma 7 and Theorem 3 of Chapter 1

Lemma 47. $Th_1(B(2))$, $Th_1(B(\eta_0))$ and $Th_1(B(\eta))$ are decidable.

Next we show that all products used to construct $\mathcal{M}_1$ preserve decidability. To do this we proceed as follows:

We enlarge the language $L_{Ba}$ in a suitable way and in this enlarged language we describe an axiomatizable and complete theory T, the theory of the product we have in mind. This theory T depends on the kind of product we have in mind and on the theories of those Boolean algebras which are factors of the product. We show that T has the following property:

(3) for each $\varphi \in Sent(L_{Ba}(Q_1))$, $T \models \varphi$ or $T \models \neg\varphi$.

Then $\{\varphi \in Sent(L_{Ba}(Q_1)) : T \models \varphi\}$ is a complete and axiomatizable theory of an axiomatizable logic and thus decidable.

We make the following

Convention. Let L' be any language with $L' \supseteq L_{Ba}$. For any structure $\mathfrak{A}$ for L', $\tilde{\mathfrak{A}}$ denotes the reduct of $\mathfrak{A}$ to $L_{Ba}$.

To prove (3) we show that player II has a winning strategy in the game $H^2(n, \tilde{\mathfrak{A}}, \tilde{\mathfrak{B}}, 1, 1)$ for any $\mathfrak{A}, \mathfrak{B} \in Mod(T)$ and any $n \in \omega$.

Lemma 48. Let $\mathfrak{C}, \mathfrak{D} \in Ba$ such that $Th_1(\mathfrak{C})$ and $Th_1(\mathfrak{D})$ are decidable. Then $Th_1(\mathfrak{C} \times \mathfrak{D})$ is decidable, too.

Remark. This lemma follows immediately from results of [Wojciechowska, 1969]. For the proof given here we use ideas which are needed for the other products, too. It is easier to understand the main ideas here than in case of more complicated products.

Proof. Let L' be the language $L_{Ba}$ enlarged by two unary predicate symbols $P_1$ and $P_2$. Let $T_0$ be the theory given by the following axioms:

(i) the axioms of Th(Ba),

(ii) $\exists x(P_1(x) \wedge P_2(\overline{\overline{x}}))$,

(iii) $\forall xy((P_1(x) \wedge P_1(y) \rightarrow x = y) \wedge (P_2(x) \wedge P_2(y) \rightarrow x = y))$,

(iv) the theory $Th_1(\mathcal{L})$ relativized to $P_1$,

(v) the theory $Th_1(\vartheta)$ relativized to $P_2$.

Let $n \in \omega$, $\mathfrak{A}, \mathfrak{B} \in Mod(T_0)$. Let $a^+ \in |\mathfrak{A}|$ and $b^+ \in |\mathfrak{B}|$ be such that $\mathfrak{A} \models P_1(a^+)$ and $\mathfrak{B} \models P_1(b^+)$. Then we have

$$\tilde{\mathfrak{A}} = \tilde{\mathfrak{A}}[a^+] \times \tilde{\mathfrak{A}}[\overline{\overline{a^+}}], \quad \tilde{\mathfrak{B}} = \tilde{\mathfrak{B}}[b^+] \times \tilde{\mathfrak{B}}[\overline{\overline{b^+}}],$$

$$\tilde{\mathfrak{A}}[a^+], \tilde{\mathfrak{B}}[b^+] \in Mod(Th_1(\mathcal{L})), \quad \tilde{\mathfrak{A}}[\overline{\overline{a^+}}], \tilde{\mathfrak{B}}[\overline{\overline{b^+}}] \in Mod(Th_1(\vartheta)).$$

Now $Th_1(\mathcal{L})$ and $Th_1(\vartheta)$ are complete theories, thus

$$\tilde{\mathfrak{A}}[a^+] \;{}^1{\equiv}^1\; \tilde{\mathfrak{B}}[b^+], \quad \tilde{\mathfrak{A}}[\overline{\overline{a^+}}] \;{}^1{\equiv}^1\; \tilde{\mathfrak{B}}[\overline{\overline{b^+}}].$$

Now we describe a winning strategy for player II in the game $H^2(n, \tilde{\mathfrak{A}}, \tilde{\mathfrak{B}}, 1, 1)$. We show that player II can ensure that we have for each $k \leq n$:

(4) if $\langle a_k, b_k \rangle$ is the result of the game before the k-th round, then $\tilde{\mathfrak{A}}[a^+ \cap a_k] \;{}^1{\equiv}^1_{n-k}\; \tilde{\mathfrak{B}}[b^+ \cap b_k]$ and $\tilde{\mathfrak{A}}[\overline{\overline{a^+}} \cap a_k] \;{}^1{\equiv}^1_{n-k}\; \tilde{\mathfrak{B}}[\overline{\overline{b^+}} \cap b_k]$.

It is clear that player II wins if he ensures (4).

For $k = 0$ everything is clear as $a_0 = 1^{\tilde{\mathfrak{A}}}$ and $b_0 = 1^{\tilde{\mathfrak{B}}}$.

Let $i < n$ and let the result $\langle a_i, b_i \rangle$ before the i-th round satisfy (4) for $k = i$. We show that player II can play in such a way that also the result $\langle a_{i+1}, b_{i+1}\rangle$ after the i-th round satisfies (4) for $k = i + 1$.

Assume player I decides to play an $\exists$-round. Without loss of generality he chooses the structure $\tilde{\mathfrak{A}}$ and an element $a \in |\tilde{\mathfrak{A}}[\overline{a_i}]|$. Now we have $\tilde{\mathfrak{A}}[a^+ \cap a_i] \; {}^1\equiv^1_{n-i} \; \mathfrak{B}[b^+ \cap b_i]$ and $\tilde{\mathfrak{A}}[\overline{a^+} \cap a_i] \; {}^1\equiv^1_{n-i} \; \tilde{\mathfrak{B}}[\overline{b^+} \cap \overline{b_i}]$, thus there are elements $b^0, b^1 \in |\tilde{\mathfrak{B}}|$ with $b^0 \leq b_i \cap b^+$, $b^1 \leq b_i \cap \overline{b^+}$,

$$\langle \tilde{\mathfrak{A}}[a^+ \cap a_i], a \cap a^+\rangle \; {}^1\equiv^1_{n-i-1} \langle \tilde{\mathfrak{B}}[b^+ \cap b_i], b^0\rangle,$$

$$\langle \tilde{\mathfrak{A}}[\overline{a^+} \cap a_i], a \cap \overline{a^+}\rangle \; {}^1\equiv^1_{n-i-1} \langle \tilde{\mathfrak{B}}[\overline{b^+} \cap b_i], b^1\rangle.$$

Choosing the element $b^0 \cup b^1$ player II ensures that the result $\langle a_{i+1}, b_{i+1}\rangle$ after the i-th round satisfies (4) for $k = i + 1$.

Assume now that player I decides to play a Q-round. Without loss of generality he chooses the structure $\tilde{\mathfrak{B}}$ and a set $B \subseteq |\tilde{\mathfrak{B}}[b_i]|$ with $\mathrm{card}(B) = \omega_1$. Then there is a set $B^* \subseteq B$ with $\mathrm{card}(B^*) = \omega_1$ so that we have for all $b^0, b^1 \in B^*$

$$\langle \tilde{\mathfrak{B}}[b_i \cap b^+], b^0 \cap b^+\rangle \; {}^1\equiv^1_{n-i-1} \langle \tilde{\mathfrak{B}}[b_i \cap b^+], b^1 \cap b^+\rangle$$

and

$$\langle \tilde{\mathfrak{B}}[b_i \cap \overline{b^+}], b^0 \cap \overline{b^+}\rangle \; {}^1\equiv^1_{n-i-1} \langle \tilde{\mathfrak{B}}[b_i \cap \overline{b^+}], b^1 \cap \overline{b^+}\rangle.$$

Let $B^0 = \{b \cap b^+ : b \in B^*\}$, $B^1 = \{b \cap \overline{b^+} : b \in B^*\}$. Then $\mathrm{card}(B^0) = \omega_1$ or $\mathrm{card}(B^1) = \omega_1$; let us assume $\mathrm{card}(B^0) = \omega_1$. Thus there is a set $A \subseteq |\mathfrak{A}[a_i \cap a^+]|$ so that for each $a \in A$ and each $b \in B^0$,

$$\langle \tilde{\mathfrak{A}}[a_i \cap a^+], a\rangle \; {}^1\equiv^1_{n-i-1} \langle \tilde{\mathfrak{B}}[b_i \cap b^+], b\rangle.$$

Furthermore, there is an element $a^*$ with $a^* \leq a_i \cap \overline{a^+}$ so that we have for each $b \in B^1$

$$\langle \tilde{\mathfrak{A}}[a_i \cap \overline{a^+}], a^*\rangle \; {}^1\equiv^1_{n-i-1} \langle \tilde{\mathfrak{B}}[b_i \cap \overline{b^+}], b\rangle.$$

Let $A^* = \{a^* \cup a : a \in A\}$. Then $\mathrm{card}(A^*) = \omega_1$. Choosing the set $A^*$ and choosing for each $a \in A^*$ an element $b \in B^*$ player II ensures that the result $\langle a_{i+1}, b_{i+1}\rangle$ after the i-th round satisfies (4) for $k = i + 1$.

Q.E.D.

Lemma 49. Let $\mathcal{L} \in$ Ba so that $Th_1(\mathcal{L})$ is decidable. Then
(i) $Th_1(\prod^{w}_{\omega} \mathcal{L})$ is decidable,
(ii) $Th_1(\prod^{\mathcal{I}_1(\mathcal{H}_1)} \mathcal{L})$ is decidable.

Proof. Ad (i): Let L' be the language $L_{Ba}$ enlarged by two unary predicate symbols P and S. Let $T_1$ be the theory given by the following axioms:

(i) the axioms of Th(Ba),

(ii) $\forall x(x \neq 0 \rightarrow \exists y(P(y) \wedge x \cap y \neq 0))$,

(iii) $\forall xy(P(x) \wedge P(y) \wedge x \neq y \rightarrow x \cap y = 0)$,

(iv) $\forall xy(S(x) \wedge S(y) \rightarrow S(x \cap y))$,

(v) $\forall xy(S(x) \wedge x \leq y \rightarrow S(y))$,

(vi) $\neg\exists xy(S(x) \wedge S(y) \wedge x \cap y = 0)$,

(vii) $\forall x(S(x) \vee S(\bar{\bar{x}}))$,

(viii) $\forall x(P(x) \rightarrow S(\bar{\bar{x}}))$,

(ix) $\forall x\exists y\forall z(P(z) \rightarrow ((z \leq x \rightarrow z \leq y) \wedge (z \cap x \neq z \rightarrow z \cap y = 0)))$,

(x) for each $\varphi(x) \in Form(L_{Ba}(Q_1))$ and any variables y, z not occuring in $\varphi(x)$,
$\forall x\exists y\forall z(P(z) \rightarrow ((z \cap y = 0 \leftrightarrow \neg\varphi^{z}(x)) \wedge (z \cap y = z \leftrightarrow \varphi^{z}(x))))$,

(xi) for each $\varphi(x) \in Form(L_{Ba}(Q_1))$ with $\neg Q_1 x\varphi(x) \in Th_1(\mathcal{L})$,
$\neg Q_1 x\forall z(P(z) \rightarrow \varphi^{z}(x))$,

(xii) for each $\varphi(x) \in Form(L_{Ba}(Q_1))$ and any variable z not occuring in $\varphi(x)$,
$\forall x(\neg S(x) \rightarrow \exists y(y \leq x \wedge \forall z(P(z) \rightarrow ((\varphi^{z}(x) \rightarrow y \cap z = x \cap z)$
$\wedge (\neg\varphi^{z}(x) \rightarrow y \cap z = 0)))))$,

(xiii) for each $\varphi(x), \psi(x,y) \in Form(L_{Ba}(Q_1))$ with
$\forall x(\varphi(x) \rightarrow \exists y(y \leq x \wedge \psi(x,y)) \in Th_1(\mathcal{L})$,
$\forall x(\neg S(x) \wedge \forall y(P(y) \wedge y \cap x \neq 0 \rightarrow \varphi^{y}(x)) \rightarrow \exists y(y \leq x \wedge$
$\forall z(P(z) \wedge z \cap x \neq 0 \rightarrow \psi^{z}(x,y))))$,

(xiv) the theory $Th_1(\mathcal{L})$ relativized to P.

We set

$$P^* = \{a \in |\prod_{\omega}^{w}\mathcal{L}| : \text{there is an } i \in \omega \text{ with } a(i) = 1 \text{ and } a(k) = 0 \text{ for } k \neq i\},$$

$$S^* = \{a \in |\prod_{\omega}^{w}\mathcal{L}| : \operatorname{card}(\{i \in \omega : a(i) = 1\}) = \omega_0\}.$$

Then

$$\left\langle \prod_{\omega}^{w}\mathcal{L}, P^*, S^* \right\rangle \in \operatorname{Mod}(T_1).$$

Let $\varphi(x) = (x = x)$. Then $\varphi^z(x) = (x \cap z = x \cap z)$ and (xi) implies

$$\text{if } \operatorname{card}(|\mathcal{L}|) < \omega_1, \text{ then } \neg Q_1 x(x = x) \in T_1.$$

Let $\mathfrak{A}, \mathfrak{B} \in \operatorname{Mod}(T_1)$. By (iv), (v), (vi), and (vii) it follows that $S^{\mathfrak{A}}$ and $S^{\mathfrak{B}}$ are ultrafilters. (viii) implies that each $a \in S^{\mathfrak{A}}$ (each $b \in S^{\mathfrak{B}}$) contains infinitely many elements c with $c \in P^{\mathfrak{A}}$ ($c \in P^{\mathfrak{B}}$). For each $k \leq n$ let

$$\operatorname{typ}(k,1,Ba) = \{\Phi_0^k, \ldots, \Phi_{g(k)-1}^k\}.$$

For $j < g(k)$ we set

$$\varphi_j^k = \bigwedge \Phi_j^k.$$

Then $\varphi_j^k \in \operatorname{Form}(L_{Ba}(Q_1))$ since there are only finitely many sentences $\varphi$ with $qr(\varphi) \leq k$ which are different with respect to $Th_1(Ba)$.

Now we describe the winning strategy for player II in the game $H^2(n, \tilde{\mathfrak{A}}, \tilde{\mathfrak{B}}, 1, 1)$. We show that player II can ensure that we have for each $k \leq n$:

If $\langle a_k, b_k \rangle$ is the result of the game before the k-th round, then either

$a_k \in S^{\mathfrak{A}}$ and $b_k \in S^{\mathfrak{B}}$

or

(5) $a_k \notin S^{\mathfrak{A}}$ and $b_k \notin S^{\mathfrak{B}}$ and for each $j < g(n-k)$,
$\mathrm{card}(\{a \in |\tilde{\mathfrak{A}}| : a \in P^{\mathfrak{A}} \text{ and } \tilde{\mathfrak{A}}[a \cap a_k] \models \varphi_j^{n-k}\}) \geq g(n)^{2(n-k)}$
and $\mathrm{card}(\{b \in |\tilde{\mathfrak{B}}| : b \in P^{\mathfrak{B}} \text{ and } \tilde{\mathfrak{B}}[b \cap b_k] \models \varphi_j^{n-k}\}) \geq g(n)^{2(n-k)}$
or
$\mathrm{card}(\{a \in |\tilde{\mathfrak{A}}| : a \in P^{\mathfrak{A}} \text{ and } \tilde{\mathfrak{A}}[a \cap a_k] \models \varphi_j^{n-k}\}) =$
$\mathrm{card}(\{b \in |\tilde{\mathfrak{B}}| : b \in P^{\mathfrak{B}} \text{ and } \tilde{\mathfrak{B}}[b \cap b_k] \models \varphi_j^{n-k}\})$.

It is clear that player II wins if he ensures (5).

For $k = 0$ everything is clear.

Let $i < n$ and let the result $\langle a_i, b_i \rangle$ before the i-th round satisfy (5) for $k = i$.

First assume that $a_i \in S^{\mathfrak{A}}$ (and thus also $b_i \in S^{\mathfrak{B}}$).

Assume player I decides to play an $\exists$-round. Without loss of generality he chooses the structure $\tilde{\mathfrak{A}}$ and an element $a \in |\tilde{\mathfrak{A}}[a_i]|$ with $a \notin S^{\mathfrak{A}}$ (if $a \in S^{\mathfrak{A}}$, we simply consider the element $\bar{\bar{a}} \cap a_i$). Now $b_i \in S^{\mathfrak{B}}$, thus there are elements $b^0, \ldots, b^{(g(n))^{2(n-i)}-1}$ with

if $j,k < (g(n))^{2(n-i)}$, then $b^j, b^k \leq b_i$, $b^j, b^k \in P^{\mathfrak{B}}$ and $b^j \cap b^k = 0$ for $j \neq k$.

For $j < g(n-i-1)$ we set

$$k_j = \min\{(g(n))^{2(n-i-1)}, \mathrm{card}(\{c \in |\tilde{\mathfrak{A}}| : c \in P^{\mathfrak{A}} \text{ and } \mathfrak{A}[a \cap c] \models \varphi_j^{n-i-1}\})\}.$$

For each $j < (g(n))^{2(n-i)}$ let $d_j \in |\tilde{\mathfrak{B}}|$ be any element with the following properties:

$d_j \leq b^j$,

if $j = l_0 \cdot (g(n))^{2(n-i-1)} + l_1$ with $l_1 < (g(n))^{2(n-i-1)}$,
then $d_j = 0$ for $l_0 \geq g(n-i-1)$ or $l_1 \geq k_{l_0}$,
and $d_j$ is such that $\mathcal{Z}[d_j] \models \varphi^{n-i-1}_{l_0}$ for $l_0 < g(n-i-1)$ and $l_1 < k_{l_0}$.

(xiv) implies the existence of such an element $d_j$. Let

$$b = \bigcup[d_j : j < (g(n))^{(n-i-1)}].$$

Choosing the element b player II ensures that the result $\langle a_{i+1}, b_{i+1} \rangle$ after the i-th round satisfies (5) for $k = i + 1$.

Assume player I decides to play a Q-round. Without loss of generality he chooses the structure $\tilde{\mathcal{Z}}$ and a set $B \subseteq |\tilde{\mathcal{Z}}[b_i]|$ with $\mathrm{card}(B) = \omega_1$. Without loss of generality we assume that there is a set $B^* \subseteq B$ with $\mathrm{card}(B^*) = \omega_1$ and $B^* \cap S^{\mathcal{Z}} = \emptyset$.

For each $b \in B^*$ and each $j < g(n-i-1)$ let $b(j)$ be the element $b^* \in |\tilde{\mathcal{Z}}|$ with:

$b^* \leq b$ and for each $d \in P^{\mathcal{Z}}$ we have $b^* \cap d = b \cap d$ for $\tilde{\mathcal{Z}}[b \cap d] \models \varphi^{n-i-1}_j$ and $b^* \cap d = 0$ for $\tilde{\mathcal{Z}}[b \cap d] \models \neg\varphi^{n-i-1}_j$.

(xii) implies the existence of $b(j)$.

For each $j < g(n-i-1)$ let

$$B_j = \{b(j) : b \in B^*\}.$$

Then there is a $j^* < g(n-i-1)$ so that $\mathrm{card}(B_{j^*}) = \omega_1$. (xi) implies $\mathcal{L} \models Q_1 x (\varphi^{n-i-1}_j)^x$, where x is a variable not occuring in $\varphi^{n-i-1}_j$.
$a_i \in S^{\mathfrak{A}}$ implies that there are elements $a^0, \ldots, a^{(g(n))2^{(n-i)}-1}$ with:

if $j,k < (g(n))^{2(n-i)}$, then $a^j, a^k \leq a_i$, $a^j, a^k \in P^{\mathfrak{A}}$, and $a^j \cap a^k = 0$ for $j \neq k$.

Let $b^* \in B^*$ be such that $b^*(j^*) \neq 0$. For each $j < g(n-i-1)$ let

$$k_j = \min\{(g(n))^{2(n-i-1)}, \mathrm{card}(\{d \in |\tilde{\mathcal{Z}}| : d \in P^{\mathcal{Z}} \text{ and } \tilde{\mathcal{Z}}[b^* \cap d] \models \varphi^{n-i-1}_j\})\}.$$

For each $j < (g(n))^{2(n-i)}$ let $c_j \in |\tilde{\mathfrak{A}}|$ be any element with the following properties:

$c_j \leq a^j$,

if $j = l_0 \cdot (g(n))^{2(n-i-1)} + l_1$ with $l_1 < (g(n))^{2(n-i-1)}$, then $c_j = 0$ for $l_0 \geq g(n-i-1)$ or $l_1 \geq k_{l_0}$, and $c_j$ is such that $\tilde{\mathfrak{B}}[c_j] \models \varphi^{n-i-1}_{l_0}$ for $l_0 < g(n-i-1)$ and $l_1 < k_{l_0}$.

Let $A^* \subseteq |\tilde{\mathfrak{A}}|$ be such that $\mathrm{card}(A^*) = \omega_1$ and for each $a \in A^*$ we have $a \leq a^{j^* \cdot (g(n))^{2(n-i-1)}}$ and $\tilde{\mathfrak{A}}[a] \models \varphi^{n-i-1}_{j^*}$. Let

$$a^* = \bigcup[c_j\colon j < (g(n))^{2(n-i)} \text{ and } j \neq j^* \cdot (g(n))^{2(n-i-1)}].$$

We set $A = \{a \cup a^*\colon a \in A^*\}$. Choosing the set A and for each $a \in A$ choosing the element $b^*$ player II ensures that also the result $\langle a_{i+1}, b_{i+1} \rangle$ after the i-th round satisfies (5) for $k = i + 1$.

Assume now $a_i \notin S^{\mathfrak{A}}$ (and thus also $b_i \notin S^{\mathfrak{B}}$).

Assume player I decides to play an $\exists$-round. Without loss of generality he chooses the structure $\tilde{\mathfrak{A}}$ and an element $a \in |\tilde{\mathfrak{A}}[a_i]|$. For each $j < g(n-i)$ we set

$$A^j = [c \cap a_i\colon c \in P^{\mathfrak{A}} \text{ and } \tilde{\mathfrak{A}}[c \cap a_i] \models \varphi^{n-i}_j\ ,$$

$$B^j = [d \cap b_i\colon d \in P^{\mathfrak{B}} \text{ and } \tilde{\mathfrak{B}}[d \cap b_i] \models \varphi^{n-i}_j\ ,$$

$$a^j = \bigcup[a\colon a \in A^j], \qquad b^j = \bigcup[b\colon b \in B^j].$$

(x) implies the existence of $a^j$ and $b^j$.

We consider now some fixed $j^* < g(n-i)$.

Assume first $\mathrm{card}(A^{j^*}) < (g(n))^{2(n-i)}$ and thus $\mathrm{card}(A^{j^*}) = \mathrm{card}(B^{j^*})$. Let $A^{j^*} = \{a^{j^*}_0, \ldots, a^{j^*}_{m_0-1}\}$, $B^{j^*} = \{b^{j^*}_0, \ldots, b^{j^*}_{m_0-1}\}$. For each $k < m_0$ there is a $d_k \in |\tilde{\mathfrak{B}}[b^{j^*}_k]|$ with $\langle \tilde{\mathfrak{A}}[a^{j^*}_k], a^{j^*}_k \cap a \rangle \ {}^1{\equiv}^1_{n-i-1} \langle \tilde{\mathfrak{B}}[b^{j^*}_k], d_k \rangle$. We set

$$d^{j^*} = \bigcup\{d_k\colon k < m_0\}.$$

Assume now that $\mathrm{card}(A^{j^*}) \geq (g(n))^{2(n-i)}$. For each pair $(j_0,j_1)$ with $j_0,j_1 < g(n-i-1)$ let

$$k(j_0,j_1) = \min\{(g(n))^{2(n-i-1)}, \mathrm{card}(\{c \in A^{j^*} : \tilde{\mathfrak{A}}[a \cap c] \models \varphi^{n-i-1}_{j_0} \text{ and } \tilde{\mathfrak{A}}[\bar{\bar{a}} \cap c] \models \varphi^{n-i-1}_{j_1}\}\}.$$

$\mathrm{card}(A^{j^*}) \geq (g(n))^{2(n-i)}$ implies that there is a pair $(k_0,k_1)$ with

$$\mathrm{card}(\{c \in A^{j^*} : \tilde{\mathfrak{A}}[a \cap c] \models \varphi^{n-i-1}_{k_0} \text{ and } \tilde{\mathfrak{A}}[\bar{\bar{a}} \cap c] \models \varphi^{n-i-1}_{k_1}\}) \geq (g(n))^{2(n-i-1)}.$$

Let $\{D(j_0,j_1): j_0,j_1 < g(n-i-1)\}$ be a sytem of disjoint subsets of $B^{j^*}$ with

$$\bigcup\{D(j_0,j_1): j_0,j_1 < g(n-i-1)\} = B^{j^*},$$

$$\text{if } (j_0,j_1) \neq (k_0,k_1), \text{ then } \mathrm{card}(D(j_0,j_1)) = k(j_0,j_1).$$

For each pair $(j_0,j_1)$ with $j_0,j_1 < g(n-i-1)$ we set

$$d_{j_0,j_1} = \bigcup D(j_0,j_1).$$

(xiii) implies that for any pair $(j_0,j_1)$ with $j_0,j_1 < g(n-i-1)$ there exists a $b_{j_0,j_1} \leq d_{j_0,j_1}$ so that for any $d \in D(j_0,j_1)$ the following holds true:

$$\tilde{\mathfrak{L}}[d \cap b_{j_0,j_1}] \models \varphi^{n-i-1}_{j_0} \quad \text{and} \quad \tilde{\mathfrak{L}}[d \cap \overline{b_{j_0,j_1}}] \models \varphi^{n-i-1}_{j_1}.$$

We set

$$d^{j^*} = \bigcup[b_{j_0,j_1} : j_0,j_1 < g(n-i-1)].$$

Dealing such as for $j^*$ for each $j < g(n-i)$ we get an element $d^j$ for each $j < g(n-i)$. Let

$$b = \bigcup[d^j : j < g(n-i)].$$

Choosing the element b player II ensures that also the result $\langle a_{i+1}, b_{i+1}\rangle$ after the i-th round satisfies (5) for $k = i + 1$.

Assume now that player I decides to play a Q-round. Without loss of generality he chooses the structure $\mathfrak{B}$ and a set $B \subseteq |\tilde{\mathfrak{B}}[b_i]|$ with $card(B) = \omega_1$.

For each $b \in B$ and each $(j,k,l)$ with $j < g(n-i)$, $k,l < g(n-i-1)$ let

$$B_{j,k,l}(b) = \{d \in |\tilde{\mathfrak{B}}|: d \in P^{\mathfrak{B}}, \tilde{\mathfrak{B}}[d \cap b_i] \models \varphi_j^{n-i},$$
$$\tilde{\mathfrak{B}}[d \cap b] \models \varphi_k^{n-i-1} \text{ and } \tilde{\mathfrak{B}}[d \cap \bar{b} \cap b_i] \models \varphi_l^{n-i-1}\}.$$

Let $B^* \subseteq B$ be such that $card(B^*) = \omega_1$, and for each $(j,k,l)$ with $j < g(n-i)$, $k,l < g(n-i-1)$ the following holds true:

if $b^0, b^1 \in B^*$, then
$card(B_{j,k,l}(b^0)) = card(B_{j,k,l}(b^1))$
or
$card(B_{j,k,l}(b^0)), card(B_{j,k,l}(b^1)) \geq (g(n))^{2(n-i-1)}$.

(xi) implies that there are $j_0 < g(n-i)$ and $k_0, l_0 < g(n-i-1)$ so that for each $b \in B^*$,

$$B_{j_0,k_0,l_0}(b) \neq \emptyset$$

and

$$\mathcal{L} \models \forall x((\varphi_{j_0}^{n-i})^x \rightarrow Qy(y \leq x \wedge (\varphi_{k_0}^{n-i-1})^y \wedge (\varphi_{l_0}^{n-i-1})^{x \cap \bar{y}})).$$

Let $a^* \in |\tilde{\mathfrak{A}}[a_i]|$ be such that for each $j < g(n-i)$ and each $k,l < g(n-i-1)$, the following holds true:

For each $b \in B^*$,
$card(B_{j,k,l}(b)) = card(\{c \in |\tilde{\mathfrak{A}}[a_j]|: c \in P^{\mathfrak{A}}, \tilde{\mathfrak{A}}[c \cap a_i] \models \varphi_j^{n-i},$
$\tilde{\mathfrak{A}}[c \cap a^*] \models \varphi_k^{n-i-1} \text{ and } \tilde{\mathfrak{A}}[c \cap \bar{a}^* \cap a_i] \models \varphi_l^{n-i-1}\})$
or
$card(B_{j,k,l}(b)) \geq (g(n))^{2(n-i-1)}$ and $card(\{c \in |\mathfrak{A}[a_i]|: c \in P^{\mathfrak{A}},$
$\tilde{\mathfrak{A}}[c \cap a_i] \models \varphi_j^{n-i}, \tilde{\mathfrak{A}}[c \cap a^*] \models \varphi_k^{n-i-1} \text{ and } \tilde{\mathfrak{A}}[c \cap \bar{a}^* \cap a_i] \models \varphi_l^{n-i-1}\})$
$\geq (g(n))^{2(n-i-1)}$.

Let $c^* \in |\tilde{\mathfrak{A}}|$ with $c^* \in P^{\mathfrak{A}}$ and $\tilde{\mathfrak{A}}[c^* \cap a^*] \models \varphi_{j_0}^{n-i}$. Then there i a set $A^* \subseteq |\tilde{\mathfrak{A}}|$ with $\mathrm{card}(A^*) = \omega_1$ so that for each $a \in A^*$,

$$a \leq c^* \cap a^*, \quad \tilde{\mathfrak{A}}[a] \models \varphi^{n-i-1}_{k_0} \quad \text{and} \quad \tilde{\mathfrak{A}}[\bar{\bar{a}} \cap c^* \cap a^*] \models \varphi^{n-i-1}_{l_0}.$$

Let $A = \{(a^* \cap \bar{\bar{c^*}}) \cup a\colon\ a \in A^*\}$. Choosing the set A and for each $a \in A$ choosing an element from $B^*$ player II ensures that also the result $\langle a_{i+1}, b_{i+1}\rangle$ after the i-th round satisfies (5) for $k = i + 1$.

Thus the proof of (i) is finished.

Ad (ii): Let $T_2$ be the theory arising from $T_1$ by replacing the axioms (iv) and (vi) by

(iv') $\forall x(P(x) \rightarrow \neg S(x))$,

(vi') $\forall xyz(x = y \cup z \wedge S(x) \rightarrow S(y) \vee S(z))$
$\wedge \forall x(S(x) \rightarrow \exists yz(y \cap z = 0 \wedge y \cup z \leq x \wedge S(y) \wedge S(z)))$,

respectively. We set

$$P^* = \{a \in |\prod^{\mathcal{I}_1(\mathcal{R}^+)} \mathcal{L}|\colon \text{there is an } \alpha \in R^+ \text{ with } a(\alpha) = 1 \text{ and } a(\beta) = 0 \text{ for } \beta \neq \alpha\},$$

$$S^* = \{a \in |\prod^{\mathcal{I}_1(\mathcal{R}^+)} \mathcal{L}|\colon \mathrm{card}(\{\alpha \in R^+\colon a(\alpha) = 1\}) = \omega_0\ .$$

Then, of course,

$$\langle \prod^{\mathcal{I}_1(\mathcal{R}^+)} \mathcal{L}, P^*, S^*\rangle \in \mathrm{Mod}(T_2).$$

Let $\mathfrak{A}, \mathfrak{B} \in \mathrm{Mod}(T_2)$. Similar to (i) it can be shown that player II can play in such a way in the game $H^2(n, \tilde{\mathfrak{A}}, \tilde{\mathfrak{B}}, 1, 1)$ that (5) is satisfied for each $k \leq n$.

Q.E.D.

Lemma 50. Let $\mathcal{L}, \vartheta \in \mathrm{Ba}$ be such that $\mathrm{Th}_1(\mathcal{L})$ and $\mathrm{Th}_1(\vartheta)$ are decidable. Let $\mathcal{L}_\alpha$ $(\alpha < \omega_1)$ be defined as follows:

$$\mathcal{L}_\alpha = \begin{cases} \mathcal{L} \text{ for } \alpha < \omega, \\ \vartheta \text{ for } \omega \leq \alpha < \omega_1. \end{cases}$$

Then

(i) $Th_1(\prod^{w}\{\mathfrak{L}_\alpha : \alpha < \omega_1\})$ is decidable,

(ii) $Th_1(\prod^{B(f_1)}\{\mathfrak{L}_\alpha : \alpha < \omega_1\})$ is decidable.

Proof. Ad (i): Let L' be the language $L_{Ba}$ enlarged by three unary predicate symbols $P_1$, $P_2$ and S. Let $T_3$ be the theory given by the following axioms:

(i) the axioms of Th(Ba),

(ii) $\forall x(x \neq 0 \rightarrow \exists y(P_1(y) \vee P_2(y)) \wedge x \cap y \neq 0))$,

(iii) $\forall xy((P_1(x) \vee P_2(x)) \wedge (P_1(y) \vee P_2(y)) \wedge x \neq y \rightarrow x \cap y = 0)$,

(iv) $\forall xy(S(x) \wedge S(y) \rightarrow S(x \cap y))$,

(v) $\forall xy(S(x) \wedge x \leq y \rightarrow S(y))$,

(vi) $\neg\exists xy(S(x) \wedge S(y) \wedge x \cap y = 0)$,

(vii) $\forall x(S(x) \vee S(\bar{x}))$,

(viii) $\forall x(P_1(x) \vee P_2(x) \rightarrow S(\bar{x}))$,

(ix) $\forall x(S(x) \rightarrow Q_1 y(y \leq x \wedge P_2(y)))$,

(x) $\forall x\exists y\forall z((P_1(z) \vee P_2(z)) \rightarrow ((z \leq x \rightarrow z \leq y) \wedge (z \cap x \neq z \rightarrow z \cap y = 0)))$,

(xi) for each $\varphi(x) \in Form(L_{Ba}(Q_1))$ and any variables y,z not occuring in $\varphi(x)$,
$\forall x(\neg S(x) \rightarrow \exists y\forall z(P_1(z) \vee P_2(z) \rightarrow ((z \cap y = 0 \leftrightarrow \neg\varphi^z(x)) \wedge (z \cap y = z \leftrightarrow \varphi^z(x)))))$,

(xii) $\forall x(\neg S(x) \rightarrow \exists y_1 y_2(y_1 \cap y_2 = 0 \wedge \forall z((P_1(z) \rightarrow z \cap y_1 = z \cap x) \wedge (P_2(z) \rightarrow z \cap y_2 = z \cap x))))$,

(xiii) for each $\varphi(x) \in Form(L_{Ba}(Q_1))$ with $\neg Q_1 x\varphi(x) \in Th_1(\mathfrak{L})$,
$\neg Q_1 x\forall z((P_1(z) \rightarrow \varphi^z(x)) \wedge (P_2(z) \rightarrow x \cap z = 0))$,

(xiv) for each $\varphi(x) \in Form(L_{Ba}(Q_1))$ and any variable z not occurring in $\varphi(x)$,
$\forall x(\neg S(x) \rightarrow \exists y(y \leq x \wedge \forall z(P_1(z) \vee P_2(z) \rightarrow ((\varphi^z(x) \rightarrow y \cap z = x \cap z) \wedge (\neg\varphi^z(x) \rightarrow y \cap z = 0)))))$,

(xv) for any $\varphi(x), \psi(x,y) \in Form(L_{Ba}(Q_1))$ with
$\forall x(\varphi(x) \rightarrow \exists y(y \leq x \wedge \psi(x,y)) \in Th_1(\mathcal{L})$,
$\forall x(\neg S(x) \wedge \forall y(P_1(y) \wedge y \cap x \neq 0 \rightarrow \varphi^y(x)) \rightarrow \exists y(y \leq x \wedge \forall z(P_1(z) \wedge z \cap x \neq 0 \rightarrow \psi^z(x,y))))$,

(xvi) for any $\varphi(x), \psi(x,y) \in Form(L_{Ba}(Q_1))$ with
$\forall x(\varphi(x) \rightarrow \exists y(y \leq x \wedge \psi(x,y)) \in Th_1(\mathfrak{B})$,
$\forall x(\neg S(x) \wedge \forall y(P_2(y) \wedge y \cap x \neq 0 \rightarrow \varphi^y(x)) \rightarrow \exists y(y \leq x \wedge \forall z(P_2(z) \wedge z \cap x \neq 0 \rightarrow \psi^z(x,y))))$,

(xvii) the theory $Th_1(\mathcal{L})$ relativized to $P_1$,

(xviii) the theory $Th_1(\mathfrak{B})$ relativized to $P_2$.

We set

$$\mathcal{E} = \prod^{w}\{\mathcal{L}_\alpha : \alpha < \omega_1\},$$

$$P_1^* = \{a \in |\mathcal{E}| : \text{there is an } \alpha < \omega \text{ with } a(\alpha) = 1 \text{ and } a(\beta) = 0 \text{ for } \beta \neq \alpha\},$$

$$P_2^* = \{a \in |\mathcal{E}| : \text{there is an } \alpha \text{ with } \omega \leq \alpha < \omega_1,\ a(\alpha) = 1 \text{ and } a(\beta) = 0 \text{ for } \beta \neq \alpha\},$$

$$S^* = \{a \in |\mathcal{E}| : card(\{\alpha < \omega_1 : a(\alpha) = 1\}) \geq \omega_0\}.$$

Then it is easily seen that

$$\langle \mathcal{E}, P_1^*, P_2^*, S^* \rangle \in Mod(T_3).$$

Let $\mathfrak{A}, \mathfrak{B} \in Mod(T_3)$. It follows from (iv), (v), (vi), and (vii) that $S^{\mathfrak{A}}$ and $S^{\mathfrak{B}}$ are ultrafilters. From (xiii) with $\varphi(x) = x = 1$ it follows that $\neg Q_1 x(P_1(x)) \in T_3$.

Let $n \in \omega$. Similar to the proof of Lemma 49 it can be shown that player II can play in such a way in the game $H^2(n, \tilde{\mathfrak{A}}, \tilde{\mathfrak{B}}, 1, 1)$ that for each $k \leq n$ the following holds true:

If $\langle a_k, b_k \rangle$ is the result of the game before the k-th round, then either

$a_k \in S^{\mathfrak{A}}$ and $b_k \in S^{\mathfrak{B}}$

or

(6) $a_k \notin S^{\mathfrak{A}}$ and $b_k \notin S^{\mathfrak{B}}$ and for each $j < g(n-k)$,
$\mathrm{card}(\{a \in |\tilde{\mathfrak{A}}|: a \in P_1^{\mathfrak{A}} \cup P_2^{\mathfrak{A}} \text{ and } \tilde{\mathfrak{A}}[a \cap a_k] \models \varphi_j^{n-k}\}) =$
$\mathrm{card}(\{b \in |\tilde{\mathfrak{B}}|: b \in P_1^{\mathfrak{B}} \cup P_2^{\mathfrak{B}} \text{ and } \tilde{\mathfrak{B}}[b \cap b_k] \models \varphi_j^{n-k}\})$
or
$\mathrm{card}(\{a \in |\tilde{\mathfrak{A}}|: a \in P_1^{\mathfrak{A}} \cup P_2^{\mathfrak{B}} \text{ and } \tilde{\mathfrak{A}}[a \cap a_k] \models \varphi_j^{n-k}\}) \geq (g(n))^{2(n-k)}$ and
$\mathrm{card}(\{b \in |\tilde{\mathfrak{B}}|: b \in P_1^{\mathfrak{B}} \cup P_2^{\mathfrak{B}} \text{ and } \tilde{\mathfrak{B}}[b \cap b_k] \models \varphi_j^{n-k}\}) \geq (g(n))^{2(n-k)}$.

<u>Ad (ii):</u> Let $T_4$ be the theory arising from $T_3$ by replacing (iv) and (vi) by

(iv') $\forall x(P_1(x) \vee P_2(x) \rightarrow \neg S(x))$,

(vi') $\forall xyz(x = y \cup z \wedge S(x) \rightarrow S(y) \vee S(z)) \wedge \forall x(S(x) \rightarrow$
$\exists yz(y \cap z = 0 \wedge y \cup z \leq x \wedge S(y) \wedge S(z)))$,

respectively. We set

$$\mathcal{E} = \prod^{B(f_1)} \{\mathcal{L}_i : i < \omega_1\},$$

$P_1^* = \{a \in |\mathcal{E}|:$ there is an $\alpha < \omega$ with $a(\alpha) = 1$ and $a(\beta) = 0$ for $\beta \neq \alpha\}$,

$P_2^* = \{a \in |\mathcal{E}|:$ there is an $\alpha$ with $\omega \leq \alpha < \omega_1$, $a(\alpha) = 1$ and $a(\beta) = 0$ for $\beta \neq \alpha\}$,

$S^* = \{a \in |\mathcal{E}|: \mathrm{card}(\{\alpha < \omega_1 : a(\alpha) = 1\}) \geq \omega_0\}$.

Then, of course,

$$\langle \mathcal{E}, P_1^*, P_2^*, S^* \rangle \in \mathrm{Mod}(T_4).$$

Let $\mathfrak{A}, \mathfrak{B} \in \mathrm{Mod}(T_4)$. (viii) implies that each $a \in S$ contains infinitely many elements c with $c \in P_1^{\mathfrak{A}}$ and (ix) implies that each $a \in S^{\mathfrak{A}}$

contains $\geq \omega_1$ elements c with $c \in P_2^{\mathfrak{A}}$. Similarly it follows that each $b \in S^{\mathfrak{B}}$ contains infinitely many d with $d \in P_1^{\mathfrak{B}}$ and $\geq \omega_1$ d with $d \in P_2^{\mathfrak{B}}$.

Similar to 5.3 (i) it can be shown that player I can play in such a way in the game $H^2(n,\tilde{\mathfrak{A}},\tilde{\mathfrak{B}},1,1)$ that for each $k \leq n$ (6) is satisfies.

Q.E.D.

Lemma 51. Let $m_0,m_1,m_2 < \omega$ with $m_0 \geq 1$, $m_2 \geq m_1$; for each $j < m_0$ let $\vartheta_j \in$ Ba be such that $Th_1(\vartheta_j)$ is decidable and for each $j < m_2$ let $X(j) \in pow(m_0)$, $X(j) \neq 0$. For $\alpha < \omega^\omega$ let $\mathfrak{L}_\alpha$ be defined as follows:

$$\mathfrak{L}_{\beta\cdot\omega+j} = \begin{cases} \vartheta_k \text{ if } j \equiv k(\mathrm{mod}(m_0+m_2)) \text{ and } 0 \leq k < m_0, \\ \prod\{\mathfrak{L}_i : i \in X(k-m_0)\} \text{ if } j \equiv k(\mathrm{mod}(m_0+m_2)) \text{ and} \\ \qquad m_0 \leq k < m_0 + m_2. \end{cases}$$

Let

$$\mathcal{E}_1 = B_{\omega_0}[m_0,m_1,m_2], \quad \mathcal{E}_2 = B^+_{\omega_0}[m_0,m_1,m_2].$$

Then

(i) $Th_1(\prod^{\mathcal{E}_1}\{\mathfrak{L}_\alpha : \alpha < \omega^\omega\})$ is decidable,

(ii) $Th_1(\prod^{\mathcal{E}_2}\{\mathfrak{L}_\alpha : \alpha < \omega^\omega\})$ is decidable.

Proof. Ad (i): Let L' be the language $L_{Ba}$ enlarged by the following unary predicate symbols:

$$P_0,\ldots,P_{m_0-1},\ R_0,\ldots,R_{m_2-1},\ S_0,\ldots,S_{m_2-1},\ S.$$

Let $T_5$ be the theory given by the following axioms:

(i) the axioms of Th(Ba),

(ii) $\forall x(x \neq 0 \rightarrow \exists y((\bigvee_{j<m_0} P_j(y)) \wedge x \cap y \neq 0))$,

(iii) $\forall xy((\bigvee_{j<m_0} P_j(x)) \wedge (\bigvee_{j<m_0} P_j(y)) \wedge x \neq y \rightarrow x \cap y = 0)$,

(iv) $\forall xy((\bigvee_{j<m_0} P_j(x)) \wedge (\bigvee_{j<m_2} R_j(y)) \rightarrow x \leq y \vee x \cap y = 0)$,

(v) for each $j < m_2$ and each $k \in X(j)$,
$\forall x(R_j(x) \rightarrow (\exists y(y \leq x \wedge P_k(y)) \wedge \forall y_0 y_1(y_0 \cup y_1 \leq x \wedge P_k(y_0) \wedge P_k(y_1) \rightarrow y_0 = y_1)))$,

(vi) for each $j < m_2$ and each $k \in m_0 \setminus X(j)$,
$\forall x(R_j(x) \rightarrow \neg \exists y(y \leq x \wedge P_k(y)))$,

(vii) for each $k < m_2$,
$\forall xy((\bigvee_{j<m_2} R_j(x) \wedge S_k(y) \rightarrow (x \cap y = 0 \vee (x \leq y \wedge R_k(x))))$,

(viii) for each $j < m_2$,
$\forall x(S_j(x) \rightarrow \exists y(y \leq x \wedge R_j(y))) \wedge \forall x(R_j(x) \rightarrow \neg S_j(x))$,

(ix) for each $j < m_2$,
$\forall xy(R_j(x) \wedge S_j(y) \rightarrow S_j(x \cup y) \wedge S_j(\bar{x} \cap y))$,

(x) for each $j < m_2$,
$\forall xy(S_j(x) \wedge y \leq x \wedge y \neq 0 \rightarrow \exists z(z \leq x \wedge z \cap y \neq 0 \wedge R_j(z)))$,

(xi) for each $j < m_2$,
$\forall xy(S_j(x) \wedge y \leq x \rightarrow \exists z(S_j(z) \wedge (z \leq y \vee z \leq x \cap \bar{y})))$,

(xii) for each $j,k < m_2$ with $j \neq k$,
$\forall x(S_j(x) \rightarrow \neg \exists y(y \leq x \wedge S_k(y)))$,

(xiii) for each $j < m_1$,
$\forall xy(S_j(x) \wedge S_j(y) \rightarrow (\neg \exists z(z \leq x \cap y \wedge S_j(z)) \vee \neg \exists z(z \leq (x \cap \bar{y}) \cup (\bar{x} \cap y) \wedge S_j(z))))$,

(xiv) for each $j$ with $m_1 \leq j < m_2$,
$\forall x(S_j(x) \rightarrow \exists y_0 y_1(y_0 \cap y_1 = 0 \wedge y_0 \cup y_1 \leq x \wedge S_j(y_0) \wedge S_j(y_1)))$,

(xv) for each $j < m_2$,
$\forall xy(S_j(x) \wedge y \leq x \rightarrow \exists z((z \leq y \wedge S_j(z)) \vee (z \leq x \cap \bar{y} \wedge S_j(z))))$,

(xvi) $\forall xy(S(x) \wedge x \leq y \rightarrow S(y))$,

(xvii) $\forall xy(S(x) \wedge S(y) \rightarrow S(x \cap y))$,

(xviii) $\neg \exists xy(x \cap y = 0 \wedge S(x) \wedge S(y))$,

(xix) $\forall x(S(x) \vee S(\bar{x}))$,

(xx) $\forall x((\bigvee_{j<m_0} P_j(x)) \vee (\bigvee_{j<m_2} S_j(x)) \rightarrow S(\bar{x}),$

(xxi) $\forall x \exists y \forall z((\bigvee_{j<m_0} P_j(z)) \rightarrow ((z \leq x \rightarrow z \leq y) \wedge (z \cap x \neq z$
$\rightarrow z \cap y = 0))),$

(xxii) for each $j < m_0$,
$\forall x(S(x) \rightarrow \exists y(y \leq x \wedge P_j(y))),$

(xxiii) for each $j < m_2$,
$\forall x(S(x) \rightarrow Q_1 y(y \leq x \wedge S_j(y))),$

(xxiv) for each $\varphi(x) \in Form(L_{Ba}(Q_1))$ and each variables $y$, $z$ not occuring in $\varphi(x)$,
$\forall x(\neg\exists y(y \leq x \wedge (\bigvee_{j<m_2} S_j(y))) \rightarrow \exists y \forall z((\bigvee_{j<m_0} P_j(z))$
$\rightarrow ((z \cap y = 0 \leftrightarrow \neg\varphi^z(x)) \wedge (z \cap y = z \leftrightarrow \varphi^z(x))))),$

(xxv) $\forall x(\neg\exists y(y \leq x \wedge (\bigvee_{j<m_2} S_j(y)) \rightarrow \exists y_0 \ldots y_{m_0-1}((\bigwedge_{\substack{j,k<m_0 \\ j \neq k}} y_j \cap y_k = 0)$
$\wedge (\bigwedge_{j<m_0} \forall z(P_j(z) \rightarrow z \cap y_j = z \cap x)))),$

(xxvi) $\forall x(\neg S(x) \rightarrow \exists y_0 \ldots y_{m_0-1} z_0 \ldots z_{m_2-1}((\bigwedge_{\substack{j,k<m_0 \\ j \neq k}} y_j \cap y_k = 0)$
$\wedge (\bigwedge_{\substack{j,k<m_2 \\ j \neq k}} z_j \cap z_k = 0) \wedge (\bigwedge_{\substack{j<m_0 \\ k<m_2}} y_j \cap z_k = 0) \wedge$
$x = \bigcup[y_j : j < m_0] \cup \bigcup[z_j : j < m_2] \wedge$
$(\bigwedge_{j<m_0} \forall y(y \cap y_j \neq 0 \rightarrow \exists z(P_j(z) \wedge z \cap y \neq 0)) \wedge$
$(\bigwedge_{j<m_2} \forall y(y \cap z_j \neq 0 \rightarrow \exists z(z \leq z_j \wedge S_j(z) \wedge z \cap y \neq 0))))),$

(xxvii) for each $j < m_0$ and each $\varphi(x) \in Form(L_{Ba}(Q_1))$ with $\neg Q_1 x\varphi(x) \in Th_1(\vartheta_j)$,
$\neg Q_1 x \forall z((P_j(z) \rightarrow \varphi^z(x)) \wedge ((\bigvee_{\substack{k<m_0 \\ k \neq j}} P_k(z)) \rightarrow x \cap z = 0)),$

(xxviii) for each $j < m_0$, each $\varphi(x), \psi(x,y) \in Form(L_{Ba}(Q_1))$ with
$\forall x(\varphi(x) \rightarrow \exists y(y \leq x \wedge \psi(x,y))) \in Th_1(\vartheta_j)$,
$\forall x(\neg \exists y(y \leq x \wedge (\bigvee_{k<m_2} S_k(y))) \wedge \forall y(P_j(y) \wedge y \cap x \neq 0 \rightarrow \varphi^y(x))$
$\rightarrow \exists y(y \leq x \wedge \forall z(P_j(z) \wedge z \cap x \neq 0 \rightarrow \psi^z(x,y)))$,

(xxix) for each $j < m_0$,
the theory $Th_1(\vartheta_j)$ relativized to $P_j$.

We set: for each $i < m_0$,

$P_i^* = \{a \in |\prod^{\mathcal{E}_1}\{\mathcal{L}_\alpha: \alpha < \omega^\omega\}|$: there are $\beta < \omega^\omega$, $k < \omega$ so that
$a(\gamma) = 0$ for $\gamma \neq \beta \cdot \omega + k$ and so that $i \equiv k(\mathrm{mod}(m_0+m_2))$
and $a(\beta \cdot \omega + k) = 1$ or there is an $l < m_0$ with $i \in X(l)$,
$k \equiv m_0 + l(\mathrm{mod}(m_0+m_2))$ and $a(\beta \cdot \omega + k)(i) = 1$,
$a(\beta \cdot \omega + k)(j) = 0$ for $j \in X(l) \setminus \{i\}\}$,

for each $i < m_2$,

$R_i^* = \{a \in |\prod^{\mathcal{E}_1}\{\mathcal{L}_\alpha: \alpha < \omega^\omega\}|$: there are $\beta < \omega^\omega$, $k < \omega$ so that
$i \equiv k + m_0(\mathrm{mod}(m_0+m_2))$, $a(\beta \cdot \omega + k) = 1$ and $a(\gamma) = 0$
for $\gamma \neq \beta \cdot \omega + k\}$,

for each $i < m_1$,

$S_i^* = \{a \in |\prod^{\mathcal{E}_1}\{\mathcal{L}_\alpha: \alpha < \omega^\omega\}|$: for each $\gamma < \omega^\omega$, $a(\gamma) = 0$ or $a(\gamma) = 1$,
and if $a(\gamma) = 1$, then there are $\beta < \omega^\omega$ and $k < \omega$ with
$\gamma = \beta \cdot \omega + k$, $i \equiv k + m_0(\mathrm{mod}(m_0+m_2))$ and there is an
$X \in E_{\omega_0}[i+m_0, m_0+m_2]$ so that $(X \setminus \{\delta < \omega^\omega: a(\delta) = 1\})$
$\cup (\{\delta < \omega^\omega: a(\delta) = 0\} \setminus X)$ is finite$\}$,

for each $i$ with $m_1 \leq i < m_2$,

$S_i^* = \{a \in |\prod^{\mathcal{E}_1}\{\mathcal{L}_\alpha: \alpha < \omega^\omega\}|$: for each $\gamma < \omega^\omega$, $a(\gamma) = 0$ or $a(\gamma) = 1$,
and if $a(\gamma) = 1$, then there are $\beta < \omega^\omega$ and $k < \omega$ with
$\gamma = \beta \cdot \omega + k$, $i \equiv k + m_0(\mathrm{mod}(m_0+m_2))$ and there is an
$X \in F_{\omega_0}[i+m_0, m_0+m_2]$ so that $(X \setminus \{\delta < \omega^\omega: a(\delta) = 1\})$
$\cup$ $(\{\delta < \omega^\omega: a(\delta) = 0\} \setminus X$ is finite$\}$.

Let $D^*$ be the ideal generated by the sets $\bigcup\{P_i^*: i < m_0\}$

and $\bigcup\{S_i^*: i < m_2\}$. We set

$$S^* = |\Pi^{\mathcal{E}_1}\{\mathcal{L}_\alpha : \alpha < \omega^\omega\}| \setminus D^*.$$

Then it is easily seen that

$$\langle \Pi^{\mathcal{E}_1}\{\mathcal{L}_\alpha : \alpha < \omega^\omega\}, P_0^*, \ldots, P_{m_0-1}^*, R_0^*, \ldots, R_{m_2-1}^*, S_0^*, \ldots, S_{m_2-1}^*, S^* \rangle \in \mathrm{Mod}(T_5).$$

Let $\mathfrak{A}, \mathfrak{B} \in \mathrm{Mod}(T_5)$. It follows from (xvi), (xvii), (xviii), and (xix) that $S^{\mathfrak{A}}$ and $S^{\mathfrak{B}}$ are ultrafilters. (xxvii) implies with $\varphi(x) = (x = 1)$ that $\neg Q_1 x(\bigvee_{j<m_0} P_j(x)) \in T_5$, and thus $\neg Q_1 x(\bigvee_{j<m_2} R_j(x)) \in T_5$, too.

Let $j < m_2$. It follows from (viii) and (ix) that each $a \in S_j^{\mathfrak{A}}$ contains infinitely many c with $c \in R_j^{\mathfrak{A}}$ and that each $b \in S_j^{\mathfrak{B}}$ contains infinitely many d with $d \in R_j^{\mathfrak{B}}$. From (xvii) and (xx) it follows that for each $j < m_2$ and each $a \in S^{\mathfrak{A}}$ there is a set $\{c_i : i < \omega\}$ of pairwise disjoint elements with $c_i \le a$ and $c_i \in S_j^{\mathfrak{A}}$ for each $i < \omega$; and for each $j < m_2$ and each $b \in S^{\mathfrak{B}}$ there is a set $\{d_i : i < \omega\}$ of pairwise disjoint elements $d_i \le b$ and $d_i \in S_j^{\mathfrak{B}}$ for each $i < \omega$.

Let $n < \omega$. Then player II can ensure in the game $H^2(n, \tilde{\mathfrak{A}}, \tilde{\mathfrak{B}}, 1, 1)$ that for each $k \le n$ the following is valid:

(7)

> if $\langle a_k, b_k \rangle$ is the result of the game before the k-th round, then either
>
> $a_k \in S^{\mathfrak{A}}$ and $b_k \in S^{\mathfrak{B}}$
>
> or
>
> $a_k \notin S^{\mathfrak{A}}$ and $b_k \notin S^{\mathfrak{B}}$ and
>
> (i) for each $j < g(n-k)$ and each $l < m_0$, $\mathrm{card}(\{a \in |\tilde{\mathfrak{A}}| : a \in P_l^{\mathfrak{A}}, \tilde{\mathfrak{A}}[a \cap a_k] \models \varphi_j^{n-k}\}) = \mathrm{card}(\{b \in |\tilde{\mathfrak{B}}| : b \in P_l^{\mathfrak{B}}, \tilde{\mathfrak{B}}[b \cap b_k] \models \varphi_l^{n-k}\})$ or $\mathrm{card}(\{a \in |\tilde{\mathfrak{A}}| : a \in P_l^{\mathfrak{A}}, \tilde{\mathfrak{A}}[a \cap a_k] \models \varphi_j^{n-k}\}) \ge (g(n))^{2(n-k)}$ and $\mathrm{card}(\{b \in |\tilde{\mathfrak{B}}| : b \in P_l^{\mathfrak{B}}, \tilde{\mathfrak{B}}[b \cap b_k] \models \varphi_l^{n-k}\}) \ge (g(n))^{2(n-k)}$,
>
> (ii) for each $j < (g(n))^{2(n-k)}$ and each $l < m_2$ there are disjoint elements $a^0, \ldots, a^{j-1} \in S_l^{\mathfrak{A}}$ with $\bigcup[a^i : i < j] \le a_k$ iff there are disjoint elements $b^0, \ldots, b^{j-1} \in S_l^{\mathfrak{B}}$ with $\bigcup[b^i : i < j] \le b_k$.

The proof of (7) is similar to the proof of Lemma 49.

Ad (ii): Let $T_6$ be the theory arising from $T_5$ by replacing the axioms (xvii) and (xviii) by

$$\text{(xvii')} \quad \forall x((\bigvee_{j<m_0} P_j(x)) \vee (\bigvee_{j<m_2} S_j(x)) \rightarrow \neg S(x)),$$

$$\text{(xviii')} \quad \forall xyz(x = y \cup z \wedge S(x) \rightarrow S(y) \vee S(z)) \wedge \forall x(S(x) \rightarrow \exists yz(y \cap z = 0 \wedge y \cup z \leq x \wedge S(y) \wedge S(z))),$$

respectively. All the other things are similar to (i).

Q.E.D.

Now it follows from Lemmas 47 - 51 that Theorem 7 of Chapter 1 can be applied; hence

Theorem 52. $Th^1$ is decidable.

Now we investigate the theories $Th^0$ and $Th^\omega$.

We assume for the following that GCH holds true. Then in order to show the decidability of $Th^0$ and $Th^\omega$ we have to show that $Th_0(\mathcal{M}_0)$ and $Th_\omega(\mathcal{M}_\omega)$ are decidable.

Let $h : Term_1 \longrightarrow \omega$ be defined as follows:

$$h(2) = h(\eta) = h(\eta_0) = 0,$$

$$h(p_1(t_1)) = h(p_2(t_1)) = h(t_1) + 1,$$

$$h(p_3(t_1,t_2)) = h(p_4(t_1,t_2)) = \max\{h(t_1),h(t_2)\} + 1,$$

$$h(p_5(t_0,\ldots,t_{m_0-1};n,X)) = h(p_6(t_0,\ldots,t_{m_0-1};n,X)) = \max\{h(t_i): i < m_0\} + 1.$$

Lemma 53. Assume GCH. Then
(i) $\mathcal{B}_1(t) \ {}^1{\equiv}^0 \ \mathcal{B}_0(t)$ for each $t \in Term_0$,
(ii) $\mathcal{B}_1(t) \ {}^1{\equiv}^\omega \ \mathcal{B}_\omega(t)$ for each $t \in Term_\omega$.

Proof. The proof is by induction on $h(t)$. Using $H^2$-games it is immediately seen that

$$B(2) \ {}^1{\equiv}^0 \ B(2), \quad B(2) \ {}^1{\equiv}^\omega \ B(2),$$

$$\mathcal{B}_1(\eta) \;{}^{1}\!\equiv^{0}\; B(\eta_0), \quad \mathcal{B}_1(\eta) \;{}^{1}\!\equiv^{\omega}\; \mathcal{B}_\omega(\eta),$$

$$B(\eta_0) \;{}^{1}\!\equiv^{\omega}\; B(\eta_0).$$

Thus the case $h(t) = 0$ is shown.

The induction step is also carried out with the help of $H^2$-games and is similar to the proofs of Lemmas 48 - 51.

We only give the proof of one special case which is representative for all cases:

Let $t_1, t_2 \in \mathrm{Term}_\omega$, $t = p_3(t_1,t_2)$. Thus

$$\mathcal{B}_1(t) = \prod^{w}\{\mathcal{B}_1(t^\alpha): \alpha < \omega_1\}$$

with

$$t_\alpha = \begin{cases} t_1 & \text{for } \alpha < \omega, \\ t_2 & \text{for } \omega \leq \alpha < \omega_1, \end{cases}$$

$$\mathcal{B}_\omega(t) = \prod^{w}\{\mathcal{B}_\omega(t^\alpha): \alpha < \omega_\omega\}$$

with

$$t^\alpha = \begin{cases} t_1 & \text{for } \alpha < \omega, \\ t_2 & \text{for } \omega \leq \alpha < \omega_\omega. \end{cases}$$

We set

$P_1^1 = \{a \in |\mathcal{B}_1(t)|$: there is an $\alpha < \omega$ with $a(\alpha) = 1$ and $a(\beta) = 0$ for $\alpha \neq \beta\}$,

$P_2^1 = \{a \in |\mathcal{B}_1(t)|$: there is an $\alpha$ with $\omega \leq \alpha < \omega_1$ and $a(\alpha) = 1$ so that $a(\beta) = 0$ for $\alpha \neq \beta\}$,

$S^1 = \{a \in |\mathcal{B}_1(t)|$: $\mathrm{card}(\{\alpha < \omega_1 : a(\alpha) = 1\}) \geq \omega_0\}$,

$P_1^\omega = \{a \in |\mathcal{B}_\omega(t)|$: there is an $i < \omega$ with $a(i) = 1$ and $a(\alpha) = 0$ for $\alpha \neq i\}$,

$P_2^{\omega} = \{a \in |\mathcal{B}_{\omega}(t)|$: there is an $\alpha$ with $\omega \leq \alpha < \omega_{\omega}$ and $a(\alpha) = 1$ so that $a(\beta) = 0$ for $\alpha \neq \beta\}$,

$S^{\omega} = \{a \in |\mathcal{B}_{\omega}(t)|: \text{card}(\{\alpha < \omega_{\omega}: a(\alpha) = 1\}) \geq \omega_0\}$.

Let $m,n \in \omega$ be such that

there are at most m sentences $\varphi \in \text{Sent}(L_{Ba}(Q_1))$ with $qr(\varphi) \leq n$ which are different with respect to $Th_1(Ba)$ and there are at most m sentences $\varphi \in \text{Sent}(L_{Ba}(Q_{\omega}))$ with $qr(\varphi) \leq n$ which are different with respect to $Th_{\omega}(Ba)$.

Then player II can ensure in the game $H^2(n,\mathcal{B}_1(t),\mathcal{B}_{\omega}(t),1,\omega)$ that for each $k \leq n$ the following is valid:

if $\langle a_k,b_k\rangle$ is the result of the game before the k-th round, then either

$a_k \in S^1$ and $b_k \in S^{\omega}$

or

(8) $a_k \notin S^1$ and $b_k \notin S^{\omega}$ and for each $\alpha < \omega_{\omega}$,

(i) $\text{card}(\{\beta < \omega_1: \mathcal{B}_1(t)[a_k(\beta)^{red}] \; {}^1\!\equiv^{\omega}_{n-k} \mathcal{B}_{\omega}(t)[b_k(\alpha)^{red}]\}) =$
$\text{card}(\{\beta < \omega_{\omega}: \mathcal{B}_{\omega}(t)[b_k(\beta)^{red}] \; {}^{\omega}\!\equiv^{\omega}_{n-k} \mathcal{B}_{\omega}(t)[b_k(\alpha)^{red}]\})$ or

(ii) $\text{card}(\{\beta < \omega_1: \mathcal{B}_1(t)[a_k(\beta)^{red}] \; {}^1\!\equiv^{\omega}_{n-k} \mathcal{B}_{\omega}(t)[b_k(\alpha)^{red}]\}) \geq m^{2(n-k)}$
and $\text{card}(\{\beta < \omega_{\omega}: \mathcal{B}_{\omega}(t)[b_k(\beta)^{red}] \; {}^{\omega}\!\equiv^{\omega}_{n-k} \mathcal{B}_{\omega}(t)[b_k(\alpha)^{red}]\}) \geq m^{2(n-k)}$.

Q.E.D.

Lemma 54. Let $n \in \omega$, $\mathfrak{A},\mathfrak{A}^*,\mathfrak{B},\mathfrak{B}^* \in Ba$.

(i) If $\mathfrak{A} \; {}^1\!\equiv^{\omega}_n \mathfrak{A}^*$, $\mathfrak{B} \; {}^1\!\equiv^{\omega}_n \mathfrak{B}^*$, then $\mathfrak{A} \times \mathfrak{B} \; {}^1\!\equiv^{\omega}_n \mathfrak{A}^* \times \mathfrak{B}^*$.

(ii) If $\mathfrak{A} \; {}^1\!\equiv^{0}_n \mathfrak{A}^*$, $\mathfrak{B} \; {}^1\!\equiv^{0}_n \mathfrak{B}^*$, then $\mathfrak{A} \times \mathfrak{B} \; {}^1\!\equiv^{0}_n \mathfrak{A}^* \times \mathfrak{B}^*$.

Proof. The proof is similar to the proof of Lemma 48. We only show (i). Let L' be the language $L_{Ba}(Q_1)$ enlarged by two binary predicate symbols $P_1$ and $P_2$. Let T be the theory given by the following axioms:

(i) the axioms of Th(Ba),

(ii) $\exists x(P_1(x) \wedge P_2(\bar{\bar{x}})$,

(iii) $\forall xy((P_1(x) \wedge P_1(y) \rightarrow x = y) \wedge (P_2(x) \wedge P_2(y) \rightarrow x = y))$,

(iv) $typ(n,1,\mathfrak{A})$ relativized to $P_1$,

(v) $typ(n',1,\mathfrak{B})$ relativized to $P_2$.

Now $\mathfrak{A} \;{}^1{\equiv}_n^\omega\; \mathfrak{A}^*$ and $\mathfrak{B} \;{}^1{\equiv}_n^\omega\; \mathfrak{B}^*$; thus

$$.typ(n,1,\mathfrak{A}). = .typ(n,\omega,\mathfrak{A}^*). \quad \text{and} \quad .typ(n,1,\mathfrak{B}). = .typ(n,\omega,\mathfrak{B}^*)..$$

We set

$$P_1^1 = \{\langle 1^{\mathfrak{A}},0\rangle\},\quad P_2^1 = \{\langle 0,1^{\mathfrak{B}}\rangle\},\quad P_1 = \{\langle 1^{\mathfrak{A}^*},0\rangle\},\; P_2^\omega = \{\langle 0,1^{\mathfrak{B}^*}\rangle\}.$$

Then $\langle \mathfrak{A}\times\mathfrak{B}, P_1^1, P_2^1\rangle \in Mod(T^1)$ and $\langle \mathfrak{A}^*\times\mathfrak{B}^*, P_1^\omega, P_2^\omega\rangle \in Mod(T^\omega)$. Let $\mathfrak{C} \in Mod(T^1)$, $\mathfrak{D} \in Mod(T^\omega)$, $P_1^{\mathfrak{C}} = \{c^+\}$, $P_1^{\mathfrak{D}} = \{d^+\}$. Then player II can ensure in the game $H^2(n,\tilde{\mathfrak{C}},\tilde{\mathfrak{D}},1,\omega)$ that for each $k \leq n$ the following holds true:

(9) if $\langle a_k,b_k\rangle$ is the result of the game before the k-th round, then $\tilde{\mathfrak{C}}[c^+ \cap a_k] \;{}^1{\equiv}_{n-k}^\omega\; \tilde{\mathfrak{D}}[d^+ \cap b_k]$ and $\tilde{\mathfrak{C}}[\overline{\overline{c^+}} \cap a_k] \;{}^1{\equiv}_{n-k}^\omega\; \tilde{\mathfrak{D}}[\overline{\overline{d^+}} \cap b_k]$.

This immediately implies the statement of the lemma.

Q.E.D.

Similar to Lemmas 49 and 50 we obtain

<u>Lemma 55.</u> Let $n \in \omega$, $\mathfrak{A},\mathfrak{A}^* \in Ba$.

(i) If $\mathfrak{A} \;{}^1{\equiv}_n^\omega\; \mathfrak{A}^*$, then $\prod_{\omega_1}^{W}\mathfrak{A} \;{}^1{\equiv}_n^\omega\; \prod_{\omega_\omega}^{W}\mathfrak{A}^*$, $\prod^{\mathcal{J}_1(\mathcal{H}_1)}\mathfrak{A} \;{}^1{\equiv}_n^\omega\; \prod^{\mathcal{J}_1(\mathcal{H}_\omega)}\mathfrak{A}^*$ and $\prod^{\mathcal{J}_1(\mathcal{H}_0)}\mathfrak{A} \;{}^1{\equiv}_n^\omega\; \prod^{\mathcal{J}_1(\mathcal{H}_0)}\mathfrak{A}^*$.

(ii) If $\mathfrak{A} \;{}^1{\equiv}_n^0\; \mathfrak{A}^*$, then $\prod_{\omega_1}^{W}\mathfrak{A} \;{}^1{\equiv}_n^0\; \prod_{\omega}^{W}\mathfrak{A}^*$ and $\prod^{\mathcal{J}_1(\mathcal{H}_1)}\mathfrak{A} \;{}^1{\equiv}_n^0\; \prod^{\mathcal{J}_1(\mathcal{H}_0)}\mathfrak{A}^*$.

<u>Lemma 56.</u> Let $n \in \omega$, $\mathfrak{A},\mathfrak{A}^*,\mathfrak{B},\mathfrak{B}^* \in Ba$ with $\mathfrak{A} \;{}^1{\equiv}_n^\omega\; \mathfrak{A}^*$ and $\mathfrak{B} \;{}^1{\equiv}_n^\omega\; \mathfrak{B}^*$. Let

$$\mathfrak{C}_\alpha = \begin{cases} \mathfrak{A} \text{ for } \alpha < \omega, \\ \mathfrak{B} \text{ for } \omega \leq \alpha < \omega_1, \end{cases}$$

$$\mathfrak{C}_\alpha^* = \begin{cases} \mathfrak{A}^* \text{ for } \alpha < \omega, \\ \mathfrak{B}^* \text{ for } \omega \leq \alpha < \omega_\omega. \end{cases}$$

Then

$$\prod^{W}\{\mathcal{L}_\alpha : \alpha < \omega_1\} \ {}^1\!\equiv^{\omega}_{n} \prod^{W}\{\mathcal{L}^{*}_\alpha : \alpha < \omega_\omega\}$$

and

$$\prod^{B(f_1)}\{\mathcal{L}_\alpha : \alpha < \omega_1\} \ {}^1\!\equiv^{\omega}_{n} \prod^{B(f_\omega)}\{\mathcal{L}^{*}_\alpha : \alpha < \omega_\omega\}.$$

Let $n \in \omega$. We set

$$\mathcal{F}^{\omega}_{n}(i) = \mathrm{card}(\{\mathrm{typ}(n,\omega,\mathcal{B}_\omega(t)) : t \in \mathrm{Term}_\omega \text{ and } h(t) \leq i\}),$$

$$\mathcal{F}^{0}_{n}(i) = \mathrm{card}(\{\mathrm{typ}(n,0,\mathcal{B}_0(t)) : t \in \mathrm{Term}_0 \text{ and } h(t) \leq i\}).$$

Thus $\langle\mathcal{F}^{\omega}_{n}(i) : i < \omega\rangle$ and $\langle\mathcal{F}^{0}_{n}(i) : i < \omega\rangle$ are monotone sequences. From Lemmas 54, 55, and 56 it follows that:

If $\mathcal{F}^{\omega}_{n}(i) = \mathcal{F}^{\omega}_{n}(i+1)$, then $\mathcal{F}^{\omega}_{n}(i) = \mathcal{F}^{\omega}_{n}(k)$ for each $k \geq i$.

Correspondingly we get:

If $\mathcal{F}^{0}_{n}(i) = \mathcal{F}^{0}_{n}(i+1)$, then $\mathcal{F}^{0}_{n}(i) = \mathcal{F}^{0}_{n}(k)$ for each $k \geq i$.

Now we have $\mathrm{card}(\mathrm{typ}(n,\alpha,\mathrm{Ba})) < \omega_0$, consequently the sequences $\langle\mathcal{F}^{\omega}_{n}(i) : i < \omega\rangle$ and $\langle\mathcal{F}^{0}_{n}(i) : i < \omega\rangle$ have to become constant for all i being sufficiently large.

Let $g : \omega \longrightarrow \omega$ be a recursive function so that for each $k \in \omega$,

$$g(k) \geq \max\{\mathrm{card}(\mathrm{typ}(n,\omega,\mathrm{Ba})), \mathrm{card}(\mathrm{typ}(n,0,\mathrm{Ba}))\}.$$

(The existence of such a function was shown in § 1.) Then

Lemma 57. For each $k,n \in \omega$, if $k \geq g(n)$, then $\mathcal{F}^{\omega}_{n}(k) = \mathcal{F}^{\omega}_{n}(g(n))$ and $\mathcal{F}^{0}_{n}(k) = \mathcal{F}^{0}_{n}(g(n))$.

Let $\langle t^i : i < \omega\rangle$ be an enumeration of $\mathrm{Term}_\omega$ so that $i < k$ implies $h(t^i) \leq h(t^k)$. Then we can apply Theorem 8 of Chapter 1 to $\{\mathcal{B}_\omega(t^i) : i < \omega\}$ and get

Theorem 58. $\mathrm{Th}^{\omega}$ is decidable.

Similar to Theorem 58 we get

<u>Theorem 59.</u> $Th^0$ is decidable.

Remember that $Term_0 \subseteq Term_\omega \subseteq Term_1$; thus

$$Th^1 \subseteq Th^\omega \subseteq Th^0.$$

We show now that the three theories are different. Let

$$\varphi_1 \equiv \forall x(\eta(x) \longrightarrow Qy(y \leq x)),$$

$$\varphi_2 \equiv \forall x(at(x) \wedge Qy(y \leq x) \longrightarrow Qy(y \leq x \wedge At(x))).$$

Then $\varphi_1 \in Th^0 \setminus Th^\omega$ and $\varphi_2 \in Th^\omega \setminus Th^1$. Hence

<u>Theorem 60.</u> $Th^1 \subseteq Th^\omega \subseteq Th^0$ and $Th^1 \neq Th^\omega \neq Th^0$.

From this lemma it especially follows that all axioms of the logic with $Q_1$ are valid in each Boolean algebra. For instance, let $\varphi(x,y) \in Form(L_{Ba}(Q_\omega))$. Then

$$\psi \equiv Q_\omega x \exists y \varphi(x,y) \longrightarrow \exists y Q_\omega x \varphi(x,y) \vee Q_\omega y \exists x \varphi(x,y)$$

is valid in each Boolean algebra, whereas this is not the case for linear orderings.

# CHAPTER 7

# PSEUDO-TREES

A partially ordered set $\langle A,\leq\rangle$ is said to be a pseudo-tree if for each $a \in A$ the set $\{b: b \leq a\}$ is linearly ordered by $\leq$. In the first paragraph a theorem on limits of n-equivalent structures is proved. This result is applied in the following paragraphs to construct a dense system of models of the $Q_1$-theory of a special class of pseudo-trees. From this we get the decidability of this theory. In the last paragraph some generalizations and problems are given.

## § 1. Ehrenfeucht's game and limits

If we proved in the preceding chapters that two structures $\mathfrak{A}$ and $\mathfrak{B}$ are n-equivalent ($\mathfrak{A} \sim_n \mathfrak{B}$), then the winning strategy for player II was given by a verbal description only. In the next paragraph we need a more refined construction. For this reason we shall give here a more special definition of a winning strategy.

Let $\mathfrak{A}$ and $\mathfrak{B}$ be structures of the same signature. In this chapter we shall use the following notion of a winning strategy W for player II in the game corresponding to $\mathfrak{A} \sim_n \mathfrak{B}$.

$$\mathrm{dom}(W) := \{\bar{w}: \bar{w} = ((\vartheta_0,d_0),\ldots,(\vartheta_{2k},d_{2k})) \text{ for certain } k < n, \text{ where for all } 0 \leq i \leq 2k\ (\vartheta_i = \mathfrak{A} \text{ or } \vartheta_i = \mathfrak{B}) \text{ and } d_i \in |\vartheta_i|\},$$

$$W : \mathrm{dom}(W) \longrightarrow |\mathfrak{A}| \cup |\mathfrak{B}|.$$

For $\bar{w} \in \mathrm{dom}(W)$ of odd length we define by induction the notion "$\bar{w}$ is chosen according to W":

$\bar{w} = ((\vartheta_0,d_0))$ is chosen according to W if $\bar{w} \in \mathrm{dom}(W)$,

$\bar{w} = ((\vartheta_0,d_0),\dots,(\vartheta_i,d_i),(\vartheta_{i+1},d_{i+1}),(\vartheta_{i+2},d_{i+2}))$ is chosen according to W <u>if</u> $((\vartheta_0,d_0),\dots,(\vartheta_i,d_i))$ is chosen according to W and ($\vartheta_{i+1} = \mathfrak{A}$ if $\vartheta_i = \mathfrak{B}$) and ($\vartheta_{i+1} = \mathfrak{B}$ if $\vartheta_i = \mathfrak{A}$) and $d_{i+1} = W(((\vartheta_0,d_0),\dots,(\vartheta_i,d_i)))$.

A function as above is said to be a winning strategy for player II in the game corresponding to $\mathfrak{A} \sim_n \mathfrak{B}$ <u>iff</u> for every $\bar{w} = ((\vartheta_0,d_0),\dots,(\vartheta_{2n-2},d_{2n-2})) \in \operatorname{dom} W$ which is chosen according to W it holds that

$$\langle \mathfrak{A}, a^0,\dots,a^{n-1}\rangle \equiv_0 \langle \mathfrak{B}, b^0,\dots,b^{n-1}\rangle,$$

where

$$a^i := \begin{cases} d_{2i} & \text{if } \vartheta_{2i} = \mathfrak{A}, \\ W(((\vartheta_0,d_0),\dots,(\vartheta_{2i},d_{2i}))) & \text{if } \vartheta_{2i} = \mathfrak{B} \end{cases}$$

and

$$b^i := \begin{cases} d_{2i} & \text{if } \vartheta_{2i} = \mathfrak{B}, \\ W(((\vartheta_0,d_0),\dots,(\vartheta_{2i},d_{2i}))) & \text{if } \vartheta_{2i} = \mathfrak{A} \end{cases}$$

for $0 \le i \le n - 1$.

Since the property of being a winning strategy only depends on the values of W for all $\bar{w}$ chosen according to W, we will sometimes define the winning strategy for these $\bar{w}$ only. Sometimes it is useful to define a winning strategy W also for the empty sequence $\varnothing$ which is by definition assumed to be chosen according to W.

Let $\mathfrak{A}$ be a relational structure of finite signature which may contain finitely many constants. For a set $C \subseteq |\mathfrak{A}|$ which contains the constants of $\mathfrak{A}$ $\mathfrak{A}\upharpoonright C$ denotes the usual restriction of $\mathfrak{A}$ to C.

Let $(\mathfrak{A}_i)_{i\in\omega}$ be a sequence of relational structures of the same finite signature and let $(C_i)_{i\in\omega}$ be a sequence of sets so that the following conditions are fulfilled:

(*) $C_i \subseteq C_j$ for $i < j \in \omega$,
every constant of $\mathfrak{A}_0$ is contained in $C_0$,
$C_i \subseteq |\mathfrak{A}_i|$ for $i \in \omega$ and
$\mathfrak{A}_i \upharpoonright C_i = \mathfrak{A}_j \upharpoonright C_i$ for $i < j \in \omega$.

Let $\mathfrak{C}$ be the structure $\bigcup_{i\in\omega} \mathfrak{A}_i \upharpoonright C_i$.

We are now interested in the following problem. Let $\mathfrak{A}_i \sim_n \mathfrak{A}_j$ for

$i < j \in \omega$. Under which conditions do we get $\mathfrak{A}_0 \sim_n \mathfrak{L}$? To find a sufficient condition which is suitable for some applications we have to regard the winning strategies in the corresponding games.

Let $(\mathfrak{A}_i)_{i\in\omega}$, $(C_i)_{i\in\omega}$ and $\mathfrak{L}$ be as above. Furthermore, let $\mathfrak{A}_0 \sim_n \mathfrak{A}_i$ for $i > 0$ and let $W_i$ be a winning strategy for player II in the corresponding game. For $i > 0$ and for every sequence $\bar{w} = ((\vartheta_0, d_0), \dots, (\vartheta_{2k}, d_{2k}))$ we define:

$\bar{w}$ fulfils ($*$i) <u>iff</u> $\bar{w} \in \mathrm{dom}(W)$ and w is chosen according to $W_i$ and for every $j \leq 2k$, if $\vartheta_j = \mathfrak{A}_i$, then $d_j \in C_i$.

Let $\bar{w} = ((\vartheta_0, d_0), \dots, (\vartheta_{2k}, d_{2k}))$ fulfil ($*$i) and let j be a natural number with $j \geq i$. Then we define:

$$\bar{w}^{(j)} := ((\vartheta'_0, d_0), \dots, (\vartheta'_{2k}, d_{2k})),$$

where

$$\vartheta'_s := \begin{cases} \mathfrak{A}_j & \text{for } \vartheta_s = \mathfrak{A}_i, \\ \mathfrak{A}_0 & \text{for } \vartheta_s = \mathfrak{A}_0. \end{cases}$$

We get $\bar{w}^{(j)} \in \mathrm{dom}(W_j)$ since $C_i \subseteq C_j$. Now we are able to state a sufficient condition for $\mathfrak{A}_0 \sim_n \mathfrak{L}$.

<u>Theorem 1.</u> Let n be a natural number and let $(\mathfrak{A}_i)_{i\in\omega}$, $(C_i)_{i\in\omega}$ and $\mathfrak{L}$ be as in ($*$) above. For every $i > 0$ we suppose $\mathfrak{A}_0 \sim_n \mathfrak{A}_i$. Let $W_i$ be a winning strategy for player II in the corresponding game. For every $i > 0$ and every sequence $\bar{w} = ((\vartheta_0, d_0), \dots, (\vartheta_{2k}, d_{2k}))$ (with $k < n$) let us suppose that:

(a) $\bar{w}$ fulfils ($*$i) and $\vartheta_{2k} = \mathfrak{A}_0$, <u>then</u> there exists a $j \geq 1$ with $W_j(\bar{w}^{(j)}) \in C_j$,

(b) $\bar{w}$ fulfils ($*$i) and $\vartheta_{2k} = \mathfrak{A}_0$ and $W_i(\bar{w}) \in C_i$, <u>then</u> for every $j \geq i$, $W_i(\bar{w}) = W_j(\bar{w}^{(j)})$,

(c) $\bar{w}$ fulfils ($*$i) and $\vartheta_{2k} = \mathfrak{A}_i$ and $d_{2k} \in C_i$, <u>then</u> for every $j \geq i$, $W_i(\bar{w}) = W_j(\bar{w}^{(j)})$.

On this suppositions $\mathfrak{A}_0 \sim_n \mathfrak{L}$ is fulfilled.

<u>Proof.</u> We prove this theorem by induction on n. In the case of $n = 0$ it is obvious since every constant of $\mathfrak{A}_0$ is contained in $C_0$ and $\mathfrak{A}_0 \upharpoonright C_0 = \mathfrak{L} \upharpoonright C_0$.

Let us assume that the theorem is true for n - 1. Presupposing that the suppositions of the theorem are fulfilled for n by $(\mathfrak{A}_i)_{i\in\omega}$, $(C_i)_{i\in\omega}$ and $\mathcal{L}$ we have to prove $\mathfrak{A}_0 \sim_n \mathcal{L}$. We shall do this by giving a winning strategy for player II in the corresponding game.

Case 1: Suppose player I chooses an element $d_0 \in |\mathfrak{A}_0|$ in the first round. It is obvious that $((\mathfrak{A}_0,d_0))$ fulfils $(*i)$. Using (a) we get the existence of a $j_0 \geq 1$ with $W_{j_0}(((\mathfrak{A}_0,d_0))) \in C_{j_0}$. Let us denote $W_{j_0}(((\mathfrak{A}_0,d_0)))$ by $d_1$. Player II has to choose this element $d_1$. What we have to show is that $\langle\mathfrak{A}_0,d_0\rangle \sim_{n-1} \langle\mathcal{L},d_1\rangle$. Since $\langle\mathfrak{A}_0 d_0\rangle \sim_{n-1} \langle\mathfrak{A}_{j_0},d_1\rangle$ holds, we only have to show that $\langle\mathfrak{A}_{j_0},d_1\rangle \sim_{n-1} \langle\mathcal{L},d_1\rangle$ holds. To do this we define:

$$\mathfrak{A}_i' := \langle\mathfrak{A}_{j_0+1},d_1\rangle \text{ for } i\in\omega;$$

$$C_i' := C_{i+j_0} \text{ for } i\in\omega$$

and

$$\mathcal{L}' := \langle\mathcal{L},d_1\rangle .$$

All $\mathfrak{A}_i'$ are relational structures of the same finite signature. Every constant of $\mathfrak{A}_0'$ is contained in $C_0'$, $\mathfrak{A}_i' \upharpoonright C_i = \mathfrak{A}_j' \upharpoonright C_i$ and $\mathcal{L}' = \bigcup_{i\in\omega} \mathfrak{A}_i' \upharpoonright C_i$ holds for $i < j \in \omega$. Furthermore, we have $\mathfrak{A}_0' \sim_{n-1} \mathfrak{A}_i'$ (see fig. 1).

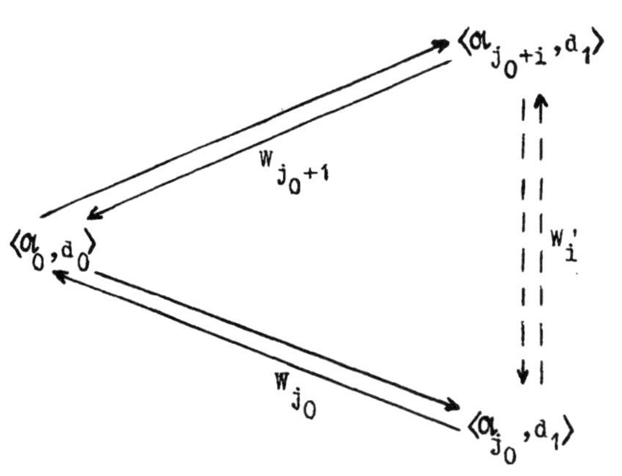


fig. 1

We get a winning strategy $W_i'$ for player II in the game corresponding to $\mathfrak{A}_0' \sim_{n-1} \mathfrak{A}_i'$ as a natural combination of $W_{j_0}$ and $W_{j_0+i}$. We shall give an exact definition of $W_i'$, since we have to show that $(\mathfrak{A}_i')_{i\in\omega}$, $(C_i')_{i\in\omega}$, $(W_i')_{i\in\omega}$ and $\mathcal{L}$ fulfil also (a), (b) and (c) of the theorem for n - 1. For distinction we shall write in this case (a'), (b'), (c'), $(*i)'$ and $\bar{w}^{(j)'}$ instead of (a), (b), (c), $(*i)$ and $\bar{w}^{(j)}$, respectively.

First of all we define some functions $W_{i,k}$ by induction on $k < n-1$. The function $W_i'$ will then be the union of all $W_{i,k}$ for $k < n-1$.

$$\mathrm{dom}(W_{i,0}) := \{((\vartheta_0,e_0)): \vartheta_0 \in \{\alpha_0',\alpha_i'\} \text{ and } e_0 \in |\vartheta_0|\},$$

$$((\vartheta_0,e_0))^{(+i)} := \begin{cases} ((\alpha_0,d_0),(\alpha_{j_0},d_1)(\alpha_{j_0},e_0),(\alpha_0,W_{j_0}(((\alpha_0,d_0),(\alpha_{j_0},d_1),(\alpha_{j_0},e_0))))) \\ \qquad \text{if } \vartheta_0 = \alpha_0', \\ ((\alpha_0,d_0),(\alpha_{j_0},d_1),(\alpha_0,W_{i+j_0}(((\alpha_0,d_0),(\alpha_{i+j_0},d_1),(\vartheta_0,e_0))))) \\ \qquad \text{if } \vartheta_0 = \alpha_i', \end{cases}$$

$$((\vartheta_0,e_0))^{(++i)} := \begin{cases} ((\alpha_0,d_0),(\alpha_{i+j_0},d_1),(\alpha_0,W_{j_0}(((\alpha_0,d_0),(\alpha_{j_0},d_1),(\vartheta_0,e_0))))) \\ \qquad \text{if } \vartheta_0 = \alpha_0', \\ ((\alpha_0,d_0),(\alpha_{i+j_0},d_1),(\alpha_{i+j_0},e_0),(\alpha_0,W_{i+j_0}(((\alpha_0,d_0),(\alpha_{i+j_0},d_1), \\ \qquad (\alpha_{i+j_0},e_0))))) \text{ if } \vartheta_0 = \alpha_i'. \end{cases}$$

Let $\bar{w} = ((\vartheta_0,e_0))$ be an element of $\mathrm{dom}(W_{i,0})$.

$$W_{i,0} := \begin{cases} W_{i+j_0}(\bar{w}^{(++i)}) \text{ if } \vartheta_0 = \alpha_0', \\ W_{j_0}(\bar{w}^{(+i)}) \text{ if } \vartheta_0 = \alpha_i'. \end{cases}$$

Now we suppose $k + 1 < n - 1$.

$$\begin{aligned} \mathrm{dom}(W_{i,k+1}) := \{ & ((\vartheta_0,e_0),\dots,(\vartheta_{2k+2},e_{2k+2})): \\ & ((\vartheta_0,e_0),\dots,(\vartheta_{2k},e_{2k})) \in \mathrm{dom}(W_{i,k}) \text{ and } \vartheta_{2k+2} \in \\ & \{\alpha_0',\alpha_i'\} \text{ and (if } \vartheta_{2k} = \alpha_i' \text{ then } \vartheta_{2k+1} = \alpha_0') \text{ and} \\ & \text{(if } \vartheta_{2k} = \alpha_0' \text{ then } \vartheta_{2k+1} = \alpha_i') \text{ and } e_{2k+1} = \\ & W_{i,k}(((\vartheta_0,e_0),\dots,(\vartheta_{2k},e_{2k})))\}. \end{aligned}$$

Let $\bar{u} = (\bar{w},(\vartheta_{2k+1},e_{2k+1}),(\vartheta_{2k+2},e_{2k+2}))$ be an element of $\mathrm{dom}(W_{i,k+1})$.

(1) If $\vartheta_{2k+1} = \alpha_i'$ is fulfilled, we define $\bar{u}^{(+i)}$ and $\bar{u}^{(++i)}$ as follows:

$$\bar{u}^{(+i)} := \begin{cases} (\bar{w}^{(+i)}, (\alpha_{j_0}, e_{2k+2}), (\alpha_0, W_{j_0}((\bar{w}^{(+i)}, (\alpha_{j_0}, e_{2k+2}))))) \text{ if } \vartheta_{2k+2} = \alpha_0', \\ (\bar{w}^{(+i)}, (\alpha_0, W_{i+j_0}((\bar{w}^{(++i)}, (\vartheta_{2k+1}, e_{2k+1}), (\vartheta_{2k+2}, e_{2k+2}))))) \\ \qquad \text{if } \vartheta_{2k+2} = \alpha_i', \end{cases}$$

$$\bar{u}^{(++i)} := \begin{cases} (\bar{w}^{(++i)}, (\vartheta_{2k+1}, e_{2k+1}), (\alpha_0, W_{j_0}((\bar{w}^{(+i)}, (\alpha_{j_0}, e_{2k+2}))))) \\ \qquad \text{if } \vartheta_{2k+2} = \alpha_0', \\ (w^{-(++i)}, (\vartheta_{2k+1}, e_{2k+1}), (\vartheta_{2k+2}, e_{2k+2}), \\ (\alpha_0, W_{i+j_0}((\bar{w}^{(++i)}, (\vartheta_{2k+1}, e_{2k+1}), (\vartheta_{2k+2}, e_{2k+2}))))) \text{ if } \vartheta_{2k+2} = \alpha_i'. \end{cases}$$

(2) If $\vartheta_{2k+1} = \alpha_0'$ is fulfilled, we define $\bar{u}^{(+i)}$ and $\bar{u}^{(++i)}$ as follows:

$$\bar{u}^{(+i)} := \begin{cases} (\bar{w}^{(+i)}, (\vartheta_{2k+1}, e_{2k+1}), (\vartheta_{2k+2}, e_{2k+2}), \\ (\alpha_0, W_{j_0}((\bar{w}^{(+i)}, (\vartheta_{2k+1}, e_{2k+1}), (\vartheta_{2k+2}, e_{2k+2}))))) \quad \text{if } \vartheta_{2k+2} = \alpha_0', \\ (\bar{w}^{(+i)}, (\vartheta_{2k+1}, e_{2k+1}), (\alpha_0, W_{i+j_0}((\bar{w}^{(++i)}, (\vartheta_{2k+2}, e_{2k+2}))))) \\ \qquad \text{if } \vartheta_{2k+2} = \alpha_i', \end{cases}$$

$$\bar{u}^{(++i)} := \begin{cases} (\bar{w}^{(++i)}, (\alpha_0, W_{j_0}((\bar{w}^{(+i)}, (\vartheta_{2k+1}, e_{2k+1}), (\vartheta_{2k+2}, e_{2k+2}))))) \\ \qquad \text{if } \vartheta_{2k+2} = \alpha_0', \\ (\bar{w}^{(++i)}, (\vartheta_{2k+2}, e_{2k+2}), (\alpha_0, W_{i+j_0}((\bar{w}^{(++i)}, (\vartheta_{2k+2}, e_{2k+2}))))) \\ \qquad \text{if } \vartheta_{2k+2} = \alpha_i'. \end{cases}$$

Let $\bar{u} = ((\vartheta_0, e_0), \ldots, (\vartheta_{2k+2}, e_{2k+2}))$ be an element of $\mathrm{dom}(W_{i,k+1})$.

$$W_{i,k+1}(\bar{u}) := \begin{cases} W_{i+j_0}(\bar{u}^{(++i)}) \text{ if } \vartheta_{2k+2} = \alpha_0', \\ W_{j_0}(\bar{u}^{(+i)}) \text{ if } \vartheta_{2k+2} = \alpha_i'. \end{cases}$$

Now we define

$$W_i' := \bigcup_{0 \le k < n-1} W_{i,k}.$$

It is an immediate consequence of the above definitions that $W_i'$ is a winning strategy for player II in the game corresponding to $\mathfrak{A}_0' \sim_{n-1} \mathfrak{A}_i'$.

Let $\bar{w} = ((\vartheta_0,e_0),\ldots,(\vartheta_{2k},e_{2k}))$ $(k < n-1)$ fulfil $(*i)'$ and $\vartheta_{2k} = \mathfrak{A}_0'$. To prove (a') we have to find a $j \geq i$ with $W_j'(\bar{w}^{(j)}{}') \in C_j'$. First of all we observe that $\bar{w}^{(++i)}$ fulfils the premise of (a) for $i + j$. Therefore it exists a $j \geq i + j_0$ with $W_j(\bar{w}^{(++i)})^{(j)}) \in C_j$.

The next step is to show that $(\bar{w}^{(j-j_0)}{}')^{(+(j-j_0))} = \bar{w}^{(+i)}$ and $(\bar{w}^{(j-j_0)}{}')^{(++(j-j_0))} = (\bar{w}^{(++i)})^{(j)}$ are fulfilled. Using (b) and (c) this can be done by induction following the definitions of $\bar{w}^{(+i)}$ and $\bar{w}^{(++i)}$. We shall omit the details here. But we have

$$\begin{aligned} W_{j-j_0}'(\bar{w}^{(j-j_0)}{}') &= W_{(j-j_0)+j_0}((\bar{w}^{(j-j_0)}{}')^{(++(j-j_0))}) \\ &= W_j((\bar{w}^{(++i)})^{(j)}) \in C_j = C_{j-j_0}'. \end{aligned}$$

This proves (a').

The proofs of (b') and (c') are similar and are left to the reader.

Using the hypothesis of our induction we get $\mathfrak{A}_0' \sim_{n-1} \mathfrak{L}'$. But this is nothing but $\langle \mathfrak{A}_{j_0},d_1\rangle \sim_{n-1} \langle \mathfrak{L},d_1\rangle$.

Case 2: Suppose player I chooses an element $d_1 \in |\mathfrak{L}|$ in the first round. Then there exists a $j_0 \in \omega$ with $d_1 \in C_{j_0}$. Player II has to choose $W_{j_0}(((\mathfrak{A}_{j_0},d_1)))$. Let us denote $W_{j_0}(((\mathfrak{A}_{j_0},d_1)))$ by $d_0$. Since $\langle \mathfrak{A}_0,d_0\rangle \sim_{n-1} \langle \mathfrak{A}_{j_0},d_1\rangle$ holds, we only have to show that $\langle \mathfrak{A}_{j_0},d_1\rangle \sim_{n-1} \langle \mathfrak{L},d_1\rangle$ holds. But this we can do as in case 1.

Q.E.D.

## § 2. $\omega$-trees and n-equivalence for $L(Q_1)$

A partially ordered set $\langle A,\leq\rangle$ is an $\omega$-tree if for each $a \in A$ the set $\hat{a} := \{b: b \in A \text{ and } b < a\}$ is linearly ordered and finite. An $\omega$-tree $\mathfrak{A} = \langle A,\leq\rangle$ is said to be connected if $\mathfrak{A} \models \exists x\forall y(x \leq y)$. The class of all (connected) $\omega$-trees is denoted by $(\omega\text{-T})$ ($(\omega\text{-CT})$, respectively). For every element $a \in A$ we define $h_{\mathfrak{A}}(a) := \text{card}(\hat{a})$. $h_{\mathfrak{A}}(a)$ is said to be the height of a in $\mathfrak{A}$. If it is clear which $\omega$-tree is meant, we omit the subscript $\mathfrak{A}$. For a subset B of A we define

$$h_{\mathfrak{A}}(B) := \bigcup_{a\in B} h_{\mathfrak{A}}(a).$$

There is a connection between $\omega$-trees and forests. Let $\mathfrak{A} = \langle A,\leq\rangle$ be an $\omega$-tree and let R be the following binary relation

$$\{(a,b)\colon a \in A \text{ and } b \in A \text{ and } (\mathfrak{A} \models \varphi[a,b] \text{ or } \mathfrak{A} \models \varphi[b,a]), \text{ where}$$
$$\varphi \text{ is the formula } v_1 < v_2 \wedge \neg\exists v_3(v_1 < v_3 \wedge v_3 < v_2) \text{ .}$$

Then $\mathfrak{A}' := \langle A,R\rangle$ is a forest in the sense of Chapter 4.

Now let $\mathfrak{B} = \langle A,R\rangle$ be a tree in the sense of Chapter 4 and let a be an element of A. We define a partial ordering $\leq$ on A by

b $\leq$ c <u>iff</u> every path which connects a and c contains b.

It is easy to see that the structure $\mathfrak{A} := \langle A,\leq\rangle$ is a connected $\omega$-tree. Moreover, we can construct for this $\mathfrak{A}$ the above forest $\mathfrak{A}'$. Then we have $\mathfrak{A}' = \mathfrak{B}$. Since each forest is the union of its components, which are trees, we can also connect each forest with a corresponding $\omega$-tree. Using these constructions we get the following

<u>Lemma 2.</u> Th(F) is interpretable in Th(($\omega$-T)).
Th(CF) is interpretable in Th(($\omega$-CT)).

In this lemma we can replace "interpretable" also by "interpretable in the sense of [Rabin, 1965]". But it is well known that Th(($\omega$-T)) is not interpretable in the sense of [Rabin, 1965] in the theory Th(F).

A subset B of A is said to be bounded in $\mathfrak{A}$ if $h_{\mathfrak{A}}(B) < \omega$. $P_b(\mathfrak{A})$ denotes the set of all bounded subsets of $|\mathfrak{A}|$. $P_{b1}(\mathfrak{A})$ is defined to be the set $\{c\colon c \in P_b(\mathfrak{A}) \text{ and } \mathrm{card}(c) \geq \omega_1\}$.

Let L be an elementary language with the binary relation symbol $\leq$ and let L' be the monadic second order language which expands L. For all $\omega$-trees $\mathfrak{A}$ and all sentences $\varphi$ of L' we define $\mathfrak{A} \models_b \varphi$ in the same way as $\mathfrak{A} \models \varphi$, but the set variables vary over elements of $P_b(\mathfrak{A})$ only. Similarly, we define $\mathfrak{A} \models_{b1} \varphi$. Here the set variables vary over elements of $P_{b1}(\mathfrak{A})$ only. We define:

$$Th_b(\mathfrak{A}) := \{\varphi : \varphi \text{ is a sentence of } L' \text{ and } \mathfrak{A} \models_b \varphi\},$$

$$Th_{b1}(\mathfrak{A}) := \{\varphi : \varphi \text{ is a sentence of } L' \text{ and } \mathfrak{A} \models_{b1} \varphi\}.$$

For each class K of $\omega$-trees we define:

$$Th_b(K) := \bigcap_{\mathfrak{A}\in K} Th_b(\mathfrak{A}),$$

$$Th_{b1}(K) := \bigcap_{\mathfrak{A}\in K} Th_{b1}(\mathfrak{A}).$$

Lemma 3.
(i) Th(($\omega$-T)) is interpretable in Th(($\omega$-CT)).
(ii) $Th_1(\mathfrak{A})$ is interpretable in $Th_{b1}(\mathfrak{A})$ for all $\omega$-trees $\mathfrak{A}$.
(iii) $Th_1((\omega\text{-CT}))$ is interpretable in $Th_{b1}((\omega\text{-CT}))$.
(iv) $Th_b((\omega\text{-CT}))$ is interpretable in $Th_{b1}((\omega\text{-CT}))$.
All interpretations are in the sense of [Rabin, 1965].

Proof. (i) We define for each formula $\varphi$ of L' a formula $\varphi^+$ by induction on the complexity of the formulas:

$$\varphi^+ := \varphi \text{ if } \varphi \text{ is atomic},$$

$$(\exists x\varphi(x))^+ := \exists xy\forall z(y \leq z \wedge x \neq y \wedge \varphi^+(x)),$$

$$(\neg\varphi)^+ := \neg\varphi^+,$$

$$(\varphi \wedge \psi)^+ := \varphi^+ \wedge \psi^+,$$

$$(\exists X\varphi(X))^+ := \exists xX\forall z(x \leq z \wedge x \notin X \wedge \varphi^+(X)).$$

For each $\omega$-tree $\mathfrak{A}$ let a not be an element of $|\mathfrak{A}|$. We define a new structure $\mathfrak{L}(\mathfrak{A})$ as it is indicated in fig. 2. Here $\mathfrak{A}_c$ ($c \in C$) are the components of $\mathfrak{A}$ and a is the minimal element of $\mathfrak{L}(\mathfrak{A})$.

It is easy to see that for all $\mathfrak{A} \in (\omega\text{-T})$ and all sentences $\varphi$ of L' the following holds: $\mathfrak{A} \models \varphi$ iff $\mathfrak{L}(\mathfrak{A}) \models \varphi^+$. Let K be the class of all models $\mathfrak{L}(\mathfrak{A})$ with $\mathfrak{A} \in (\omega\text{-T})$. Obviously we have:

$$K = \{\mathfrak{L} : \mathfrak{L} \models \exists v_1v_2\forall v_3(v_1 \leq v_3 \wedge v_1 \neq v_2) \in (\omega\text{-CT}).$$

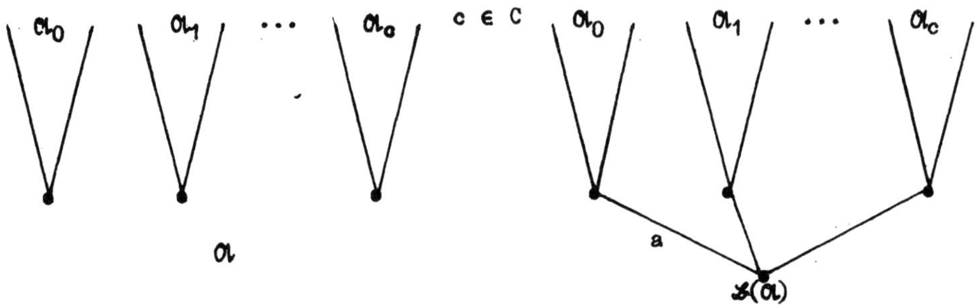


fig. 2

But this gives us:

$$(\omega\text{-T}) \models \varphi \quad \text{iff} \quad (\omega\text{-CT}) \models (\exists v_1 v_2 \forall v_3 (v_1 \leq v_3 \wedge v_1 \neq v_2)) \longrightarrow \varphi^+$$

for all sentences $\varphi$ of L'. This proves (i).

(ii) For each formula $\varphi$ of $L(Q_1)$ we define a formula $\varphi^*$ for L':

$\varphi^* := \varphi$ if $\varphi$ is atomic,

$(\varphi \wedge \psi)^* := \varphi^* \wedge \psi^*$,

$(\neg\varphi)^* := \neg\varphi^*$,

$(\exists x\varphi(x))^* := \exists x\varphi^*(x)$,

$(Q_1 x\varphi(x))^* := \exists X \forall x(x \in X \longrightarrow \varphi^*(x))$.

For each $\omega$-tree $\mathfrak{A}$ we have to prove: $\mathfrak{A} \models \varphi$ iff $\mathfrak{A} \models_{b1} \varphi^*$. We do this by induction on the complexity of $\varphi$. The only non trivial step is that for $Q_1$. Here we use the simple fact that for each subset $B \subseteq |\mathfrak{A}|$ of cardinality $\geq \omega_1$ there exists an n so that

$$\text{card}(\{a: h_{\mathfrak{A}}(a) = n \text{ and } a \in B\}) \geq \omega_1$$

(hint: use $\text{cf}(\omega_1) > \omega$).

(iii) This immediately follows from (ii).

(iv) Let $\mathfrak{A}$ be a connected $\omega$-tree and let $a,b,c,d,e_\gamma$ (with $\gamma < \omega_1$) be new elements (i.e. $(\{a,b,c,d\} \cup \{c_\gamma: \gamma < \omega_1\}) \cap |\mathfrak{A}| = \emptyset$). We define a new connected $\omega$-tree as it is indicated in fig. 3.

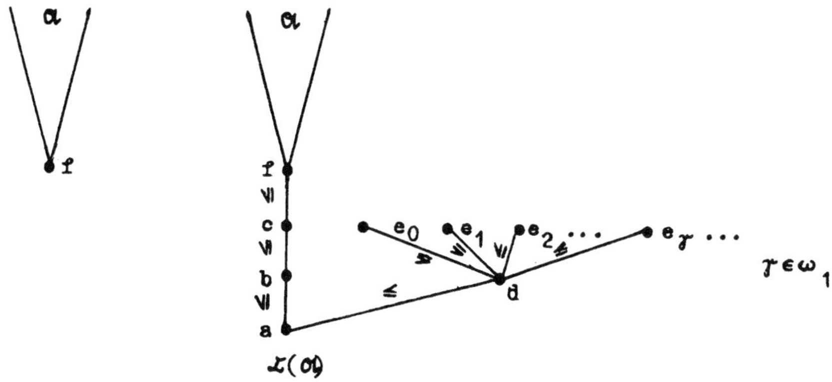


fig. 3

a is the minimal element of $\mathcal{L}(\mathfrak{A})$, while f is the minimal element of $\mathfrak{A}$. Let C be the set $\{a,b,c,d\} \cup \{c_\gamma: \gamma \in \omega_1\}$. Obviously this set is definable in $\mathcal{L}(\mathfrak{A})$ by using bounded sets of cardinality $\geq \omega_1$. Moreover, it is easy to see that f is definable in $\mathcal{L}(\mathfrak{A})$. Both definitions are independent of $\mathfrak{A}$. This justifies the following definitions:

$\varphi' := \varphi$ if $\varphi$ is atomic,

$(\neg\varphi)' := \neg\varphi'$,

$(\varphi \wedge \psi)' := \varphi' \wedge \psi'$,

$(\exists x\varphi(x))' := \exists x\varphi(x) \wedge f \leq x$,

$(\exists X\varphi(X))' := \exists X(C \subseteq X \wedge \varphi'(X \setminus C))$.

Then it is easy to see that $\mathfrak{A} \models_b \varphi$ iff $\mathcal{L}(\mathfrak{A}) \models_{b1} \varphi'$. We left it as an exercise to find a sentence $\psi$ of L' so that

$\{\mathcal{B}: \mathcal{B} \models_{b1} \psi$ and $\mathcal{B} \in (\omega\text{-CT})\} = \{\mathcal{L}(\mathfrak{A}): \mathfrak{A} \in (\omega\text{-CT})\}$.

Q.E.D.

In [Seese, 1978] it was proved that $Th_b((\omega\text{-CT}))$ is decidable. In the next paragraph we shall prove the decidability of $Th_{b1}((\omega\text{-CT}))$. Before we can prove some basic technical lemmas we need the following definition.

<u>Definition.</u> Let $\mathfrak{A} = \langle A, \leq \rangle$ be an $\omega$-tree. $\mathfrak{A}^+$ is defined to be the structure $\langle (S \cup P_{b1}(\mathfrak{A})), S, E, \leq^+ \rangle$, where

$$S := \{a\colon a \subseteq P(\mathfrak{A}) \text{ and } card(a) = 1\},$$

$$E := \{(a,b)\colon b \in P_{b1}(\mathfrak{A}) \text{ and } a \in S \text{ and } a \subseteq b\}$$

and

$$\leq^+ := \{(a,b)\colon a \in S \text{ and } b \in S \text{ and there exist } c,d \text{ with } \{c\} = a \text{ and } \{d\} = b \text{ and } c \leq d\}.$$

A similar notion can be introduced if $\mathfrak{A} = \langle \mathfrak{B}, \ldots \rangle$ is the expansion of an $\omega$-tree $\mathfrak{B}$ by certain sets, constants, and other relations. We shall use the corresponding notion for these structures.

Let $\mathfrak{A}$ and $\mathfrak{B}$ be $\omega$-trees and let n be a natural number. We define the relations $\equiv_n^{b1}$ and $\equiv_n^b$ by:

$\mathfrak{A} \equiv_n^{b1} \mathfrak{B}$ <u>iff</u> for all sentences $\varphi$ of L' with $qr(\varphi) \leq n$ we have

$$\mathfrak{A} \models_{b1} \varphi \quad \text{iff} \quad \mathfrak{B} \models_{b1} \varphi.$$

$\mathfrak{A} \equiv_n^{b} \mathfrak{B}$ <u>iff</u> for all sentences $\varphi$ of L' with $qr(\varphi) \leq n$ we have

$$\mathfrak{A} \models_{b} \varphi \quad \text{iff} \quad \mathfrak{B} \models_{b} \varphi.$$

Here qr is defined (for the monadic second order language L') in the same way as for an elementary language; we only have to add the clause

$$qr(\exists X\varphi) := qr(\varphi) + 1.$$

Now we define the relation $\sim_n^{b1}$ for expanded $\omega$-trees.

$$\mathfrak{A} \sim_n^{b1} \mathfrak{B} \text{ iff } \mathfrak{A}^+ \sim_n \mathfrak{B}^+.$$

From the definitions and the results of Chapter 1 we immediately get the following two results.

<u>Lemma 4.</u> $\mathfrak{A}^+ \equiv_n \mathfrak{B}^+$ implies $\mathfrak{A} \equiv_n^{b1} \mathfrak{B}$.

<u>Lemma 5.</u> Let $\mathfrak{A}, \mathfrak{B}$ be expansions of $\omega$-trees by certain sets, constants, and relations. For every natural number n:

$$\text{if } \mathfrak{A} \sim_n^{b1} \mathfrak{B}, \text{ then } \mathfrak{A} \equiv_n^{b1} \mathfrak{B}.$$

Let $\sigma(i,j)$ be an extension of the signature for $\omega$-trees by i individual constants and j set constants. As a corollary of Lemma 5 we get the following

<u>Lemma 6.</u> Let n,i,j be natural numbers. The number of equivalence classes of $\sim_n^{b1}$ for $\omega$-trees of the signature $\sigma(i,j)$ is finite.

Let $\varkappa^{\sigma(i,j)}(n)$ be the number of equivalence classes of $\sim_n^{b1}$ for $\omega$-trees of signature $\sigma(i,j)$. We define

$$\varkappa(l,m) := \sum_{i+j=l} \varkappa^{\sigma(i,j)}(m).$$

For every $\omega$-tree $\mathfrak{A} = \langle A, \leq \rangle$ and every element a of $|\mathfrak{A}|$ we define the $\omega$-tree $\mathfrak{A}_a$ by

$$|\mathfrak{A}_a| := \{b\colon a \leq b \text{ and } b \in |\mathfrak{A}|\} \qquad \mathfrak{A}_a := \langle |\mathfrak{A}_a|, \leq \restriction |\mathfrak{A}_a| \rangle.$$

We illustrate this definition by fig. 4.

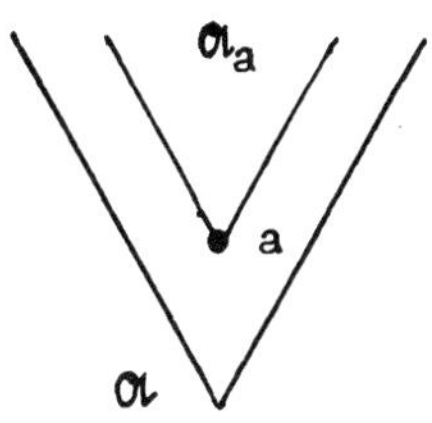


fig. 4

Let $\mathfrak{A}_1$, $\mathfrak{A}_2$ be connected $\omega$-trees with minimal elements $a_1, a_2$, respectively, and let n be a natural number. We assume $|\mathfrak{A}_1| \cap |\mathfrak{A}_2| = \emptyset$ and define

$$S_i := \{a\colon a \in |\mathfrak{A}_i| \text{ and } h_{\mathfrak{A}_i}(a) \neq 1\} \quad \text{for } i = 1,2.$$

Let $\bullet$ be the equivalence relation on $S_1 \cup S_2$ defined by

$$\begin{aligned} a \bullet c \text{ \underline{iff}} \quad & (a \in S_1 \text{ and } c \in S_1 \text{ and } (\mathfrak{A}_1)_a \sim_n^{b1} (\mathfrak{A}_1)_c) \\ \text{or } & (a \in S_2 \text{ and } c \in S_2 \text{ and } (\mathfrak{A}_2)_a \sim_n^{b1} (\mathfrak{A}_2)_c) \\ \text{or } & (a \in S_1 \text{ and } c \in S_2 \text{ and } (\mathfrak{A}_1)_a \sim_n^{b1} (\mathfrak{A}_2)_c) \\ \text{or } & (a \in S_2 \text{ and } c \in S_1 \text{ and } (\mathfrak{A}_2)_a \sim_n^{b1} (\mathfrak{A}_1)_c). \end{aligned}$$

For every element a of $S_1 \cup S_2$ let $\overset{\bullet}{a}$ denote its equivalence class under $\bullet$.

<u>Lemma 7.</u> Let n, $\mathfrak{A}_1, \mathfrak{A}_2$, $a_1, a_2$, $S_1$, and $S_2$ be as above. Moreover, we assume that for all $b \in S_1 \cup S_2$ the following conditions are fulfilled:

(i) $\prod_{i=0}^{n} \varkappa(i,n-i) < \operatorname{card}(\overset{\bullet}{b} \cap S_1) < \omega_1$ <u>iff</u>

$\prod_{i=0}^{n} \varkappa(i,n-i) < \operatorname{card}(\overset{\bullet}{b} \cap S_2) < \omega_1$.

(ii) <u>If</u> $\operatorname{card}(\overset{\bullet}{b} \cap S_1) \geq \omega_1$ $(\operatorname{card}(\overset{\bullet}{b} \cap S_2) \geq \omega_1)$, <u>then</u> $\operatorname{card}(\overset{\bullet}{b} \cap S_2) \geq \omega_1$ $(\operatorname{card}(\overset{\bullet}{b} \cap S_1) \geq \omega_1$, respectively).

(iii) <u>If</u> $(\operatorname{card}(\overset{\bullet}{b} \cap S_1) \leq \prod_{i=0}^{n} \varkappa(i,n-i)$ or $\operatorname{card}(\overset{\bullet}{b} \cap S_2) \leq \prod_{i=0}^{n} \varkappa(i,n-i))$, <u>then</u> $\operatorname{card}(\overset{\bullet}{b} \cap S_1) = \operatorname{card}(\overset{\bullet}{b} \cap S_2)$.

Then $\mathfrak{A}_1 \sim_n^{b1} \mathfrak{A}_2$ holds.

This lemma can also be generalized to $\omega$-trees which are expanded by a finite number of individuals and sets (structures of signature $\sigma(i,j)$ whose reduction to $\sigma(0,0)$ is an $\omega$-tree). The proof of this lemma is left as an exercise to the reader.

<u>Lemma 8.</u> Let $\mathfrak{A}_1 = \langle A_1, \leq_1 \rangle$, $\mathfrak{A}_2 = \langle A_2, \leq_2 \rangle$ be connected $\omega$-trees and let $B_1, B_2$ be subsets of $A_1, A_2$, respectively. We assume that any two elements a,b $(a \neq b)$ of $B_i$ (for i = 1,2) are incomparable by $\leq_i$ (that is $a \nleq_i b$ and $b \nleq_i a$) and we define $C_i$ to be the set $A_i \setminus \{b\colon$ there exists an $a \in B_i$ so that $a <_i b\}$ (for i = 1,2). Moreover, let $\mathfrak{A}_1 \upharpoonright C_1$ be isomorphic to $\mathfrak{A}_2 \upharpoonright C_2$ and let f be a corresponding isomorphism. If for all $a \in B_1$, $(\mathfrak{A}_1)_a \sim_n^{b1} (\mathfrak{A}_2)_{f(a)}$ and $\operatorname{rng}(f \upharpoonright B_1) = B_2$ are fulfilled, then $\mathfrak{A}_1 \sim_n^{b1} \mathfrak{A}_2$ holds.

The proof is obvious.

Let $\mathfrak{N}$ be a connected $\omega$-tree and n > 0. Let b be an element of $|\mathfrak{N}|$ $(B \in P_b(\mathfrak{N}))$ and let m be a natural number. We say that b is bounded in $\mathfrak{N}$ by m (B is bounded in $\mathfrak{N}$ by m) iff $h_{\mathfrak{N}}(b) < m$ (every $c \in B$ is bounded in $\mathfrak{N}$ by m). We need the following

<u>Lemma 9.</u> For every n > 0 there exists a natural number $m_n$ with the following property. For every connected $\omega$-tree $\mathfrak{N}$ there exists a connected $\omega$-tree $\mathfrak{N}'$ so that the following holds: For every $b \in |\mathfrak{N}|$ and every $B \in P_{b1}(\mathfrak{N})$ there exist a $b' \in |\mathfrak{N}'|$ and a $B' \in P_{b1}(\mathfrak{N}')$

so that b' and B' are bounded in $\mathfrak{N}'$ by $m_n$ and $\langle\mathfrak{N},b\rangle \sim^{b1}_{n-1} \langle\mathfrak{N}',b'\rangle$ and $\langle\mathfrak{N},B\rangle \sim^{b1}_{n-1} \langle\mathfrak{N}',B'\rangle$ and $\mathfrak{N} \sim^{b1}_{n} \mathfrak{N}'$.

<u>Proof.</u> Let $n > 2$ and $\mathfrak{N}$ be given. We suppose that $\{a_0,\dots,a_{k-1}\}$ is a finite subset of $|\mathfrak{N}|$ with the following properties:

for every $i,j \in k$, $a_i = a_j$ holds iff $\langle\mathfrak{N},a_i\rangle \sim^{b1}_{n-1} \langle\mathfrak{N},a_j\rangle$,

for every $a \in |\mathfrak{N}|$ there exists an $i \in k$ so that $\langle\mathfrak{N},a\rangle \sim^{b1}_{n-1} \langle\mathfrak{N},a_i\rangle$ holds.

Now we extend the signature $\langle\leq\rangle$ by an individual constant and call the resulting signature $\sigma(1,0) = \sigma$. Let $\varkappa^{\sigma}(n-1)$ be the number of equivalence classes of $\sim^{b1}_{n-1}$ for this signature.

If $h_{\mathfrak{N}}(a_0) > \varkappa^{\sigma}(n-1)$, then there exist two elements $a < b$ from $\hat{a}_0$ so that $\langle\mathfrak{N}_a,a_0\rangle \sim^{b1}_{n-1} \langle\mathfrak{N}_b,a_0\rangle$ holds. We assume that

$$\operatorname{card}(\hat{b}) - \operatorname{card}(\hat{a}) = \max\{\operatorname{card}(\hat{c}) - \operatorname{card}(\hat{d}) : c,d \in \hat{a}_0 \text{ and } \langle\mathfrak{N}_c,a_0\rangle \sim^{b1}_{n-1} \langle\mathfrak{N}_d,a_0\rangle\}.$$

Then we define $\mathfrak{N}_0$ to be the following tree:

$$|\mathfrak{N}_0| := (|\mathfrak{N}| \setminus |\mathfrak{N}_a|) \cup |\mathfrak{N}_b|,\ \leq_0 := \leq \upharpoonright |\mathfrak{N}_0| \text{ and } \mathfrak{N}_0 := \langle|\mathfrak{N}_0|,\leq_0\rangle.$$

As an illustration use fig. 5.

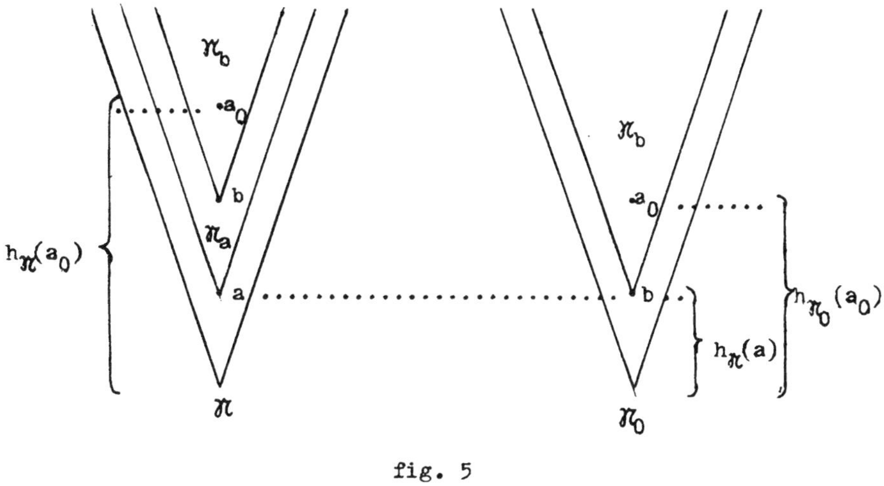


fig. 5

It is obvious that $\mathfrak{X} \sim_n^{b1} \mathfrak{X}_0$ and $\langle \mathfrak{X}, a_0\rangle \sim_{n-1}^{b1} \langle \mathfrak{X}_0, a_0\rangle$ hold. Moreover, we have $h_{\mathfrak{X}_0}(a_0) \leq \varkappa^{\sigma}(n-1)$. Let $a_0'$ be $a_0$.

Now let $k' \in k$. We assume that $\mathfrak{X}_{k'-1}$ and $a_j'$ are defined for every $j \in k'$ and have the following properties:

(i) $\mathfrak{X} \sim_n^{b1} \mathfrak{X}_{k'-1}$,

(ii) $a_j' \in |\mathfrak{X}_{k'-1}|$,

(iii) $\langle \mathfrak{X}, a_j\rangle \sim_{n-1}^{b1} \langle \mathfrak{X}_{k'-1}, a_j'\rangle$,

(iv) $h_{\mathfrak{X}_{k'-1}}(a_j') \leq k'(\varkappa^{\sigma}(n-1) + 1)$.

Since (i), we find an $a_{k'}'$ so that $\langle \mathfrak{X}, a_{k'}\rangle \sim_{n-1}^{b1} \langle \mathfrak{X}_{k'-1}, a_{k'}'\rangle$ holds. If $h_{\mathfrak{X}_{k'-1}}(a_{k'}') \leq (k'+1)(\varkappa^{\sigma}(n-1)+1)$ holds, we define $\mathfrak{X}_{k'} := \mathfrak{X}_{k'-1}$. If this is not the case, we proceed as follows: There exists an $a \in \overset{k'-1}{\widehat{a_{k'}'}}$, with $\mathrm{card}(\overset{k'-1}{\widehat{a}}) = (k'+1)(\varkappa^{\sigma}(n-1)+1)$. Here $\overset{k'-1}{\widehat{\ }}$ denotes the operation ^ for the tree $\mathfrak{X}_{k'-1}$. Moreover, there exist two elements b,c with $a \leq_{k'-1} b <_{k'-1} c \leq_{k'-1} a'$ so that

$$\langle (\mathfrak{X}_{k'-1})_b, a_{k'}'\rangle \sim_{n-1}^{b1} \langle (\mathfrak{X}_{k'-1})_c, a_{k'}'\rangle.$$

We assume that

$$h_{\mathfrak{X}_{k'-1}}(c) - h_{\mathfrak{X}_{k'-1}}(b) = \max\{h_{\mathfrak{X}_{k'-1}}(d) - h_{\mathfrak{X}_{k'-1}}(e) : a \leq_{k'-1} e <_{k'-1} d \leq_{k'-1} a_{k'}', \text{ and } \langle (\mathfrak{X}_{k'-1})_e, a_{k'}'\rangle \sim_{n-1}^{b1} \langle (\mathfrak{X}_{k'-1})_d, a_{k'}'\rangle\}.$$

We define $\mathfrak{X}_{k'}$ in the following way:

$$|\mathfrak{X}_{k'}| := (|\mathfrak{X}_{k'-1}| \setminus |(\mathfrak{X}_{k'-1})_b|) \cup |(\mathfrak{X}_{k'-1})_c|,$$

$$\leq_{k'} := \leq_{k'-1} \upharpoonright |\mathfrak{X}_{k'}|,$$

$$\mathfrak{X}_{k'} := \langle |\mathfrak{X}_{k'}|, \leq_{k'}\rangle.$$

Using the above induction we find a tree $\mathfrak{X}_{k-1}$ and elements $a_j'$ (for $j \in k$) with the properties (i),...,(iv) (change k' by k).

Let $\{A_0, \dots, A_{l-1}\}$ be a finite subset of $P_{b1}(\mathfrak{N})$ with the following properties:

for every $i, j \in l$ it holds: $A_i = A_j$ iff $\langle \mathfrak{N}, A_i \rangle \sim^{b1}_{n-1} \langle \mathfrak{N}, A_j \rangle$,

for every $A \in P_{b1}(\mathfrak{N})$ there exists an $i \in l$ so that $\langle \mathfrak{N}, A \rangle \sim^{b1}_{n-1} \langle \mathfrak{N}, A_i \rangle$ holds.

Now we extend the signature $\langle \leq \rangle$ by a set constant and call the resulting signature $\sigma(0,1) = \nu$. Let $\varkappa^{\nu}(n-1)$ be the number of equivalence classes of $\sim^{b1}_{n-1}$ for this signature.

Since $\mathfrak{N} \sim^{b1}_{n} \mathfrak{N}_{k-1}$, we find a set B from $P_{b1}(\mathfrak{N}_{k-1})$ so that $\langle \mathfrak{N}, A_0 \rangle \sim^{b1}_{n-1} \langle \mathfrak{N}_{k-1}, B \rangle$. If for all elements a of B

$$h_{\mathfrak{N}_{k-1}}(a) \leq k(\varkappa^{\sigma}(n-1)+1) + \varkappa^{\nu}(n-1) + 1$$

holds, we define $\mathfrak{N}_k := \mathfrak{N}_{k-1}$ and $A_0' := B$. If this is not the case, we proceed as follows. Let $a_0$ be an element of B for which

$$h_{\mathfrak{N}_{k-1}}(a_0) = \max\{h_{\mathfrak{N}_{k-1}}(a) : a \in B\}$$

holds. We find elements $b_0^0, c_0^0, d_0^0$ of $\widehat{a_0}^{k-1}$ which have the following properties:

$$b_0^0 \leq_{k-1} c_0^0 <_{k-1} d_0^0,$$

$$h_{\mathfrak{N}_{k-1}}(b_0^0) = k(\varkappa^{\sigma}(n-1)+1) + 1,$$

$$\langle (\mathfrak{N}_{k-1})_{c_0^0}, B \cap |(\mathfrak{N}_{k-1})_{c_0^0}| \rangle \sim^{b1}_{n-1} \langle (\mathfrak{N}_{k-1})_{d_0^0}, B \cap |(\mathfrak{N}_{k-1})_{d_0^0}| \rangle.$$

Let $\varrho_0$ be 1 and let $\mu'$ be some ordinal. Suppose $\varrho_\mu$ (for all $\mu \leq \mu'$) is some ordinal and we defined $b_\tau^\mu$, $c_\tau^\mu$, $d_\tau^\mu$ for all $\tau \in \varrho_\mu$ so that the following condition is fulfilled:

$$(*)\begin{cases} c_\tau^\mu \not\equiv_{k-1} c_{\tau'}^\mu, \text{ for all } \tau' \in \varrho_\mu \text{ with } \tau \neq \tau', \\ b_\tau^\mu \leq_{k-1} c_\tau^\mu <_{k-1} d_\tau^\mu, \\ h_{\mathfrak{N}_{k-1}}(b_\tau^\mu) = k(\varkappa^{\sigma}(n-1)+1) + 1, \\ \langle (\mathfrak{N}_{k-1})_{c_\tau^\mu}, B \cap |(\mathfrak{N}_{k-1})_{c_\tau^\mu}| \rangle \sim^{b1}_{n-1} \langle (\mathfrak{N}_{k-1})_{d_\tau^\mu}, B \cap |(\mathfrak{N}_{k-1})_{d_\tau^\mu}| \rangle, \\ f_\mu := \{(c_\tau^\mu, d_\tau^\mu) : \tau \in \varrho_\mu\} \text{ is a 1-1 function.} \end{cases}$$

Now we define

$$B_{\mu'} := |\mathfrak{A}_{k-1}| \quad (\bigcup_{\tau\in\rho_{\mu'}} |(\mathfrak{A}_{k-1})_{c^{\mu'}_\tau}|) \text{ and } \mathfrak{A}'_{\mu'} := \mathfrak{A}_{k-1}\upharpoonright B_{\mu'}.$$

If there is an element $a_{\mu'} \in B_{\mu'}$ for which $h_{\mathfrak{A}_{k-1}}(a_{\mu'}) = h_{\mathfrak{A}_{k-1}}(a_0)$ holds, then we find elements $b^{\mu'}, c^{\mu'}, d^{\mu'}$ of $\widehat{a_{\mu'}}^{k-1}$ which have the following properties:

$$b^{\mu'} \leq_{k-1} c^{\mu'} <_{k-1} d^{\mu'},$$

$$h_{\mathfrak{A}_{k-1}}(b^{\mu'}) = k(\varkappa^{\sigma}(n-1)+1) + 1,$$

$$\langle(\mathfrak{A}_{k-1})_{c^{\mu'}}, B \cap |(\mathfrak{A}_{k-1})_{c^{\mu'}}|\rangle \sim^{b1}_{n-1} \langle(\mathfrak{A}_{k-1})_{d^{\mu'}}, B \cap |(\mathfrak{A}_{k-1})_{d^{\mu'}}|\rangle.$$

Let $C_{\mu'}$ be the set $\{c^{\mu'}_\tau : \tau \in \rho_{\mu'}$ and $c^{\mu'} \not\leq_{k-1} c^{\mu'}_\tau\} \cup \{c^{\mu'}\}$. We define $\rho_{\mu+1} := \mathrm{card}(C_{\mu'})$. Let $c^{\mu'+1}_0, \ldots, c^{\mu'+1}_\tau$ $(\tau < \rho_{\mu'+1})$ be an enumeration of $C_{\mu'}$. Moreover, we define $b^{\mu'+1}$ ($d^{\mu'+1}$, respectively) to be that element $b^{\mu'}_{\tau'}$ ($d^{\mu'}_{\tau'}$, respectively) for which $c^{\mu'+1}_\tau := c^{\mu'}_{\tau'}$. Obviously, $b^{\mu'+1}_\tau, c^{\mu'+1}_\tau, d^{\mu'+1}_\tau$ have the property (*) for all $\tau \in \rho_{\mu'+1}$. Moreover, the function $f_{\mu'+1} := \{(c^{\mu'+1}_\tau, d^{\mu'+1}_\tau) : \tau \in \rho^{\mu'+1}\}$ is a 1-1 function which is compatible with $f_\mu$ for all $\mu \leq \mu'$. This means that $f_{\mu'+1}(c) = f_\mu(c)$ holds for every c from the set $\{c^{\mu'+1}_\tau : \tau \in \rho^{\mu'+1}\}$ $\cap$ $\{c^{\mu}_\tau : \tau \in \rho_\mu\}$. If for all elements $a \in B_{\mu'}$, $h_{\mathfrak{A}_{k-1}}(a) <$ $h_{\mathfrak{A}_{k-1}}(a_0)$ holds, we define

$$|\mathfrak{A}'_k| := B_{\mu'} \cup (\bigcup_{\tau\in\mu'} |(\mathfrak{A}_{k-1})_{d^{\mu'}_\tau}|),$$

$$\mathfrak{A}'_k := \langle|\mathfrak{A}'_k|, \leq_{k-1}\upharpoonright|\mathfrak{A}'_k|\rangle$$

and

$$B' := B \cap |\mathfrak{A}'_k|.$$

Using Lemma 8 we get $\langle\mathfrak{A}, A_0\rangle \sim^{b1}_{n-1} \langle\mathfrak{A}'_k, B'\rangle$. Moreover, we have $h_{\mathfrak{A}'_k}(a) < h_{\mathfrak{A}_{k-1}}(a_0)$ for all $a \in B'$. If for all elements a of $B'$

$$h_{\mathfrak{A}'_k}(a) \leq k(\varkappa^{\sigma}(n-1)+1) + \varkappa^{\nu}(n-1) + 1$$

holds, we define $\mathfrak{A}_k := \mathfrak{A}'_k$ and $A'_0 := B'$. If this inequality does not hold for all $a \in B'$, we repeat the above procedure. After a finite number of steps the condition is fulfilled.

If $\lambda$ is a limit ordinal and we defined $\varrho_\mu$, $f_\mu$, $b^\mu_\tau$, $c^\mu_\tau$, $d^\mu_\tau$ for all $\mu \in \lambda$ and $\tau \in \varrho_\mu$ so that (*) is fulfilled and the $f_\mu$ are pairwise compatible, then we proceed as follows. Let $C_\lambda$ be the set $\bigcup_{\gamma\in\lambda} \bigcap_{\gamma\leq\mu\leq\lambda} \{c^\mu_\tau : \tau<\mu\}$. By the above constructions $C_\lambda$ is the set

$$\{c^\mu_\tau : \mu \in \lambda \ \& \ \tau \in \mu \ \& \text{ there is no } \mu' \in \lambda \text{ and no } \tau' \in \mu' \text{ so that } c^{\mu'}_{\tau'} < c^\mu_\tau\}.$$

Since all our trees are $\omega$-trees, it is $C_\lambda \neq 0$. Moreover, for every $c, c'$ from $C_\lambda$, $c < c'$ does not hold. Let $\varrho_\lambda$ be $\mathrm{card}(C_\lambda)$ and let $c^\lambda_0, \ldots, c^\lambda_\tau, \ldots$, $\tau \in \varrho_\lambda$ be an enumeration of $C_\lambda$. If $f_\mu(c^\lambda_\tau)$ is defined for some $\mu \in \lambda$, we define $d^\lambda_\tau := f_\mu(c^\lambda_\tau)$. This definition does not depend on $\mu$ since the $f_\mu$ are compatible. Now we can proceed as in the successor case.

Now let $k < l' < k+l$. We assume that $\mathfrak{N}_{l'-1}$, $a'_j$ and $A'_i$ are defined for every $j \in k$ and every $i \in l'$ with $k \leq i$ so that the following conditions are fulfilled:

(i) $\mathfrak{N} \sim^{b1}_{n} \mathfrak{N}_{l'-1}$,

(ii) $a'_j \in |\mathfrak{N}_{l'-1}|$,

(iii) $\langle \mathfrak{N}, a_j \rangle \sim^{b1}_{n-1} \langle \mathfrak{N}_{l'-1}, a'_j \rangle$

(iv) $h_{\mathfrak{N}_{l'-1}}(a_j) \leq k(\varkappa^\sigma(n-1)+1)$,

(v) $A'_i \in P_{b1}(\mathfrak{N}_{l'-1})$,

(vi) $\langle \mathfrak{N}, A_i \rangle \sim^{b1}_{n-1} \langle \mathfrak{N}_{l'-1}, A'_i \rangle$,

(vii) for every $a \in A'_i$ it holds that
$$h_{\mathfrak{N}_{l'-1}}(a) \leq k(\varkappa^\sigma(n-1)+1) + (l'-k)(\varkappa^\nu(n-1)+1).$$

From (vi) and $n > 2$ we get that: $A_i \in P_{b1}(\mathfrak{N})$ iff $A'_i \in P_{b1}(\mathfrak{N}_{l'-1})$. Using the same technique as in case $l' = k + 1$ we prove the existence of a structure $\mathfrak{N}_{l'}$ and a set $A'_{l'}$ with the above properties ((i),..., (vii)) for $l' + 1$ instead of $l'$.

By induction we prove in this way the existence of an $\mathfrak{N}_{k+l-1}$ and of $a'_0,\ldots,a'_{k-1}$, $A'_k,\ldots,A'_{k+l-1}$ with the above properties.

Now let $n > 0$ be given. We define $n' := n + 2$ and

$$m_n := k'(\varkappa^{\sigma}(n'-1)+1) + l'(\varkappa^{\nu}(n'-1)+1).$$

Here $k'$ and $l'$ are defined in the same way as above, but with $n'$ instead of $n$. For every $\omega$-tree $\mathfrak{N}$ we define $\mathfrak{N}'$ to be $\mathfrak{N}_{k'+l'-1}$. Using the above and $n' > n$ it is clear that this $\mathfrak{N}'$ has the desired properties.

Q.E.D.

## § 3. A dense system for connected $\omega$-trees

As in Chapter 4 we introduce some helpful unary predicates $W$ (roots) and for all $n \in \omega$, $P_n$ (buds of color $n$). We shall extend the corresponding language $L'$ by these predicates. The new language is called $L_p$. Let $\mathfrak{A}$ be a model of this language, then $W(\mathfrak{A})$, $P_n(\mathfrak{A})$ and $\leq(\mathfrak{A})$ denote the corresponding interpretations of $W$, $P_n$ and $\leq$ in $\mathfrak{A}$. Sometimes we also use $\mathfrak{A}$ as a subscript to denote these relations.

Definition. A structure $\mathfrak{A} = \langle A,\leq,W,P_0,P_1,\ldots\rangle$ for $L_p$ is a spring-tree iff

(i) $\langle A,\leq\rangle$ is an $\omega$-tree,
(ii) $\mathfrak{A} \models \forall v_1 v_2 \neg(W(v_1) \wedge v_2 < v_1)$,
(iii) for every $i \neq j$, $\mathfrak{A} \models \neg\exists v_1(P_i(v_1) \wedge P_j(v_1))$,
(iv) for every $i$, $\mathfrak{A} \models \forall v_1 v_2 \neg(P_i(v_1) \wedge v_1 < v_2)$.

Spring-trees are useful to construct new $\omega$-trees from given ones. Sometimes we need a copy of a given model with different universe. For that reason we introduce as in Chapter 4 the following notion: let $p$ be any object, then $\mathfrak{A}(p)$ is the unique model with universe $\{(p,a) : a \in |\mathfrak{A}|\}$ so that the canonical map $a \longrightarrow (p,a)$ is an isomorphism.

Now we have to define some operations for $\omega$-trees. The operations which we use here are very closely connected with the corresponding operations of Chapter 4. Helpful illustrations can be found there.

Let $\mathfrak{A}$ and $\mathfrak{B}$ be spring-trees, then we define:

$$\mathfrak{A} + \mathfrak{B} := \mathfrak{A}(0) \cup \mathfrak{B}(1), \quad \Sigma_1(\mathfrak{A}) := \bigcup_{\alpha < \omega_1} \mathfrak{A}(\alpha).$$

$S\mathfrak{A}$ is the spring-tree described as follows:

$$|S\mathfrak{A}| := |\mathfrak{A}(0)| \cup W(\mathfrak{A}(1)),$$

$$\leq(S\mathfrak{A}) := \leq(\mathfrak{A}(0)) \cup \{((1,a),(0,b)) : a \in W(\mathfrak{A}) \text{ and } \mathfrak{A} \models a \leq b\},$$

$$W(S\mathfrak{A}) := W(\mathfrak{A}(1)),$$

and for all $n \in \omega$,

$$P_n(S\mathfrak{A}) := P_n(\mathfrak{A}(0)).$$

At last the spring-tree $\Pi\mathfrak{A}$ (up to isomorphism) arises from $\mathfrak{A}$ by identifying the elements of $W(\mathfrak{A}) \setminus (\bigcup_{n\in\omega} P_n(\mathfrak{A}))$.

A connected spring-tree $\mathfrak{A}$ is said to be a branch if $W(\mathfrak{A}) = \{a\}$ for the minimal element a of $\mathfrak{A}$. A tree-valuation f is a map into the class of branches. Let $\mathfrak{M}$ be a spring-tree. Then $\mathfrak{M}^-$ is the spring-tree which arises from $\mathfrak{M}$ by setting $W(\mathfrak{M}^-) = \emptyset$ and leave all the other relations unchanged. $\mathfrak{M}^+$ arises from $\mathfrak{M}$ by setting $P_n(\mathfrak{M}^+) = \emptyset$ for all n and leave all the other relations unchanged.

Suppose f is a tree-valuation with $\mathrm{dom}(f) \supseteq \bigcup_{n\in\omega} P_n(\mathfrak{M})$ ($\mathrm{dom}(f) = \omega$, respectively). We define

$$|\mathfrak{M}_f| := |\mathfrak{M}(\mathfrak{M})| \cup \bigcup_{n\in\omega} \bigcup_{a\in P_n(\mathfrak{M})} |(f(a))(a)|$$

$$(|\mathfrak{M}_f| := |\mathfrak{M}(\mathfrak{M})| \cup \bigcup_{n\in\omega} \bigcup_{a\in P_n(\mathfrak{M})} |(f(n))(a)|, \text{ respectively}),$$

$\leq(\mathfrak{M}_f) := \{((\mathfrak{M},a),(b,c))$: there exists an $n \in \omega$ with $b \in P_n(\mathfrak{M})$ and there exists a $d \in W(f(b))$ so that $\mathfrak{M} \models a \leq b$ and $f(b) \models d \leq c\}$

$(\leq(\mathfrak{M}_f) := \{((\mathfrak{M},a),(b,c))$: there exists an $n \in \omega$ with $b \in P_n(\mathfrak{M})$ and there exists a $d \in W(f(n))$ so that $\mathfrak{M} \models a \leq b$ and $f(b) \models d \leq c\}$, resp.),

$$W(\mathfrak{M}_f) := W(\mathfrak{M}(\mathfrak{M}))$$

$$(W(\mathfrak{M}_\varphi) := W(\mathfrak{M}(\mathfrak{M})), \text{ respectively}),$$

for every n,

$$P_n(\mathfrak{M}_f) := \bigcup_{m\in\omega} \bigcup_{a\in P_m(\mathfrak{M})} P_n(f(a)(a))$$

$$(P_n(\mathfrak{M}_f) := \bigcup_{m\in\omega} \bigcup_{a\in P_m(\mathfrak{M})} P_n(f(m)(a)), \text{ respectively}).$$

The resulting spring-tree $\mathfrak{M}_f$ is also denoted by $\sigma^1(\mathfrak{M},f)$ in both cases. $\sigma^1(\mathfrak{M},f)$ has a simple meaning: at $a \in \bigcup_{n\in\omega} P_n(\mathfrak{M})$ ($a \in P_n(\mathfrak{M})$) we substitute a copy of the spring-tree $f(a)^-$ ($f(n)^-$, respectively). In the second case we can iterate this process by defining:

$$\sigma^{m+1}(\mathfrak{M},f) := \sigma^1(\sigma^m(\mathfrak{M},f),f).$$

Obviously the canonical map $\varphi : |\mathfrak{M}| \longrightarrow |\sigma^1(\mathfrak{M},f)|$ defined by $\varphi(a) := (\mathfrak{M},a)$ is an embedding of $\mathfrak{M}$ into $\sigma^1(\mathfrak{M},f)$ with respect to $\leq$ and the predicate W. Let $\sigma(\mathfrak{M},f)$ be the direct limit of $\{\sigma^i(\mathfrak{M},f)^+: i \in \omega\}$. For a spring-tree $\mathfrak{M}$ and an element a of $|\mathfrak{M}|$ we define:

$$|\mathfrak{M}_a| := \{b: b \in |\mathfrak{M}| \text{ and } \mathfrak{M} \models a \leq b\},$$

$$\leq(\mathfrak{M}_a) := (\leq(\mathfrak{M}))\upharpoonright|\mathfrak{M}_a|,$$

$$W(\mathfrak{M}_a) := \{a\},$$

$$P_n(\mathfrak{M}_a) := P_n(\mathfrak{M})\upharpoonright|\mathfrak{M}_a| \quad \text{for every } n \in \omega.$$

Now let us regard $+,S,\Sigma_1$, and $\Pi$ as function symbols of a first order language. Let 1 be a constant and $\{v_i: i\in\omega\}$ be the set of variables of this language. The set T of terms is built as usual. $T^{\Pi S}$ is the set of terms of the form $\Pi S\alpha$, where $\alpha$ is a term. A finite sequence $(\alpha_0,\dots,\alpha_m)$ of terms is called an extended term iff

(1) $\alpha_1,\dots,\alpha_m$ are terms from $T^{\Pi S}$,
(2) in each term $\alpha_i$ at most $v_1,\dots,v_m$ occur.

The set of extended terms is denoted by E. Let EC be the set

$$\{(\alpha_0,\dots,\alpha_n): (\alpha_0,\dots,\alpha_n) \in E \ \&\ \alpha_0 \in T^{\Pi S}\}.$$

Now we shall see how every extended term represents a certain spring-tree. Suppose $\alpha$ is an extended term, then $\alpha^*$ denotes the spring-tree represented by $\alpha$. We identify terms with extended terms of length 1. Firstly let us define which trees are represented by 1 and $v_i$ for $i \in \omega$.

$$1^* := (\{0\},\{(0,0)\},W(1^*),P_0(1^*),\dots),$$

where $W(1^*) := \{0\}$ and for all $j < \omega$, $P_j(1^*) := \emptyset$,

$$v_i^* := (\{0\},\{(0,0)\},W(v_i^*),P_0(v_i^*),\dots),$$

where $W(v_i^*) := \{0\}$ and for all $j \neq i$, $P_j(v_i^*) := \emptyset$, while $P_i(v_i^*) = \{0\}$.

We extend the representation to terms by:

$$(\alpha + \beta)^* := \alpha^* + \beta^*, \qquad (S\alpha)^* := S(\alpha^*),$$

$$(\Pi\alpha)^* := \Pi(\alpha^*) \quad \text{and} \quad (\Sigma_1\alpha)^* := \Sigma_1\alpha^*.$$

Thus we defined the spring-tree $\alpha^*$ for every term $\alpha$. Let $\alpha$ be an extended term, i.e. $\alpha = (\alpha_0,\dots,\alpha_n)$, where $\alpha_0,\dots,\alpha_n$ are terms. Let $f_\alpha$ be a tree-valuation so that $\mathrm{dom}(f_\alpha) = \omega$ and for $1 \leq i \leq n$, $f_\alpha(i) = \alpha_i^*$. Then $\alpha^*$ is defined to be $\sigma(\alpha_0^*, f_\alpha)$.

Remark. If $W(\alpha^*) = \{a\}$ for the minimal element a of $\alpha^*$ (we are assuming here that there is a minimal element of $\alpha^*$) and all $P_k(\alpha^*)$ (for $k \in \omega$) are empty, then we shall identify $\alpha^*$ with the restriction of $\alpha^*$ to the language without the predicates W and $P_k$ ($k \in \omega$). Using this natural identification all structures $\alpha^*$ with $\alpha \in EC$ are (normal) $\omega$-trees.

Lemma 10. The sets E and EC are recursive.

The proof is obvious.

Let $\langle A,\leq\rangle$ be an $\omega$-tree and a be an element of A. An element b of A is said to be an immediate successor of a if $a < b$ and there is no $c \in A$ with $a < c < b$.

**Lemma 11.** For every $n > 0$ there exists a natural number m with the following property: For every connected $\omega$-tree $\mathfrak{N}$ which contains elements of each height there exist a term $\alpha$, a tree-valuation f, and a connected $\omega$-tree $\mathfrak{N}''$ so that:

(i) $\mathrm{dom}(f) \supseteq \bigcup_{m \in \omega} P_m(\alpha^*)$,

(ii) $\max(\{h_{\alpha^*}(a) : a \in |\alpha^*|\}) = m_n$,

(iii) $\mathfrak{N}'' = \sigma^1(\alpha^*, f)$,

(iv) for every $b \in |\mathfrak{N}|$ there exists a $b' \in |\mathfrak{N}''|$ which is bounded in $\mathfrak{N}''$ by $m_n$ so that $\langle \mathfrak{N}, b\rangle \sim^{b1}_{n-1} \langle \mathfrak{N}'', b'\rangle$,

(v) for every $B \in P_{b1}(\mathfrak{N})$ there exists a $B' \in P_{b1}(\mathfrak{N}'')$ bounded in $\mathfrak{N}''$ by $m_n$ so that $\langle \mathfrak{N}, B\rangle \sim^{b1}_{n-1} \langle \mathfrak{N}'', B'\rangle$.

**Proof.** Let $n > 0$ and $\mathfrak{N}$ be given and let $\mathfrak{N}'$ and $m_n$ be as in Lemma 9. Let a be the minimal element of $\mathfrak{N}'$. We define $S^{1}_{\mathfrak{N}'}(a)$ to be the set

$$\{b : b \in |\mathfrak{N}'| \text{ and b is the immediate successor of a in } \mathfrak{N}'\}.$$

Let $b_0, \ldots, b_{k-1}$ be elements of $\mathfrak{N}'$ and let $B_0, \ldots, B_{l-1}$ be elements of $P_{b1}(\mathfrak{N}')$ so that the following is fulfilled:

(+)
- for every $i, j \in k$ $b_i = b_j$ holds iff $\langle \mathfrak{N}', b_i\rangle \sim^{b1}_{n-1} \langle \mathfrak{N}', b_j\rangle$,
- for every $b \in |\mathfrak{N}'|$ there exists an $i \in k$ so that $\langle \mathfrak{N}', b\rangle \sim^{b1}_{n-1} \langle \mathfrak{N}', b_i\rangle$ holds,
- $h_{\mathfrak{N}'}(b_i) < m_n$ for all $i \in k$,
- for every $i, j \in l$ $B_i = B_j$ holds iff $\langle \mathfrak{N}', B_i\rangle \sim^{b1}_{n-1} \langle \mathfrak{N}', B_j\rangle$,
- for every $B \in P_{b1}(\mathfrak{N}')$ there exists an $i \in l$ so that $\langle \mathfrak{N}', B\rangle \sim^{b1}_{n-1} \langle \mathfrak{N}', B_i\rangle$ holds,
- $B_i$ is bounded by $m_n$ in $\mathfrak{N}'$ for all $i \in l$.

Obviously there exists a natural number $n_0$ so that

$$\langle \mathfrak{N}', b_0, \ldots, b_{k-1}, B_0, \ldots, B_{l-1}\rangle \sim^{b1}_{n_0} \langle \mathfrak{M}, c_0, \ldots, c_{k-1}, C_0, \ldots, C_{l-1}\rangle$$

implies that $c_0, \ldots, c_{k-1}, C_0, \ldots, C_{l-1}$ have the property (+) for $\mathfrak{M}$ for all $\omega$-trees $\mathfrak{M}$ and all $c_0, \ldots, c_{k-1} \in |\mathfrak{M}|$, $C_0, \ldots, C_{l-1} \in P_{b1}(\mathfrak{M})$.

Now we define the following binary relation E on $S^1_{\mathfrak{N}'}(a)$:

$$(b,c) \in E \text{ \underline{iff} } (c = b \text{ and } (b \leq b_0 \text{ or } \ldots \text{ or } b \leq b_{k-1} \text{ or } c \leq b_0 \text{ or } \ldots \text{ or } c \leq b_{k-1}))$$

$$\text{or}$$

$$(\langle \mathfrak{N}'_b, B_0 \cap |\mathfrak{N}'_b|, \ldots, B_{l-1} \cap |\mathfrak{N}'_b|\rangle \sim^{b1}_{n_0} \langle \mathfrak{N}'_c, B_0 \cap |\mathfrak{N}'_c|, \ldots, B_{l-1} \cap |\mathfrak{N}'_c|\rangle$$
$$\text{and } b \not\leq b_0 \text{ and } \ldots \text{ and } b \not\leq b_{k-1} \text{ and } c \not\leq b_0 \text{ and } \ldots \text{ and } c \not\leq b_{k-1}).$$

Obviously E is an equivalence relation with finitely many equivalence classes only.

Using Lemma 7 it is easy to find a subtree $\mathfrak{N}^1$ of $\mathfrak{N}'$ and a natural number s with the following properties:

(i) $b_0, \ldots, b_{k-1} \in |\mathfrak{N}^1|$,

(ii) $|\mathfrak{N}^1| \subseteq |\mathfrak{N}'|$,

(iii) $\mathfrak{N}^1 = \mathfrak{N}' \upharpoonright |\mathfrak{N}^1|$,

(iv) $\langle \mathfrak{N}^1, b_0, \ldots, b_{k-1}, B_0 \cap |\mathfrak{N}^1|, \ldots, B_{l-1} \cap |\mathfrak{N}^1|\rangle \sim^{b1}_{n_0} \langle \mathfrak{N}', b_0, \ldots, b_{k-1}, B_0, \ldots, B_{l-1}\rangle$,

(v) for each equivalence class $\mathfrak{b}$ of E, $\text{card}(\mathfrak{b} \cap |\mathfrak{N}^1|) \leq s$ or $\text{card}(\mathfrak{b} \cap |\mathfrak{N}^1|) = \omega_1$.

But from these properties (especially from (v)) we get the existence of a term $\beta^1$ and a tree-valuation $f^1$ with:

(vi) $\text{dom}(f^1) \supseteq \bigcup_{m \in \omega} P_m((\beta^1)^*)$

and

(vii) $\mathfrak{N}^1 \cong \sigma^1((\beta^1)^*, f^1)$.

Moreover, from (iv) we get that the conditions (iv) and (v) from Lemma 11 hold for $\mathfrak{N}^1$.

Now we can proceed by induction. Let t be a natural number with $1 \leq t < m_n$ and let us assume that $\pi^t$, $\beta^t$, and $f^t$ are already defined so that:

(viii) $a, b_0, \ldots, b_{k-1} \in |\pi^t|$,

(ix) $|\pi^t| \subseteq |\pi'|$,

(x) $\pi^t = \pi' \restriction |\pi^t|$,

(xi) $\langle \pi^t, b_0, \ldots, b_{k-1}, B_0 \cap |\pi^t|, \ldots, B_{l-1} \cap |\pi^t| \rangle \sim^{b1}_{n_0}$
$\langle \pi', b_0, \ldots, b_{k-1}, B_0, \ldots, B_{l-1} \rangle$,

(xii) $\max(\{h_{(\beta^t)^*}(b) : b \in |(\beta^t)^*|\}) = t$,

(xiii) $\mathrm{dom}(f^t) \supseteq \bigcup_{m \in \omega} P_m((\beta^t)^*)$,

(xiv) $\pi^t \cong \sigma^1((\beta^t)^*, f^t)$.

Now let $d \in |\pi^t|$ be given such that $\mathrm{card}(\hat{d}) = t-1$. Let $E^t$ be defined by:

$$
\begin{aligned}
(b,c) \in E^t \ \underline{\text{iff}}\ & \mathrm{card}(\hat{b}) = \mathrm{card}(\hat{c}) = t \text{ and} \\
& [(c = b \text{ and } (b \leq b_0 \text{ or } \ldots \text{ or } b \leq b_{k-1} \text{ or} \\
& \qquad\qquad c \leq b_0 \text{ or } \ldots \text{ or } c \leq b_{k-1})] \\
& \text{or} \\
& (\langle \pi^t_b, B_0 \cap |\pi^t_b|, \ldots, B_{l-1} \cap |\pi^t_b| \rangle \sim^{b1}_{n_0} \\
& \langle \pi^t_c, B_0 \cap |\pi^t_c|, \ldots, B_{l-1} \cap |\pi^t_c| \rangle \\
& \text{and } b \nleq b_0 \text{ and } \ldots \text{ and } b \nleq b_{k-1} \\
& \text{and } c \nleq b_0 \text{ and } \ldots \text{ and } c \nleq b_{k-1}).
\end{aligned}
$$

We consider $S^1_{\pi^t_d}(d)$ and the restriction of $E^t$ to this set. Then we define a subtree $(\pi^t_d)^1$ of $\pi^t_d$, a term $\alpha_d$, and a tree-valuation $f_d$ for $\pi^t_d$ as above for $\pi'$ and E. But obviously there are (up to isomorphic copies) only finitely many of such spring-trees $(\alpha_d)^*$ for $d \in |\pi^t|$ (hint: use the analogy of (v) above in this case).

Using this and the properties of $\pi^t$, $\beta^t$, $f^t$, $(\pi^t_d)^1$, $\alpha_d$, and $f_d$ it is easy to construct $\pi^{t+1}$, $\beta^{t+1}$, and $f^{t+1}$ so that (viii),...,(xiv) are fulfilled (hint: use also Lemma 8).

In case $t = m_n - 1$, we define $\mathfrak{N}'' = \sigma^{1}((\beta^{m_n})^{*}, f^{m_n})$ and proved the lemma.

Q.E.D.

Using the same technique we get the following lemma.

<u>Lemma 12.</u> For each natural number n there exists a natural number $m_n$ with the following property: For every connected $\omega$-tree $\mathfrak{N}$, which does not contain elements of any height, there exists a term $\alpha$ so that $\mathfrak{N} \sim_n^{b1} \alpha^*$ holds, and $\max(\{h_{\alpha^*}(b) : b \in |\alpha^*|\}) < m_n$.

The following strengthened version of Lemma 11 we need in the proof of the next theorem.

<u>Lemma 13.</u> For every $n > 0$ there exists a natural number $m_n$ with the following property: For every connected $\omega$-tree $\mathfrak{N}$ which contains elements of each height there exist a term $\alpha \in T^{\Pi S}$, a tree-valuation f, and a connected $\omega$-tree $\mathfrak{N}''$ so that $\mathrm{dom}(f) = \omega$ and (ii), ..., (iv) of Lemma 11 are fulfilled.

The proof of this lemma is left as an exercise to the reader. (Hint: Use a similar technique as in the proof of Lemma 11 and remember that $\sim_n^{b1}$ has only finitely many equivalence classes.)

<u>Theorem 14.</u> Let $\mathfrak{M}$ be a connected $\omega$-tree and $n \in \omega$. Then there is an extended term $\alpha$ so that $\mathfrak{M} \sim_n^{b1} \alpha^*$.

<u>Proof.</u> If there is a natural number l so that there is no point of of height greater than l, then we can choose $\alpha$ by Lemma 12.

$\sim_n^{b1}$ defines an equivalence relation on the class of all connected $\omega$-trees which are not isomorphic to $1^*$. This relation has finitely many equivalence classes only. Let $\mathfrak{M}_1^n, \ldots, \mathfrak{M}_{k(n)}^n$ be a complete system of representatives with respect to $\sim_n^{b1}$ over the class of $\omega$-trees being not isomorphic to $1^*$. If $\mathfrak{M}_i^n$ contains elements of each height, then let $(\mathfrak{M}_i^n)''$ be as in Lemma 13. In case that $\mathfrak{M}_i^n$ does not contain elements of each height, let $(\mathfrak{M}_i^n)''$ be the corresponding $\alpha^*$ as in Lemma 12. For simplicity we denote $(\mathfrak{M}_1^n)'', \ldots, (\mathfrak{M}_{k(n)}^n)''$ by $\mathfrak{N}_1, \ldots, \mathfrak{N}_{k(n)}$, respectively.

Now we define a spring-tree $\mathfrak{S}_i$ for every i $(1 \le i \le k(n))$. We use here the corresponding extended language $L_p$. In case $\mathfrak{N}_i$ does not have an element of height $m_n$ (we assume that the $m_n$ of Lemma 12 and Lemma 13 are the same, by choosing the maximum of both) we set

$$\langle |\mathfrak{L}_i|, \leq(\mathfrak{L}_i), W(\mathfrak{L}_i), P_0(\mathfrak{L}_i), P_1(\mathfrak{L}_i), \ldots\rangle := \langle |\mathfrak{N}_i|, \leq(\mathfrak{N}_i), \{a_i\}, \emptyset, \emptyset, \ldots\rangle,$$

where $a_i$ is the minimal element of $\mathfrak{N}_i$. Otherwise we define:

$$|\mathfrak{L}_i| := \{a: a \in |\mathfrak{N}_i| \text{ and } h_{\mathfrak{N}_i}(a) \leq m_n\},$$

$$\leq_{\mathfrak{L}_i} := \leq_{\mathfrak{N}_i} \upharpoonright |\mathfrak{L}_i|,$$

$$W(\mathfrak{L}_i) := \{a: a \in |\mathfrak{N}_i| \text{ and a is the minimal element of } \mathfrak{N}_i\},$$

$$P_j(\mathfrak{L}_i) := \{a: a \in |\mathfrak{N}_i| \text{ and } h_{\mathfrak{N}_i}(a) = m_n \text{ and } (\mathfrak{N}_i)_a \sim_n^{b1} \mathfrak{N}_j\} \quad \text{for } 1 \leq j \leq k(n),$$

$$P_j(\mathfrak{L}_i) := \emptyset \text{ for } k(n) < j.$$

Using Lemma 13 and Lemma 12 it is not difficult to find a term $\alpha_i \in T^{\Pi S}$ representing $\mathfrak{L}_i$ (for all i with $1 \leq i \leq k(n)$). Let f be a tree-valuation so that $f(i) := \mathfrak{L}_i$ for $1 \leq i \leq k(n)$; otherwise let f be arbitrary. Then we obviously have $\mathfrak{N}_i \sim_n^{b1} \sigma^1(\alpha_i^*, f)$. We define by induction:

$$\alpha_i^0 := \alpha_i, \quad \alpha_i^{m+1} := \alpha_i^m(\alpha_1, \ldots, \alpha_{k(n)}) \ (m \in \omega).$$

Here $\alpha_i^n(\alpha_1, \ldots, \alpha_{k(n)})$ is obtained from $\alpha_i^m$ by substituting the term $\alpha_j$ ($j = 1, \ldots, k(n)$) for the variables $v_j$. From this we get for all m

$$\mathfrak{N}_i \sim_n^{b1} \sigma^1((\alpha_i^m)^*, f).$$

We have to show that each $\mathfrak{N}_i$ can be represented by an extended term. Without loss of generality we show this for $\mathfrak{N}_1$. Let $\alpha$ be the extended term $(\alpha_1, \alpha_1, \ldots, \alpha_{k(n)})$. To use Theorem 1 we define for every $i \in \omega$: $\mathfrak{A}_i' := \sigma^1((\alpha_1^i)^*, f)$. We choose for every $i \in \omega$ an isomorphic copy $\vartheta_i$ of $\mathfrak{A}_i'$ so that for all natural numbers $k < n$

$$\{a: a \in |\vartheta_k| \text{ and } h_{\vartheta_k}(a) \leq (k+1)m_n\} \subseteq \{a: a \in |\vartheta_l| \text{ and } h_{\vartheta_l}(a) \leq (l+1)m_n\}$$

holds. Then we define for every $i \in \omega$

$$D_i := \{a: a \in |\vartheta_i| \text{ and } h_{\vartheta_i}(a) \leq (i+1)m_n\} \text{ and } \vartheta := \bigcup_{i\in\omega} \vartheta_i \upharpoonright D_i.$$

It is sufficient to show $\vartheta_0 \sim_n^{b1} \vartheta$, since $\alpha^* \cong \vartheta$.

Let $\mathfrak{A}_i = \langle A_i, S_i, E_i, \leq_i\rangle$ be $\vartheta_i^+$ and let $C_i$ be $P_{b1}(D_i) \cup \{\{a\}: a \in D_i\}$ for every $i \in \omega$. We define $\mathfrak{L} := \vartheta^+$ and get $\mathfrak{L} = \bigcup_{i\in\omega} \mathfrak{A}_i \upharpoonright C_i$. We only have to prove $\mathfrak{A}_0 \sim_n \mathfrak{L}$. We know that $\mathfrak{A}_0 \sim_n \mathfrak{A}_i$ (for $i \in \omega$) since $\vartheta_0 \sim_n^{b1} \vartheta_i$. To apply Theorem 1 we have to define the corresponding winning strategies $W_i$ for player II so that (a), (b), and (c) of this theorem are fulfilled. First of all we define for every natural number i a winning strategy $V_i$ for player II in the game corresponding to $\mathfrak{A}_i \sim_n \mathfrak{A}_{i+1}$. Let $H_i$ be the set

$$\{a: a \in |\vartheta_i| \text{ and } h_{\vartheta_i}(a) = (i+1)m_n\}$$

and let $G_i$ be the set

$$\{a: a \in |\vartheta_{i+1}| \text{ and } h_{\vartheta_{i+1}}(a) = (i+1)m_n\}.$$

Using Lemma 12 and Lemma 13 we find for every connected $\omega$-tree $\pi \neq 1^*$ an $i_0$ and a $U'_\pi$ with the following properties:

$1 \leq i_0 \leq k(n)$,

$\pi^+ \sim_n \pi_{i_0}^+$ and $U'_\pi$ is a winning strategy for player II in the corresponding game,

for every $M \in |\pi^+|$ and every $a \in U'_\pi((\pi^+, M))$ it holds that $h_{\pi_{i_0}}(a) \leq m_n$.

There is an isomorphic embedding $f'_i$ of $\vartheta_i \upharpoonright C_i$ in $\vartheta_{i+1}$ which is directly given from the construction of $\vartheta_{i+1}$ from $\vartheta_i$. Using the above we get for every $a \in H_i$ a winning strategy $U_a$ for player II in the game corresponding to $((\vartheta_i)_a)^+ \sim_n ((\vartheta_{i+1})_{f'_i(a)})^+$ which has the property that for every $M \in |((\vartheta_i)_a)^+|$ and every $b \in U_a((((\vartheta_i)_a)^+, M))$, $h_{((\vartheta_{i+1})_{f'_i(a)})^+}(b) \leq m_n$ holds.

The isomorphic embedding $f'_i$ directly leads to a winning strategy $F_i$ for player II in the game corresponding to $\mathfrak{A}_i \upharpoonright C_i \sim_n \mathfrak{A}_{i+1} \upharpoonright P(\mathrm{range}(f'_i))$. Now we define $V_i$ in an inductive way. Let us define $V_i(\varphi) := \varphi$,

$\varrho^0 := \varrho$ and $\varrho_a := \varrho$ for every $a \in H_i$. We assume $\bar{u} = (\bar{v},(\mathfrak{A}_i,M))$, where $\bar{v}$ is chosen according to (the already defined part of) $V_i$ and $\bar{v}$ is of length at most 2n - 2. Moreover, we assume that $M \in |\mathfrak{A}_i|$ and that for every $a \in H_i$, $\bar{v}_a$ is defined and chosen according to $U_a$ and that $\bar{v}^0$ is defined and chosen according to $F_i$. We define:

$$M_0 := M \cap D_i, \quad M_a := M \cap |(\vartheta_i)_a| \text{ for every } a \in H_i,$$

$$\bar{u}^0 := (\bar{v}^0,(\mathfrak{A}_i \upharpoonright C_i,M_0),(\mathfrak{A}_{i+1} \upharpoonright P(\text{range}(f_i')),F_i((\bar{v}^0,(\mathfrak{A}_i \upharpoonright C_i,M_0))))),$$

$$\bar{u}_a := \bar{v}_a \qquad \text{if } a \in H_i \text{ and } M_a = \emptyset,$$

$$\bar{u}_a := (\bar{v}_a,(((\vartheta_i)_a)^+,M_a),(((\vartheta_{i+1})_{f_i'(a)})^+,U_a((\bar{v}_a,(((\vartheta_i)_a)^+,M_a)))))$$
$$\text{if } a \in H_i \text{ and } M_a \neq \emptyset,$$

$$V_i(\bar{u}) := F_i((\bar{v}^0,(\mathfrak{A}_i \upharpoonright C_i,M_0))) \cup \bigcup_{\substack{a \in H_i \\ M_a \neq \emptyset}} U_a((\bar{v}_a,(((\vartheta_i)_a)^+,M_a))).$$

Now we assume that $\bar{u} = (\bar{v},(\mathfrak{A}_{i+1},M))$, where $\bar{v}$ is chosen according to (the already defined part of) $V_i$ and $\bar{v}$ is of length at most 2n - 2. Moreover, we assume: $M \in |\mathfrak{A}_{i+1}|$, $\bar{v}_a$ is defined and chosen according to $U_a$ for every $a \in H_i$, $\bar{v}^0$ is defined and chosen according to $F_i$. We define:

$$M_0 := M \cap \text{range}(f_i'), \quad M_a := M \cap |(\vartheta_i)_{f_i'(a)}|,$$

$$\bar{u}^0 := (\bar{v}^0,(\mathfrak{A}_{i+1} \upharpoonright P(\text{range}(f_i')),M_0),(\mathfrak{A}_i \upharpoonright C_i,F_i((\bar{v}^0,$$
$$(\mathfrak{A}_{i+1} \upharpoonright P(\text{range}(f_i')),M_0))))),$$

$$\bar{u}_a := \bar{v}_a \qquad \text{if } a \in H_i \text{ and } M_a = \emptyset,$$

$$\bar{u}_a := (\bar{v}_a,(((\vartheta_{i+1})_{f_i'(a)})^+,M_a),(((\vartheta_i)_a)^+,$$
$$U_a((\bar{v}_a,(((\vartheta_{i+1})_{f_i'(a)})^+,M_a))))) \text{ if } a \in H_i \text{ and } M_a \neq \emptyset,$$

$$V_i(\bar{u}) := F_i((\bar{v}^0,(\mathfrak{A}_{i+1} \upharpoonright P(\text{range}(f_i')),M_0)))$$
$$\cup \bigcup_{\substack{a \in H_i \\ M_a \neq \emptyset}} U_a((\bar{v}_a,(((\vartheta_{i+1})_{f_i'(a)})^+,M_a))).$$

Obviously, $V_i$ is a winning strategy for player II in the game corresponding to $\mathfrak{A}_i \sim_n \mathfrak{A}_{i+1}$. $W_i$ is defined to be the natural combination of $V_0,\dots,V_{i-1}$.

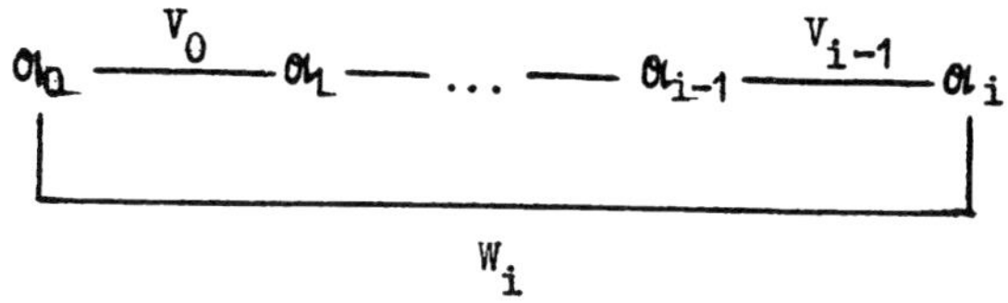


It is an immediate consequence of the above definitions that $W_i$ fulfils conditions (a), (b), and (c) from Theorem 1. This proves the theorem.

Q.E.D.

An immediate consequence of the proof of Theorem 14 is the following

Corollary 15. For every $n \in \omega$ there exists a finite set $\Delta_n$ of extended terms which has the following property: For every connected $\omega$-tree $\mathfrak{M}$ there exists an $\alpha \in \Delta_n$ with $\mathfrak{M} \overset{p1}{\sim}_n \alpha^*$. Moreover, the function $n \longrightarrow \Delta_n$ is recursive under a certain coding.

To prove the fact that "$n \longrightarrow \Delta_n$ is recursive" it should first be veryfied that $m_n$ and the set of all possible terms $\alpha$ in Lemma 11, Lemma 12, and Lemma 13 can be effectively constructed from n. The details are left to the reader.

It is possible to generalize the above results also to the extended signatures $\sigma(i,j)$. Before we give an exact formulation we have to give some definitions.

We extend the language by i individual constants $a_1,\dots,a_i$ and j set constants $A_1,\dots,A_j$. For any subset A of $\{a_1,\dots,a_i\}$ let Lj(A) be the extension of the original language L' by $A_1,\dots,A_j$ and those $a_k$ which are in A. For every such A let Lpj(A) be the extension of Lj(A) by unary predicates W and $P_n$ (for $n \in \omega$).

An A-spring-tree is an Lpj(A) structure $\mathfrak{A}$ so that the reduction of $\mathfrak{A}$ to Lp is a spring-tree.

Let A and B be disjoint subsets of $\{a_1,\dots,a_i\}$ and let $\mathfrak{A}$ be an A-spring-tree and let $\mathfrak{B}$ be a B-spring-tree. Then we define $\mathfrak{A} + \mathfrak{B}$ to be the $A \cup B$-spring-tree $\mathfrak{A}(0) \cup \mathfrak{B}(1)$.

For $A = \emptyset$ we define for all A-spring-trees $\mathfrak{A}$: $\Sigma_1 \mathfrak{A} := \bigcup_{\alpha \in \omega_1} \mathfrak{A}(\alpha)$.

Let A and B be disjoint subsets of $\{a_1,\dots,a_i\}$ and let C be a subset of $\{1,\dots,j\}$. For every connected A-spring-tree $\mathfrak{A}$ we define the

$A \cup B$-spring-tree $S_{B,C}\mathfrak{A}$ in the following way: Let $\mathfrak{A}'$ be the reduct of $\mathfrak{A}$ to Lp. There is an isomorphic embedding f of $\mathfrak{A}$ in $S\mathfrak{A}'$. For every k $(1 \leq k \leq j)$ we define

$$A_k' := \{f(a): a \in A_k\} \cup \{a: a \in W(S\mathfrak{A}') \text{ and } k \in C\}.$$

For every $a \in A \cup B$ we define:

$$a' := \begin{cases} f(a) & \text{if } a \in A, \\ b & \text{if } b \in W(S\mathfrak{A}') \text{ and } a \in B. \end{cases}$$

$S_{B,C}\mathfrak{A}$ is defined to be $\langle S\mathfrak{A}', A_1', \ldots, A_j', a' \rangle_{a \in A \cup B}$.

For an A-spring-tree $\mathfrak{A}$, $\Pi\mathfrak{A}$ arises from $\mathfrak{A}$ by identifying the elements of $W(\mathfrak{A}) \setminus (\bigcup_{n \in \omega} P_n(\mathfrak{A}))$. We have to keep attention to the constants. $\Pi\mathfrak{A}$ is then an A-spring-tree, too.

A tree-valuation f is now a map into the class of connected $\emptyset$-spring-trees. In the same way as above we define the notions $\sigma^1(\mathfrak{M}, f)$ and $\sigma(\mathfrak{M}, f)$.

<u>Remark.</u> All constants which appear in these trees lie in $\mathfrak{M}$.

Now let us regard $+$, $\Sigma_1$, $\Pi$, and $S_{A,C}$ for every $A \subseteq \{a_1, \ldots, a_i\}$ and every $C \subseteq \{1, \ldots, j\}$ as function symbols of a first order language. Let

$$\{1_{A,C}: A \subseteq \{a_1, \ldots, a_i\} \text{ and } C \subseteq \{1, \ldots, j\}\}$$

be the set of constants and

$$\{v_{l,A,C}: l \in \omega \text{ and } A \subseteq \{a_1, \ldots, a_i\} \text{ and } C \subseteq \{1, \ldots, j\}\}$$

be the set of variables of this language.

The set $T^{(i,j)}$ of terms is built as usual. $T^{(i,j)\Pi S}$ is defined to be the set

$$\{\Pi S_{A,B}\alpha: \alpha \in T^{(i,j)} \text{ and } A \subseteq \{a_1, \ldots, a_i\} \text{ and } C \subseteq \{1, \ldots, j\}\}.$$

A finite sequence $(\alpha_0, \ldots, \alpha_n)$ of terms is called an extended term iff

(1) $\alpha_1,\dots,\alpha_n$ are terms from $T^{(i,j)\Pi S}$,

(2) for each term $\alpha_i$ the set $\{l$: there exist A, C with $A \subseteq \{a_1,\dots,a_i\}$ and $C \subseteq \{1,\dots,j\}$ and $v_{l,A,C}$ occurs in $\alpha_i\}$ is contained in $\{1,\dots,n\}$,

(3) the elements of the set $\{A$: there exist l, C with $l \in \omega$ and $C \subseteq \{1,\dots,j\}$ and ($S_{A,C}$ occurs in $\alpha_0$ or $1_{A,C}$ occurs in $\alpha_0$ or $v_{l,A,C}$ occurs in $\alpha_0$)$\}$ are pairwise disjoint,

(4) the set $\{A$: there exist l, C with $l \in \omega$ and $C \subseteq \{1,\dots,j\}$ and ($S_{A,C}$ occurs in $\alpha_k$ or $1_{A,C}$ occurs in $\alpha_k$ or $v_{l,A,C}$ occurs in $\alpha_k$)$\}$ contains only the empty set for every $1 \leq k \leq n$.

The set of extended terms is denoted by $E^{(i,j)}$. Let $EC^{(i,j)}$ be the set

$$\{(\alpha_0,\dots,\alpha_n) : (\alpha_0,\dots,\alpha_n) \in E^{(i,j)} \text{ and } \alpha_0 \in T^{(i,j)\Pi S}\}.$$

As above we now define the spring-tree $\alpha^*$ represented by $\alpha$ for every $\alpha \in E^{(i,j)}$. We identify terms with extended terms of length 1. We define:

$$(1_{A,C})^* := \langle 1^*, A_1', \dots, A_j', a' \rangle_{a\in A},$$

where

$$A_l' = \begin{cases} 0 \text{ if } l \notin C, \\ 1 \text{ if } l \in C, \end{cases}$$

$a' = 0$ for all $a \in A$; and with the same notions

$$(v_{l,A,C})^* := \langle (v_l)^*, A_1', \dots, A_j', a' \rangle_{a\in A}.$$

Similarly as above we expand the representation to all terms and extended terms.

An extended term $\alpha \in EC^{(i,j)}$ is said to be i-full iff all the constants $a_1,\dots,a_i$ occur in $\alpha^*$.

<u>Lemma 16.</u> The set

$$\{(i,j,\alpha) : i \in \omega,\ j \in \omega,\ \alpha \in EC^{(i,j)} \text{ and } \alpha \text{ is i-full}\}$$

is recursive (under a certain coding).

The proof is obvious.

Theorem 17. Let $\Gamma_c$ and $\Gamma_s$, respectively, be the set

$$\{(i,j,\alpha,\beta):\ i\in\omega,\ j\in\omega,\ \alpha\in EC^{(i,j)},\ \beta\in EC^{(i+1,j)},\ \alpha \text{ is } i\text{-full},\ \beta \text{ is } (i+1)\text{-full, and } \alpha^{*}\cong\mathfrak{A}, \text{ where } \mathfrak{A} \text{ is the restriction of } \beta^{*} \text{ to the language } Lj(\{a_1,\ldots,a_i\})\}$$

and

$$\{(i,j,\alpha,\beta):\ i\in\omega,\ j\in\omega,\ \alpha\in EC^{(i,j)},\ \beta\in EC^{(i,j+1)},\ \alpha \text{ is } i\text{-full},\ \beta \text{ is } i\text{-full, and } \alpha^{*}=\mathfrak{A}, \text{ where } \mathfrak{A} \text{ is the restriction of } \beta^{*} \text{ to the language } Lj(\{a_1,\ldots,a_i\})\},$$

respectively. Then $\Gamma_c$ and $\Gamma_s$ are recursive (using certain coding).

Remark. We use here and in the following the convention of the Remark above Lemma 10.

Proof. We prove that $\Gamma_c$ is recursive. Let $i,j,\alpha,\beta$, and $\mathfrak{A}$ be given so that $i\in\omega$, $j\in\omega$, $\alpha\in EC^{(i,j)}$, $\beta\in EC^{(i+1,j)}$, $\alpha$ is i-full, $\beta$ is (i+1)-full, and $\mathfrak{A}$ is the restriction of $\beta^{*}$ to the language $Lj(\{a_1,\ldots,a_i\})$. First of all we construct a term $\beta^{0}\in EC^{(i,j)}$ so that $(\beta^{0})^{*}\cong\mathfrak{A}$. We can assume that $\alpha^{*}$ and $\beta^{*}$ are both $\omega$-trees, which contain elements of each height, since in the other case we can easily see whether $\alpha^{*}=\mathfrak{A}$ holds or not.

Let $\beta$ be $(\beta_0,\ldots,\beta_k)$. There exists a subset A of $\{a_1,\ldots,a_{i+1}\}$ which contains $a_{i+1}$ so that either $S_{A,C}$ or $1_{A,C}$ or $v_{l,A,C}$ occurs in $\beta_0$ for some l and C. Let $\bar{\beta}_0$ be the term which we get from $\beta_0$ by substituting $S_{A\setminus\{a_{i+1}\},C}$ ($1_{A\setminus\{a_{i+1}\},C}$, $v_{l,A\setminus\{a_{i+1}\},C}$, respectively) for $S_{A,C}$ ($1_{A,C}$, $v_{l,A,C}$, respectively). $(\bar{\beta}_0,\beta_1,\ldots,\beta_k)^{*}$ is isomorphic to $\mathfrak{A}$.

Let $\alpha$ be $(\alpha_0,\ldots,\alpha_n)$. Since all $\alpha_0^{*},\ldots,\alpha_n^{*}$, $\beta_0^{*},\ldots,\beta_k^{*}$ are trees which do not contain elements of any height, it is easy to construct terms $\alpha_0',\ldots,\alpha_{n'}'$, $\beta_0',\ldots,\beta_{k'}'$ so that

$$\alpha^{*}\cong(\alpha_0',\ldots,\alpha_{n'}')^{*},\quad \beta\cong(\beta_0',\ldots,\beta_{k'}')^{*}$$

and

$$h(\alpha_0'^{*})=\ldots=h(\alpha_{n'}'^{*})=h(\beta_0'^{*})=\ldots=h(\beta_{k'}'^{*}).$$

Here $h(\mathcal{T}) := \max\{h_{\mathcal{T}}(a): a \in |\mathcal{T}|\}$ for every tree $\mathcal{T}$ which does not contain elements of any height. Since $\alpha^*$ and $\beta^*$ contain elements of every height, it is easy to construct terms $\alpha_0'',\ldots,\alpha_{n''}''$, $\beta_0'',\ldots,\beta_{k''}''$ so that the following is fulfilled:

$$\alpha^* = (\alpha_0'',\ldots,\alpha_{n''}'')^*, \quad \beta^* = (\beta_0'',\ldots,\beta_{k''}'')^*,$$

$$h(\alpha_0''^*) = h(\beta_0''^*),$$

$$h(\alpha_1''^*) = \ldots = h(\alpha_{n''}''^*) = h(\beta_1''^*) = \ldots = h(\beta_{k''}''^*),$$

$$\{l\text{: there exist } A,C \text{ so that } v_{l,A,C} \text{ occurs in } \alpha_0''\}$$
$$= \{l\text{: there exist } A,C,s \text{ so that } v_{l,A,C} \text{ occurs in } \alpha_s''\}$$
$$= \{1,\ldots,n''\}$$

and

$$\{l\text{: there exist } A,C \text{ so that } v_{l,A,C} \text{ occurs in } \beta_0''\}$$
$$= \{l\text{: there exist } A,C,s \text{ so that } v_{l,A,C} \text{ occurs in } \beta_s''\}$$
$$= \{1,\ldots,k''\}.$$

Now let $f_{\alpha''}$ ($f_{\beta''}$) be a tree-valuation so that $\mathrm{dom}(f_{\alpha''}) = \omega$ ($\mathrm{dom}(f_{\beta''}) = \omega$) and for $1 \le l \le n''$ ($1 \le l \le k''$), $f_{\alpha''}(l) = \alpha_l''^*$ ($f_{\beta''}(l) = \beta_l''^*$, respectively).

Let $\alpha''$ be $(\alpha_0'',\ldots,\alpha_{n''}'')$ and let $\beta''$ be $(\beta_0'',\ldots,\beta_{k''}'')$. Moreover, let $\mathfrak{A}',\mathfrak{B}'$ be the restriction of $\sigma^{(n''\cdot k''+1)}(\alpha_0''^*, f_{\alpha''})$ and $\sigma^{(n''\cdot k''+1)}(\beta_0''^*, f_{\beta''})$, respectively, to the language $Lj(\{a_1,\ldots,a_i\})$. Using the construction of $\alpha''^*$ and $\beta''^*$ we get:

$$\alpha''^* \cong \beta''^* \quad \text{iff} \quad \mathfrak{A}' \cong \mathfrak{B}'.$$

But $\mathfrak{A}' = \mathfrak{B}'$ is decidable since $\mathfrak{A}'$ and $\mathfrak{B}'$ do not contain elements of any height. This proves that $\Gamma_c$ is recursive.

To prove that $\Gamma_s$ is recursive we can proceed in the same way.

Q.E.D.

Theorem 18. Let $i,j,n$ be natural numbers and let $\alpha$ be an extended term from $EC^{(i,j)}$.

(a) There exists a finite set $\Delta_1(\alpha,i,j,n) \subseteq EC^{(i+1,j)}$ which has the following property: For every $a \in |\alpha^*|$ there exists a $\beta \in \Delta_1(\alpha,i,j,n)$

with $(\alpha^*,a) \sim_n^{b1} \beta^*$ and $(i,j,\alpha,\beta) \in \Gamma_c$. Moreover, $\Delta_1(\alpha,i,j,n)$ can be chosen in such a way that the map $(\alpha,i,j,n) \longrightarrow \Delta_1(\alpha,i,j,n)$ is recursive (under a certain coding).

(b) There exists a finite set $\Delta_s(\alpha,i,j,n) \subseteq EC^{(i,j+1)}$ which has the following property: For every $A \in P_{b1}(\alpha^*)$ there exists a $\beta \in \Delta_s(\alpha,i,j,n)$ with $(\alpha^*,A) \sim_n^{b1} \beta^*$ and $(i,j,\alpha,\beta) \in \Gamma_s$. Moreover, $\Delta_s(\alpha,i,j,n)$ can be chosen in such a way that the map $(\alpha,i,j,n) \longrightarrow \Delta_s(\alpha,i,j,n)$ is recursive.

(c) There exists a finite set $\Delta'(\alpha,i,j,n) \subseteq EC^{(i,j+1)}$ which has the following property: For every $A \subseteq |\alpha^*|$ there exists a $\beta \in \Delta'(\alpha,i,j,n)$ so that $(\alpha^*,A) \sim_n^{b1} \beta^*$ and $(i,j,\alpha,\beta) \in \Gamma_s$. Moreover, $\Delta'(\alpha,i,j,n)$ can be chosen in such a way that the map $(\alpha,i,j,n) \longrightarrow \Delta'(\alpha,i,j,n)$ is recursive.

The proof of the existence of $\Delta_1(\alpha,i,j,n)$ and $\Delta_s(\alpha,i,j,n)$ and $\Delta'(\alpha,i,j,n)$ is a straightforward generalization of the proof of Theorem 14 and Corollary 15. We have to generalize Lemma 12 and Lemma 13, too. To get the recursivity of the corresponding functions we have to use Theorem 17. Since we need no more ideas here we, shall omit the details.

Lemma 19. The set

$$\{(\alpha,i,j,\varphi): i \in \omega,\ j \in \omega,\ \alpha \in EC^{(i,j)},\ \alpha \text{ is i-full, } \varphi \text{ is an atomic sentence in the language for the structure } \alpha^*, \text{ and } \alpha^* \models \varphi\}$$

is recursive (under certain coding).

Proof. Let i, j, $\alpha$, and $\varphi$ be given so that $i \in \omega$, $j \in \omega$, $\alpha \in EC^{(i,j)}$, $\alpha$ is i-full, and $\varphi$ is an atomic sentence in the language for the structure $\alpha^*$. From $\alpha$ and the definition of $\alpha^*$ we immediately get a recursive construction of $\alpha^*$. But this gives the decidability of $\alpha^* \models \varphi$. By Lemma 16 this proves the lemma.

Q.E.D.

A combination of the corollary to Theorem 14, Theorem 18 (a and b), and Lemma 19 immediately gives

Theorem 20. $Th_{b1}((\omega\text{-CT}))$ is decidable.

## § 4. Some consequences, pseudo-trees and trees

A corollary of Theorem 20 is the following theorem.

Theorem 21. The theories $Th_b((\omega\text{-CT}))$ and $Th_b((\omega\text{-T}))$ are decidable.

This result was proved in [Seese, 1978] analogous to Theorem 20. We get it here by Lemma 3 from this theorem. A trivial consequence of Theorem 20 is

Theorem 22. $Th_1((\omega\text{-CT}))$ and $Th_1((\omega\text{-T}))$ are decidable.

This result can also be proved in another way. To do this we extend the monadic second order language for $\omega$-trees L' by the quantifier $Q_1$ (used for individual variables only). The resulting language we denote by $L'(Q_1)$. For a class K of $\omega$-trees we define $Th_{II,Q_1}(K)$ to be the set of all sentences $\varphi$ of $L'(Q_1)$ with $K \models \varphi$.

Theorem 23. Let $(\omega\text{-CT}(\omega_1))$ be the class of all connected $\omega$-trees of cardinality at most $\omega_1$. Then $Th_{II,Q_1}((\omega\text{-CT}(\omega_1)))$ is decidable.

Proof. First of all we have to remember the following notion of [Shelah, 1975]: For a model $\mathfrak{M}$ with relations only let $\mathfrak{M}^{\#}$ be the following model:

(i) its universe is the set of finite sequences of elements of $|\mathfrak{M}|$,
(ii) its relations are
 (a) $<$, where $\bar{a} < \bar{b}$ means $\bar{a}$ is an initial segment of $\bar{b}$,
 (b) for each n-place predicate R from the language of $\mathfrak{M}$,
 $R_{\mathfrak{M}^{\#}} := \{((a_1,\dots,a_{m-1},b^1),\dots,(a_1,\dots,a_{m-1},b^n)) : m \in \omega,$ $a_i$ and $b^j$ (for $1 \le i < m$ and $1 \le j \le n$) are elements of $|\mathfrak{M}|$, and $\mathfrak{M} \models R(b^1,\dots,b^n)\}$.

In [Shelah, 1975] the following result is given: $Th_{II}(\mathfrak{M}^{\#})$ is recursive in $Th_{II}(\mathfrak{M})$.

Using this theorem and the decidability of $Th_{II}(\langle\omega_1,\le\rangle)$ (see [Büchi, 1973]) we get the decidability of $Th_{II}(\langle\omega_1,\le\rangle^{\#})$. Let $(\omega_1,\le)^{\#}$ be $\langle(\omega_1)^{<\omega},<^{\#},\le\rangle$. For each subset $A \subseteq (\omega_1)^{<\omega}$, $\langle A,<^{\#}\restriction A\rangle$ is an $\omega$-tree of cardinality at most $\omega_1$. But obviously for each $\omega$-tree $\mathfrak{A}$ of cardinality at most $\omega_1$, too, there is a subset A of $(\omega_1)^{<\omega}$ so that

$\mathfrak{A} \cong \langle A, <^{*}\upharpoonright A\rangle$. To finish the proof we only have to show that $Q_1$ is definable in the monadic second order theory of $\langle \omega_1, \leq\rangle^{*}$.

Let $\varphi(x,\bar{y},\bar{Y})$ be a formula of L'. Since $cf(\omega_1) > \omega_0$, we get

$$\langle \omega_1, \leq\rangle^{*} \models \forall\bar{y}\forall\bar{Y}[Q_1 x\varphi(x,\bar{y},\bar{Y}) \leftrightarrow \exists zZ(\text{"Z contains only immediate successors of z"}) \wedge \forall z_1(z_1 \in Z \rightarrow \exists z_2(z_1 <^{*} z_2 \vee z_1 = z_2) \wedge \varphi(z_2,\bar{y},\bar{Y})) \wedge \forall z_3 \text{"}z_3 \text{ is an immediate successor of z"} \rightarrow \exists z_4(z_3 \leq z_4 \wedge z_4 \in Z)].$$

The parts of the above formula, which are not written down formally, can obviously be formalized in L'.

Q.E.D.

The method which was developed in § 1 - § 3 is also strong enough to prove the following

Theorem 24. For every relational structure $\mathfrak{M}$ it holds that
(i) $Th(\mathfrak{M}^{*})$ is recursive in $Th(\mathfrak{M})$,
(ii) $Th_W(\mathfrak{M}^{*})$ is recursive in $Th_W(\mathfrak{M})$.

[Rabin, 1969] proved the decidability of $Th_{II}((\omega\text{-CT}))$. By interpretability this implies the decidability of $Th_W((\omega\text{-CT}))$. A trivial consequence of this is the decidability of $Th_0((\omega\text{-CT}))$. In [Hauschild, Herre, 1976] the decidability of $Th_\omega(T)$ is proved. In connection with this the following problem seems to be open.

Question. Is $Th_\omega(\omega\text{-CT})$ decidable?

Definition. A (pseudo-)tree is a partially ordered structure $\langle A, \leq\rangle$ so that for all $a \in A$ the set $\hat{a} := \{b: b \leq a\}$ is (linearly ordered) well-ordered by $\leq$. Let TR, PTR be the class of all trees and pseudo-trees, respectively.

Obviously we have $(\omega\text{-T}) \subset TR \subset PTR$.

In [Seese, 1977] it is proved that the monadic second order theory of the class of all countable pseudo-trees is decidable, which is easily proved by interpretability in the theory $Th_{II}((\omega\text{-CT}))$. This was independently proved also by Schmerl. We can conclude the following theorem.

Theorem 25. The theories $Th_w(TR)$, $Th_w(PTR)$, $Th_0(TR)$, and $Th_0(PTR)$ are all decidable.

Conjecture. The following theories are decidable: $Th_1(TR)$, $Th_\omega(TR)$, $Th_1(PTR)$, and $Th_\omega(PTR)$.

With respect to this conjecture the following lemma is of some interest. Let $2^{<\omega_2}$ be the set of all functions $f : \alpha \longrightarrow 2$ with $\alpha \in \omega_2$. Then $\langle 2^{<\omega_2}, \subseteq \rangle$ is a tree. Here $\subseteq$ denotes the usual inclusion. Let $K_1$ denote the class of all subtrees of $\langle 2^{<\omega_2}, \subseteq \rangle$.

Lemma 26. $Th_1(TR)$ and $Th_1(PTR)$ are interpretable in $Th_1(K_1)$.

This lemma is an easy consequence of the theory of $\eta_1$-sets (see e.g. [Kuratowski, Mostowski, 1976], Theorem 1, page 322 and Theorem 2, page 323), Theorem 2 of Chapter 1 and some well-known interpretability techniques.

# CHAPTER 8

# FURTHER QUANTIFIERS

The first paragraph of this chapter gives a survey on the results concerning the decidability of theories in languages with generalized quantifiers or in related languages. A table is used to present the decidable theories. The second part is used to consider the quantifier I, especially the theory of Abelian groups with this quantifier.

## § 1. $L(Q_\alpha^{<\omega})$, $L^{pos}$, $L^{neg}$, $L(aa)$

Let L be a countable first order language. For each natural number $n \geq 1$ we take a new quantifier symbol $Q^n$ which denotes an n-ary quantifier. The $\omega_\alpha$-interpretation of $Q^n$ is given by

$\mathfrak{A} \models (Q^n x_1 \ldots x_n)\varphi(x_1,\ldots,x_n)$ iff there is a set C of power $\omega_\alpha$ so that for every distinct $c_1,\ldots,c_n$ from C, $\mathfrak{A} \models \varphi(c_1,\ldots,c_n)$ holds.

If we have the $\omega_\alpha$-interpretation in mind, then we shall write $Q_\alpha^n$ instead of $Q^n$. The quantifier $Q_\alpha^1$ is the quantifier $Q_\alpha$ investigated throughout this book. It was introduced by [Mostowski, 1957]. For $n > 1$ these quantifiers go back to [Magidor, Malitz, 1977]. $L(Q_\alpha^n)$ denotes the extension of L by $Q_\alpha^n$, while $L(Q_\alpha^{<\omega})$ denotes the extension of L by all $Q_\alpha^i$ for fixed $\alpha$ and all natural numbers i. Using $\diamondsuit_{\omega_1}$ Magidor and Malitz showed the compactness and axiomatizability for $L(Q_1^{<\omega})$.

The logics $L^{pos}$ and $L^{neg}$ are defined in [Makowsky, 1978] and are generalizations of $L(Q_1^1)$ and $L(Q_1^{<\omega})$, respectively. At first we define $L^{pos}$. We add countably many set variables $X_1, X_2, \ldots$ and the "$\in$" symbol to L. Atomic formulas are the atomic formulas from L and $x_i \in X_j$. Then we take the closure under finite conjunction, negation, and existential quantification for individual variables. Existential quantification for set variables is restricted to positive occurrences of the variable, i.e. $\exists X\varphi(X)$ is a well formed formula iff X is in the scope of an even number of negation symbols. The interpretation of $L^{pos}$ is defined as follows

$$\mathfrak{A} \models \exists X\varphi(X) \quad \text{iff} \quad \text{there exists a set } C \subseteq |\mathfrak{A}| \text{ with card}(C) < \omega_1 \text{ and } \mathfrak{A} \models \varphi(C).$$

Similarly the language $L^{neg}$ is defined. We only have to replace "positive occurrence" by "negative occurrence", and in the interpretation the set variables range on uncountable subsets now.

An interesting extension of $L^{pos}$ was given by [Shelah, 1975]. Before we define it we want to give some other definitions. For a set A $P_1(A)$ denotes the set of all countable subsets of A. A set $X \in P_1(A)$ is said to be unbounded if every $B \in P_1(A)$ is a subset of some $C \in X$. X is said to be closed if $(\bigcup_{n<\omega} B_n)$ belongs to X for every increasing sequence $(B_n)_{n<\omega}$ in X. The filter cub(A) is defined to be the set

$$\{X: X \subseteq P_1(A) \text{ and there exists an } X' \subseteq X \text{ so that } X' \text{ is closed and unbounded}\}.$$

Now L is extended by countably many set variables $X_1, X_2, \ldots$ and a new second order quantifier aa. Formulas of L(aa) are formed as usual. We take closure under $\wedge$, $\vee$, $\neg$, $\exists$, and $\forall$ for individual variables. For each set variable X and each formula $\varphi$, $(aaX)\varphi(X)$ is again a formula. We allow neither $\exists X\varphi(X)$ nor $\forall X\varphi(X)$. Satisfaction is defined by induction as usual, using the clause

$$\mathfrak{A} \models (aaX)\varphi(X) \quad \text{iff} \quad \{B: B \in P_1(|\mathfrak{A}|) \text{ and } \mathfrak{A} \models \varphi(B)\} \in \text{cub}(|\mathfrak{A}|).$$

The logic L(aa) is investigated in [Barwise, Kaufmann, Makkai, 1978]. Completeness and compactness are shown there.

All notions of generalized quantifiers given here are different with respect to their expressive power. In the case of the $\omega_1$-interpretation this is illustrated by the following diagram.

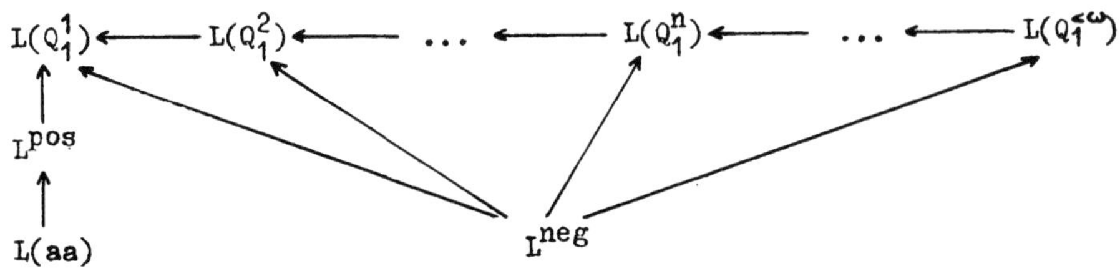


If an arrow is directed from one language L' to another language L", this indicates that L" is definable in L', but not vice versa. Proofs and examples can be found in [Barwise, Kaufmann, Makkai, 1978], [Baudisch, 1978], [Garavaglia, 1978], [Magidor, Malitz, 1977], [Makowsky, Shelah, 1976].

In the following table some interesting theories are collected, which are decidable in languages with the quantifiers mentioned above.

| theory | decidable in the language | |
|---|---|---|
| Monadic predicates without "=" | $L(Q^1_\alpha)$ for all $\alpha$ | [Mostowski, 1957] |
| Monadic predicates with "=" | $L(Q^1_\alpha)$ for all $\alpha$ | [Slomson, 1968], [Vinner, 1972] |
| | $L(Q^{<\omega}_0)$, L(aa), $L^{neg}$ | folklore |
| | with Henkin quantifier | [Krynicki, Lachlan, 1977] |
| One equivalence relation | $L(Q^1_\alpha)$ for all $\alpha$ | [Vinner, 1972] using [Rabin, 1969] |
| | $L(Q^{<\omega}_0)$, L(aa), $L^{neg}$ | folklore |
| Algebraically closed fields (of char. 0 of char. p) | $L(Q^{<\omega}_\alpha)$ for all $\alpha$ | [Cowles, 1975], see furthermore [Jensen, 1975] [Tuschik, 1976] [Vinner, 1975] |
| | L(aa) | folklore |
| Real closed fields | $L(Q^1_1)$ | [Hauschild, Wolter, 1969] |
| | $L(Q^1_\alpha)$ for all $\alpha$ | [Cowles, 1975], [Vinner, 1975] |
| | L(aa) | [Eklof, Mekler, 1978] |
| Arithmetic with "+", "<" | $L(Q^1_\alpha)$ for all $\alpha$ | [Wolter, 1975] |

| theory | decidable in the language | |
|---|---|---|
| Abelian groups | $L(Q_\alpha^{<\omega})$ for all $\alpha$ | [Baudisch, 1976], [Baudisch, 1977] |
| | $L(aa)$ | [Baudisch, 1978a], [Eklof, Mekler, 1978] |
| One unary function and forests | $L(Q_0^{<\omega})$ | follows from [Rabin, 1969] |
| | $L(Q_\alpha^1)$ for regular $\omega_\alpha, \alpha > 0$ | [Hauschild, Herre, 1976], see furthermore [Seese, Tuschik, 1977] |
| | $L^{pos}$, $L^{neg}$ | [Seese, 1977b] |
| | $L(aa)$ | [Baudisch, Tuschik, 1979] |
| Well-orderings | $L(Q_0^{<\omega})$ | follows from [Rabin, 1969] |
| | $L(Q_\alpha^{<\omega})$ for all $\alpha$ | [Slomson, 1976], earlier papers [Lipner, 1970], [Slomson, 1972] |
| | $L(Q_\alpha, Q_\beta)$ | [Wolter, 1975] |
| Linear orderings | $L(Q_0^{<\omega})$ | follows from [Rabin, 1969] |
| | $L(Q_1^1)$ | [Tuschik, 1977], [Herre, Wolter, 1977] |
| | $L(Q_\alpha^1)$ for regular $\omega_\alpha$ | [Herre, Wolter, 1978] |
| | $L(Q_\alpha^1)$ for singular $\omega_\alpha$ | [Herre, Wolter, 1978a] |
| | $L(Q_{\alpha_1}^1, \ldots, Q_{\alpha_n}^1)$, $\omega_{\alpha_i}$ regular | [Tuschik, 1977a] |
| (In [Seese, Tuschik, Weese, 1979] the undecidability of the L(aa)-theory of linear orderings is shown.) | | |
| Trees of height at most $\omega$ | $L(Q_0^{<\omega})$ | follows from [Rabin, 1969] |
| | $L^{pos}$, $L^{neg}$ | [Seese, 1977b] |
| Boolean algebras | $L(Q_0^1)$ | [Pinus, 1976], [Weese, 1977] |
| | $L(Q_\alpha^1)$ for all $\alpha$ | [Weese, 1977a] |
| (By [Rubin, 1978] $L(Q_1^2)$ and by [Seese, Tuschik, Weese, 1979] L(aa) for Boolean algebras are undecidable.) | | |

## § 2. The quantifier I

The L(aa)-theories of Boolean algebras and linear orderings and the theory of Boolean algebras with the quantifier $Q_1^2$ are the only known examples of "natural" elementary theories which become undecidable if they are extended by means of one of the logics above. The situation changes completely for the quantifier I, which was introduced in

[Härtig, 1965]. This quantifier binds one variable in each pair of formulas. Satisfaction is defined by

$$\mathfrak{A} \models (Ix)[\varphi(x),\psi(x)] \quad \text{iff} \quad \begin{array}{l}\mathrm{card}(\{a: a \in |\mathfrak{A}| \text{ and } \mathfrak{A} \models \varphi(a)\}) = \\ \mathrm{card}(\{a: a \in |\mathfrak{A}| \text{ and } \mathfrak{A} \models \psi(a)\}).\end{array}$$

Since the definition of I is very natural there are many investigations about this quantifier although its model-theoretical properties are bad. For instance, compactness fails in the logic L(I). This follows easily because $Q_0$ is definable with the help of I:

$$\mathfrak{A} \models Q_0 x\varphi(x) \quad \text{iff} \quad \mathfrak{A} \models \exists y(\varphi(y) \wedge Ix[\varphi(x), \varphi(x) \wedge x \neq y]).$$

The expressive power of I is very high. Therefore, a lot of theories become undecidable by adding the quantifier I. The undecidability of the following theories with added I can be easily proved:

> One unary function, one equivalence relation, orderings [Herre, Pinus, 1978], well-orderings [Weese, 1977a], Boolean algebras [Weese, 1976a], and even the structure $\langle \omega^2, < \rangle$ (follows from [Seese, 1972]).

However, in the case of Abelian groups the situation changes. The theory of Abelian p-groups with additional quantifier I was proved to be decidable by [Baudisch, 1979]. But this problem remains open for Abelian groups. It is reduced in [Baudisch, 1977a], [Baudisch, 1977b] to the problem of the existence of an effective method for solving a special class of systems of number-theoretical equations and inequalities, so-called I-systems.

Let $Th_I(AG)$ and $Th_I(AGp)$ be the theories of Abelian groups and Abelian p-groups, respectively, with the quantifier I. An I-system X depends on a finite set P of primes and a natural number n.

A (P,n)-term has the form

$$x^v \prod_{p \in P} p^{(\sum_{1 \le i \le n} w_{pi} x_{pi}) + v x_{pn+1}},$$

where $v \in \{0,1\}$ and $w_{pi}$ for $p \in P$ and $1 \le i \le n$ are natural numbers with $w_{pi} \le w_{pi+1}$.

An I-equation in (P,n)-terms has the form $F(x,\dots,x_{pi},\dots) = 0$, where F is a linear combination of (P,n)-terms with coefficients in the integers. Now X consists of finitely many I-equations and I-inequalities, all conditions $p \nmid x$ for $p \in P$, and, may be, additional conditions $x_{pn+1} = 0$ for some $p \in P$.

Theorem 1. $Th_I(AG)$ is decidable iff there is an effective method to solve I-systems.

Partially results concerning I-equations can be found in [Schinzel, 1967].

Theorem 2. Th(AGp) is decidable.

To prove the Theorems 1 and 2 new core sentences will be defined. They will be used to extend Szmielew's elimination procedure. For details see [Baudisch, 1977b].

# R E F E R E N C E S


Apelt, H.

1966 Axiomatische Untersuchungen über einige mit der Presburgerschen Arithmetik verwandte Systeme, ZML 12 (1966), 131 - 168

Ax, J.; Kochen, S.

1965 Diophantine problems over local fields I and II, Amer. J. Math. 87 (1965), 605 - 648

1966 Diophantine problems over local fields III, Ann. of Math. 83 (1966), 437 - 456

Baldwin, J.; Lachlan, A.

1971 On strongly minimal sets, JSL 36 (1971), 79 - 96

Barwise, J.; Kaufmann, M.; Makkai, M.

1978 Stationary logic, Annals of Math. Logic 13 (1978), 171 - 224

Baudisch, A.

1976 Elimination of the quantifier $Q_\alpha$ in the theory of Abelian groups, Bull. de l'Acad. Polon. des Sci. 24 (1976), 543 - 551

1977 Decidability of the theory of Abelian groups with Ramsey quantifiers, Bull. de l'Acad. Polon. des Sci. 25 (1977), 733 - 739

1977a The theory of Abelian groups with the quantifier ($\leq x$), ZML 23 (1977), 447 - 462

1977b Theorien von Klassen Abelscher Gruppen mit verallgemeinerten Quantoren, Dissertation B (1977), Berlin

1978 Uncountable n-cubes in models of $\omega_0$-categorical theories, to appear in Bull. de l'Acad. Polon. des Sci.

1978a The elementary theory of Abelian groups with m-chains of pure subgroups, preprint 1978

1979 The theory of Abelian p-groups with the quantifier I is decidable, to appear in Fund. Math.

Baudisch, A.; Tuschik, H. P.

1979 $L(aa) = L(Q_1)$ for trees, to appear

Baudisch, A.; Weese, M.

1977 The Lindenbaum-algebra of the theory of well-orders and Abelian groups with the quantifier $Q_\alpha$, Lecture Notes in Math. 619, Springer (1977), 59 - 75

Bell, J. L.; Slomson, A. B.

1969 Models and Ultraproducts, Amsterdam - London 1969

Brown, W.

1971 Infinitary languages, generalized quantifiers and generalized products, Ph. D. Thesis, Dartmouth

Bruce, K. B.

1978 Model constructions in stationary logic, Part I: Forcing, preprint

Büchi, J. R.

1973 The monadic second order theory of $\omega_1$, Lecture Notes in Math. 328, Springer (1973), 1 - 127

1977 Using determinacy of games to eliminate quantifiers, Lecture Notes in Computer Science 56, Springer (1977), 367 - 378

Büchi, J. R.; Siefkes, D.

1973 Decidable theories II, Lecture Notes in Math. 328, Springer (1973)

Chang, C. C.; Keisler, H. J.

1973 Model theory, Amsterdam 1973

Cowles, J. R.

1975 Abstract logic and extensions of first order logic, Ph. D. Thesis, Pennsylvania (1975)

1976 Ordered structures and logics with Ramsey quantifiers, Notices AMS 23, No. 4, 76T-E41

1978 Generalized Archimedean fields and logics with Malitz quantifiers, preprint

1978a The theory of Archimedean real closed fields in logics with Ramsey quantifiers, preprint

Doner, J. E.

1965 Decidability of the weak second-order theory or two successors (abstract), Notices AMS 12 (1965), 819

Doner, J. E.; Mostowski, A.; Tarski, A.
1978 The elementary theory of well-ordering - a metamathematical study, Logic Colloquium '77 (A. Macintyre, L. Pacholski, J. Paris (eds.)), North-Holland, Amsterdam 1978, 183 - 193

Ehrenfeucht, A.
1959 Decidability of the theory of one function (abstract), Notices AMS 6 (1959), 268
1959a Decidability of the theory of one linear ordering relation (abstract), Notices AMS 6 (1959), 268 - 269
1961 An application of games to the completeness problem for formalized theories, Fund. Math. 49 (1961), 129 - 141

Eklof, P. C.; Mekler, A. H.
1978 Stationary logic of finitely determinate structures, preprint

Ershov, Yu. L.
1964 Decidability of the theory of relatively complemented distributive lattices and the theory of filters (Russian), Algebra i Logika 3 (1964), No. 3, 17 - 38
1965 On elementary theories of local fields (Russian), Algebra i Logika 4 (1965), No. 2, 5 - 30

Ershov, Yu. L.; Lavrov, I. A.; Taimanov, A. D.; Taitslin, M. A.
1965 Elementary theories, Russian Math. Surveys, Vol. 20 (1965), 35 - 105

Feferman, S.
1960 Arithmetization of metamathematics in a general setting, Fund. Math. 49 (1960), 35 - 92

Feferman, S.; Vaught, R.
1959 The first order properties of algebraic systems, Fund. Math. 47 (1959), 57 - 105

Fraissé, R. J.
1955 Sur quelques classifications des relations basées sur des isomorphismes restreints. I. Etude générale. II. Application aux relations d'ordre, Alger-Mathématiques 2 (1955), 16 - 60, 273 - 295

Fuchs, L.
1966 Abelian groups, Budapest 1966

Fuhrken, G.
1964 Skolem type normal forms for first order language with a generalized quantifier, Fund. Math. 54 (1964), 291 - 302
1965 Languages with added quantifier "there exist at least $\aleph_\alpha$", in The Theory of Models (ed. by Addison, Henkin, and Tarski), North-Holland, Amsterdam 1965, 121 - 131

Garavaglia, S.
1978 Relative strength of Malitz quantifiers, Notre Dame Journal of Formal Logic Vol. XIX (1978), 495 - 503

Gödel, K.
1931 Über formal unentscheidbare Sätze der Principia Mathematica und verwandter Systeme I, Monatsh. Math. Phys. 38 (1931), 173 - 198

Hanf, W.
1965 Model-theoretic methods in the study of elementary logic, in The Theory of Models (ed. by Addison, Henkin, and Tarski), North-Holland, Amsterdam 1965, 132 - 145

Härtig, K.
1965 Über einen Quantifikator mit zwei Wirkungsbereichen, Colloquium on Foundations of Mathematics, Mathematical Machines and their Applications, Tihany (Hungary) 1962, Budapest - Paris (1965). 31 - 36

Hauschild, K.; Herre, H.
1976 Entscheidungsprobleme für Theorien in nichtelementaren Logiken, preprint (1976)

Hauschild, K.; Wolter, H.
1969 Über die Kategorisierbarkeit gewisser Körper in nichtelementaren Logiken, ZML 15 (1969), 157 - 162

Herre, H.; Pinus, A.
1978 Zum Entscheidungsproblem für Theorien in Logiken mit monadischen verallgemeinerten Quantoren, ZML 24 (1978), 375 - 384

Herre, H.; Wolter, H.

1975 Entscheidbarkeit von Theorien in Logiken mit verallgemeinerten Quantoren, ZML 21 (1975), 229 - 246

1977 Entscheidbarkeit der linearen Ordnung in $L_{Q_1}$, ZML 23 (1977), 273 - 282

1978 Entscheidbarkeit der Theorie der linearen Ordnung in $L_{Q_\varkappa}$ für reguläres $\varkappa$, ZML 24 (1978), 73 - 78

1979 Entscheidbarkeit der Theorie der linearen Ordnung in Logiken mit Mächtigkeitsquantoren bzw. mit Chang-Quantor, to appear in ZML

Jensen, F. V.

1975 On completeness in cardinality logics, Bull. de l'Acad. Polon. des Sci. 23 (1975), 117 - 122

Kalmar, L.

1936 Zurückführung des Entscheidungsproblems auf den Fall von Formeln mit einer einzigen binaren Funktionsvariablen, Composito Math. 4 (1936), 137 - 144

Kaufmann, M.

1978 Some results in stationary logic, Ph. D. Dissertation, University of Wisconsin

Keisler, H. J.

1968 Models with orderings, in Logic, Methodology and Philosophy of Science III, edited by B. van Rootselaar and J. F. Staal, North-Holland, Amsterdam 1968, 35 - 62

1970 Logic with the quantifier "there exist uncountably many", Annals of Math. Logic 1 (1970), 1 - 94

Korec, I.; Rautenberg, W.

1976 Model-interpretability into trees and applications, Arch. math. Logik 17 (1976), 97 - 104

Krynicki, M.; Lachlan, A. H.

1977 On the semantics of the Henkin quantifier, preprint

Lachlan, A.

1972 A property of stable theories, Fund. Math. 77 (1972), 9 - 20

Läuchli, H.
1968 A decision procedure for the weak second order theory of linear order, Contributions to Math. Logic, Proc. of the Logic Colloquium Hanover 1966, North-Holland, Amsterdam 1968

Läuchli, H.; Leonard, J.
1966 On the elementary theory of linear order, Fund. Math. 49 (1966), 109 - 116

Lipner, L. D.
1970 Some aspects of generalized quantifiers, Ph. D. Thesis, Berkeley (1970)

Magidor, M.; Malitz, J.
1977 Compact extensions of L(Q), Annals of Math. Logic 12 (1977), 217 - 261

Makowsky, J. A.
1975 Topological model theory, Proceedings of the CIME model theory session, Bressanone 1975
1977 Elementary equivalence and definability in stationary logic, preprint
1978 Quantifying over countable sets, positive vs stationary logic, Logic Colloquium '77 (A. Macintyre, L. Pacholski, J. Paris (eds.)), North-Holland, Amsterdam 1978, 183 - 193

Makowsky, J. A.; Shelah, S.
1976 The theorems of Beth and Craig in abstract model theory, preprint (1976)

Makowsky, J. A.; Shelah, S.; Stavi, J.
1976 $\Delta$-logics and generalzed quantifiers, Annals of Math. Logic 10 (1976), 155 - 192

Morley, M.
1965 Categoricity in power, Trans. Amer. Math. Soc. 114 (1965), 514 - 538

Mostowski, A.
1957 On a generalization of quantifiers, Fund. Math. 44 (1957), 12 - 36

Mostowski, A.; Tarski, A.
1949 Arithmetical classes and types of well ordered systems, preliminary report, Bull. AMS Vol. 55 (1949), 65

Pinus, A.
1976 The theory of Boolean algebras in the predicate calculus with the quantifier "there are infinitely many" (Russian), Sibirskij Math. J. 17 (1976), 1417 - 1421
1978 The theory of locally S-free algebras with the quantifier Q, preprint

Rabin, M. O.
1964 A simple method of undecidability proofs and some applications, Log. Meth. Phil. Science, Proc. Jerusalem 1964, 58 - 68
1969 Decidability of second order theories and automata on infinite trees, Trans. AMS 149 (1969), 1 - 35

Robinson, J.
1949 Definability and decision problems in arithmetic, JSL 14 (1949), 98 - 114

Rubin, M.
1978 A Boolean algebra with few subalgebras, interval Boolean algebras, and retractiveness, preprint

Schinzel, A.
1967 On two theorems of Gelfand and some of their applications, Acta Arith. 13 (1967), 177 - 236

Seese, D.
1972 Entscheidbarkeits- und Definierbarkeitsfragen der Theorie "netzartiger" Graphen - I, Wissenschaftliche Zeitschrift der Humboldt-Universität, Math.-Nat. R. 21 (1972), 513 - 517
1977 Second order logic, generalized quantifiers and decidability, to appear in Bull. de l'Acad. Polon. des Sci.
1977a Decidability of $\omega$-trees with bounded sets - a survey, Lecture Notes in Computer Science 56, Springer (1977), 511 - 515
1977b Decidability of $\omega$-trees with bounded sets, preprint (1977)
1977c A remark to "The undecidability of well-ordering with the Härtig quantifier", submitted to Bull. de l'Acad. Polon. des Sci.

1978 Decidability and generalized quantifiers, Logic Colloquium '77 (A. Macintyre, L. Pacholski, J. Paris (eds.)), North-Holland, Amsterdam 1978, 229 - 237

Seese, D. Tuschik, H. P.
1977 Construction of nice trees, Lecture Notes in Math. 619, Springer (1977), 257 - 273

Seese, D.; Tuschik, H. P.; Weese, M.
1979 The L(aa)-theories of linear orderings and Boolean algebras are undecidable, to appear

Shelah, S.
1975 The monadic theory of order, Annals of Math. 102 (1975), 379 - 419
1975a Generalized quantifiers and compact logic, Trans. AMS 204 (1975), 342 - 364

Shoenfield, J. R.
1967 Mathematical logic, Reading, Mass., Addison-Wesley 1967

Siefkes, D.
1975 Decidability of monadic second order theories of ordinals and underlying set theory, a short survey, preprint (1975)

Slomson, A. B.
1968 The monadic fragment of predicate calculus with the Chang quantifier, Lecture Notes in Math. 70 , Springer (1968), 279 - 301
1972 Generalized quantifiers and well orderings, Arch. Math. Logik 15 (1972), 57 - 73
1976 Decision problems for generalized quantifiers - a survey, Lecture Notes in Math. 537, Springer (1976), 249 - 258

Szmielew, W.
1955 Elementary properties of Abelian groups, Fund. Math. 41 (1955), 203 - 271

Tarski, A.
1949 Arithmetical classes and types of Boolean algebras, Bull. AMS 55 (1949), 64

1949a Undecidability of the theory of lattices and projective geometries, JSL 15 (1949), 77 - 78

1949b Arithmetical classes and types of mathematical systems, Bull. Amer. Math. Soc. 55 (1949), 63 - 64

Tarski, A.; McKinsey, J. C. C.

1948 A decision method for elementary algebra and geometry, 2nd ed., Berkeley - Los Angeles 1948

Tarski, A.; Mostowski, A.; Robinson, R. M.

1953 Undecidable theories, Amsterdam 1953

Le Tourneau, J. J.

1968 Decision problems related to the concept of operation, Ph. D. Thesis (1968), Berkeley

Tuschik, H. P.

1976 Elimination verallgemeinerter Quantoren in $\aleph_1$-kategorischen Theorien, to appear in Wiss. Zeitschrift Humboldt-Universität

1977 On the decidability of the theory of linear orderings in the language $L(Q_1)$, Lecture Notes in Math. 619, Springer (1977), 291 - 305

1977a On the decidability of the theory of linear orderings with generalized quantifiers, to appear in Fund. Math.

1978 Elimination of cardinality quantifiers, preprint

Väänänen, J.

1977 Remarks on generalized quantifiers and second order logics, in: Set theory and hierarchy theory, Wrocław 1977

1977a Applications of set theory to generalized quantifiers, Thesis, Manchester 1977

Vaught, R. L.

1964 The completeness of logic with added quantifier "there are uncountably many", Fund. Math. 54 (1964), 303 - 304

Vinner, S.

1972 A generalization of Ehrenfeucht's game and some applications, Israel J. Math. 12 (1972), 279 - 298

1975 Model-completeness in a first order language with a generalized quantifier, Pacific J. of Math. 56 (1975), 265 - 273

Weese, M.
1974 Entscheidbarkeit der Topologie der p-adischen Zahlkörper in Sprachen mit Mächtigkeitsquantoren, Dissertation A, Berlin 1974
1976 Ein neuer Beweis für die Entscheidbarkeit der Theorie der Booleschen Algebren, to appear in Wiss. Zeitschrift Humboldt-Universität
1976a The universality of Boolean algebras with the Härtig quantifier, Lecture Notes in Math. 537, Springer (1976), 291 - 296
1977 The decidability of the theory of Boolean algebras with the quantifier "there exist infinitely many", Proc. AMS 64 (1977), 135 - 138
1977a The undecidability of well-ordering with the Härtig quantifier, Bull. de l'Acad. Polon. des Sci. 25 (1977), 89 - 91
1977b The decidability of the theory of Boolean algebras with cardinality quantifiers, Bull. de l'Acad. Polon. des Sci. 25 (1977), 93 - 97
1977c Entscheidbarkeit der Theorie der Booleschen Algebren in Sprachen mit Mächtigkeitsquantoren, Dissertation B (1977)
1977/8 Definierbare Prädikate in Booleschen Algebren, ZML 23 (1977), 511 - 526, and ZML 24 (1978), 257 - 278
1978 Generalized Ehrenfeucht games, to appear in Fund. Math.

Wojciechowska, A.
1969 Generalized products for Q-languages, Bull. de l'Acad. Polon. des Sci. 17 (1969), 337 - 339

Wolter, H.
1973 Eine Erweiterung der elementaren Prädikatenlogik, Anwendungen in der Arithmetik und anderen mathematischen Theorien, ZML 19 (1973), 181 - 190
1975 Untersuchungen zu nicht-elementaren Logiken, Dissertation B, Berlin 1975
1975a Entscheidbarkeit der Arithmetik mit Addition und Ordnung in Logiken mit verallgemeinerten Quantoren, ZML 21 (1975), 321 - 330

# INDEX OF NOTATION

# SUBJECT INDEX

Bei Fragen zur Produktsicherheit wenden Sie sich bitte an:
If you have any questions regarding product safety,
please contact:

Walter de Gruyter GmbH
Genthiner Straße 13
10785 Berlin
productsafety@degruyterbrill.com